Lecture Notes in Bioinformatics 5750

Subseries of Lecture Notes in Computer Science

W0111598

Corrado Priami   Ralph-Johan Back
Ion Petre (Eds.)

Transactions on Computational Systems Biology XI

 Springer

Series Editors

Sorin Istrail, Brown University, Providence, RI, USA
Pavel Pevzner, University of California, San Diego, CA, USA
Michael Waterman, University of Southern California, Los Angeles, CA, USA

Editor-in-Chief

Corrado Priami
The Microsoft Research - University of Trento
Centre for Computational and Systems Biology
Piazza Manci, 17, 38050 Povo (TN), Italy
E-mail: priami@cosbi.eu

Guest Editors

Ralph-Johan Back
Ion Petre
Åbo Akademi University
Department of Information Technologies
Joukahaisenkatu 3-5, 20520 Turku, Finland
E-mail: {backrj,ipetre}@abo.fi

Library of Congress Control Number: 2009933672

CR Subject Classification (1998): J.3, F.1, F.2, I.6, I.2, C.1.3

ISSN   0302-9743 (Lecture Notes in Computer Science)
ISSN   1861-2075 (Transactions on Computational Systems Biology)
ISBN-10   3-642-04185-X Springer Berlin Heidelberg New York
ISBN-13   978-3-642-04185-3 Springer Berlin Heidelberg New York

springer.com

© Springer-Verlag Berlin Heidelberg 2009
Printed in Germany
Typesetting: Camera-ready by author, data conversion by Scientific Publishing Services, Chennai, India
Printed on acid-free paper   SPIN: 12743292   06/3180   5 4 3 2 1 0

Preface

Biology is witnessing a transformation towards a more quantitative science, based on the major technological breakthroughs of the past decade. In this transformation, biology is incorporating mathematical modeling techniques and computational approaches towards numerical simulations, model analysis, and quantitative predictions. An important goal is to formalize and analyze the ever-changing inter-connections between components (often on different time and space scales), their influence on one another, regulatory patterns, alternative pathways, etc. Formal reasoning rather than empirical observations is the main driving force in this new type of biological research. At the same time, computer science and applied mathematics are faced with considerable methodological challenges in handling an unprecedented level of concurrency, stochastic effects, a mix of large and small populations, combinatorial explosions in the state space, model refinement, and model (de)composition, etc.

This special issue of *Transactions on Computational Systems Biology* on Computational Models for Cell Processes is based on a workshop with the same name that took place in Turku, Finland, on May 27, 2008. The workshop was organized as a satellite event of *the 15th International Symposium on Formal Methods* that took place in Turku in the period May 28-31, 2008. This special issue however had an open call for paper submissions, with a separate peer-review process. The accepted papers span an interesting mix of approaches to systems biology, ranging from quantitative to qualitative techniques, from continuous to discrete mathematics, from deterministic to stochastic methods, from computational models for biology to computing paradigms inspired by biology. Overall, they give a good glimpse into some of the exciting current research avenues in computational systems biology.

This volume also contains three regular submissions that deal with the relationships between ODEs and stochastic concurrent constraint programming (by Bertolussi and Policriti), with the equilibrium points of genetic regulatory networks (by Chesi), and with probability models describing how epigenetic context affects gene expression and organismal development (by Wallace and Wallace).

July 2009

Ralph-Johan Back
Ion Petre
Corrado Priami

LNCS Transactions on
Computational Systems Biology –
Editorial Board

Table of Contents

Computational Models for Cell Processes

Process Algebra Modelling Styles for Biomolecular Processes

Muffy Calder[1] and Jane Hillston[2]

[1] Department of Computing Science, University of Glasgow,
Glasgow G12 8QQ, Scotland
[2] Laboratory for Foundations of Computer Science and
Centre for Systems Biology, Edinburgh
The University of Edinburgh, Edinburgh EHA 9AB, Scotland

Abstract. We investigate how biomolecular processes are modelled in process algebras, focussing on chemical reactions. We consider various modelling styles and how design decisions made in the definition of the process algebra have an impact on how a modelling style can be applied. Our goal is to highlight the often implicit choices that modellers make in choosing a formalism, and illustrate, through the use of examples, how this can affect expressability as well as the type and complexity of the analysis that can be performed.

1 Introduction

Much recent research has considered the problem of providing suitable abstract models to allow biologists to construct mechanistic models to enhance understanding of biomolecular processes. Process algebras, formal modelling languages originally conceived for modelling concurrent computations, have been widely applied, most notably in the area of signalling pathways [RSS01, CGH06, TK08]. This is experimental science and we are currently evaluating the hypothesis that such formal models can add value to the mathematical analysis that is already undertaken within systems biology in terms of ordinary differential equation (ODE) models or stochastic simulations directly. In exploring this goal, even within work on process algebras, several different styles of modelling have emerged. Ultimately we hope to be able to give guidance on how to choose among these modelling styles, or on how to map molecular components and their interactions to processes, process communication and process composition. However, in the first instance we investigate how design decisions made in the definition of the language have an impact on how a modelling style can be applied, and highlight the often implicit choices that modellers make in choosing a formalism.

Recent research effort on process algebras for biomolecular processes, e.g. [CGH06, CVOG06, CH08, Car08], has focussed on defining alternative semantics, such as discrete-state (stochastic) or continuous-state (ODE) semantics. These provide important links with the work where mathematical representations are used directly and establish a valid foundation for process algebra models. Based on these semantics, analysis may be carried out by model-checking,

C. Priami et al. (Eds.): Trans. on Comput. Syst. Biol. XI, LNBI 5750, pp. 1–25, 2009.

stochastic simulation based on Gillespie's algorithm or ODE simulations. Our emphasis in this paper is different. Here we consider the forms of abstraction supported by process algebra and how the abstraction and the process algebra chosen affect the expressiveness of the model with respect to the biological processes, as well as the type and complexity of the analysis that can be performed.

We focus on one of the most important types of interaction between molecular components: chemical reactions. In chemical notation, these may be first order reactions, for example A degrades to B: $A \xrightarrow{k1} B$, or second order reactions, for example A and B combine to form C or C and D: $A + B \xrightarrow{k2} C$, or $A + B \xrightarrow{k3} C + D$. Typically, $k1 \ldots k3$ are rate constants for kinetic laws (e.g. mass action).

A fundamental aspect of the abstraction used in modelling is the nature of the process mapping. In the literature on process algebras for systems biology we find predominantly the *molecule-as-process* [RSS01, Car08] abstraction, but the *species-as-process* and *reaction-as-process* mappings have also been proposed [CGH06, CH08, BP08]. The distinction between the first two can be understood by appealing to ecology: the former is essentially individuals-based, whereas the latter is population-based. We note that this distinction is less common in distributed computing system modelling, the origins of process algebra, where population-based models are rarely considered.

Further stylistic differentiation was identified in [CGH06] where the concepts of *reagent-centric* and *pathway-centric* models are introduced, in the context of population-based modelling. Reagent-centric models map all reagents in a reaction to processes, whose variation reflect decrease through consumption and increase through product formation (consumers and producers). Reagents such as modifiers that do not vary species amounts can also be modelled in this approach. Reagent-centric models provide a fine-grained, distributed view of a system. Pathway-centric models provide a more abstract view of a system, tracking serialisations of events, which are then composed concurrently. Here, processes vary according to their biological state rather than their quantity. Whereas in a reagent-centric approach the processes may be molecules or molecular species, in the pathway-centric approach the processes are molecules or sub-pathways. Thus the interactions between processes are between flows of events corresponding to producers, i.e. components on the left hand sides of a reactions.

Most modelling approaches map *chemical reactions* to *events* in a straightforward way, and map (possibly a subset of) the *chemical components* to *processes*. Bortolussi and Policriti's work on sCCP, using the *reaction-as-process* abstraction, is an exception to this. When chemical components are mapped to processes within the reagent-centric approach there is a further choice: between associating processes with *all* components or only with the reagents on the *left hand side* of equations, i.e. those reagents that are the reactants of the reaction. To distinguish these two cases, we call the former reagent-centric and the latter *reactant-centric*. This modelling choice is often influenced by the form of synchronisation available within the algebra: binary or multi-way. If we have only the former, then only the reactant-centric approach is possible and we are left with an interesting dilemma when there are fewer components on the right hand side of the equation than on the left hand side, e.g. $A + B \xrightarrow{k2} C$.

In summary, a number of factors will influence the structure of a process algebra model of a biomolecular process:

- population-based or individuals-based,
- reagent-centric, reactant-centric, pathway-centric or reaction-centric,
- the form of synchronisation available in the algebra.

In this paper we investigate the interplay between these three factors. Our motivation is to explore the extent to which we can build clear and faithful models using current algebras and analysis techniques, and how design decisions with respect to the process algebra determine the mappings available to the modeller. We consider different combinations, investigating their advantages and disadvantages.

We will use five process algebras for illustration: π-calculus, Beta-binders, PEPA, Bio-PEPA, and sCCP; these are briefly outlined in the next section. These are chosen as they represent a spectrum of different modelling style, including languages that have been adapted (π-calculus, PEPA and sCCP) and designed (Beta-binders and Bio-PEPA) for biological modelling. This is by no means a comprehensive list of process algebras used in systems biology. In particular we do not include any of the process algebras designed to consider spatial aspects of biomolecular processes [CPR+04, Car04, V07, BMMT06, CG09] as they are beyond the scope of this paper.

The remainder of the paper is organised as follows. Section 2 gives an overview of the process algebras and Section 3 describes the example pathway used throughout for illustration and comparison. In Sections 4 to 8 we consider modelling in PEPA, Bio-PEPA, π-calculus, Beta-binders and sCCP. We discuss the results in Section 9 and give our conclusions in Section 10.

2 Process Algebras

Process algebras were originally defined to give semantics to concurrent processes in a computing context and have enjoyed considerable success over the three decades since they emerged. Classical process algebras such as CCS [Mil80] and CSP [Hoa85] focus on the functional capabilities of processes and all actions are atomic with only relative timing of actions captured. Subsequently there have been many extensions of process algebras to capture more information about the system being modelled, for example the relative probability of alternative actions (probabilistic process algebras) and the expected duration of actions (stochastic process algebras).

Each of the process algebras that we consider is based on three fundamental binary operators: action prefix, choice, which is associative and commutative, and synchronous composition, which is also associative and commutative.

Note that in the following we omit the cooperation sets for composition in PEPA and Bio-PEPA and assume them to be the *intersection* of the alphabets of the processes involved (denoted $\bowtie$). We disregard quantitative aspects of actions, since the representation of kinetics is orthogonal to the expressiveness we consider here. Therefore in our examples, we will assume that the reaction rate for each considered reaction is unique and use this as the name of the corresponding reaction event, i.e. the reaction $A + B \xrightarrow{r_1} C + D$ in chemical notation maps to the process algebra event r_1.

In seminal work, Regev and Shapiro [RS01] suggested an abstraction of *cell-as-computation* and proposed that models formerly used in the study of interacting computational entities, such as Petri nets, process algebras and automata, could be usefully employed for the study of biological processes. In particular they focussed on the π-calculus [Mil99], and subsequently the stochastic π-calculus [Pri95] based on the *molecule-as-process* abstraction. This work has been hugely influential with many other authors following the same abstraction in their own work, even when the details of the process algebra differ.

However, the π-calculus has some particular characteristics that are independent of the *molecule-as-process* abstraction that also shape the style in which models are expressed. In this section we give a brief introduction to process algebras, focussing on the features which lead to different modelling paradigms.

2.1 Forms of Synchronisation

The original process algebras, CCS and CSP, differ in their interpretation of actions and consequently the meaning of synchronisation. In CCS all actions are assumed to be communications, and therefore *conjugate*, i.e. actions are paired, corresponding to an input and an output. An action cannot be carried out without its partner, and the pairing of an input and an output becomes a private τ action. This has the consequence that the interaction, or synchronisation, between processes is strictly binary as once an input has been paired with an output both become unavailable for further interaction. In contrast, in CSP no distinction is made between inputs and outputs and there is no notion of complementarity between actions. Instead action type denotes ownership of a *channel* and synchronisation is assumed to take place whenever processes undertake actions of the same type, i.e. communication over the named channel. This is termed *multiway* synchronisation as there is no restriction on the number of processes that may own a channel and thus join a synchronisation. Note that in both these cases the parallel composition operator is generic: in CCS any complementary actions which are on either side of the parallel composition may synchronise; in CSP, processes composed by the parallel operator *must* synchronise on common actions.

Synchronisation in PEPA is a subtle variation of the CSP scheme. Here the parallel composition operator, termed *cooperation*, is decorated by a set of action types (the *cooperation set*) and processes are only forced to synchronise on action types within this set, being able to act concurrently and individually on other action types. Thus the parallel composition is not generic, but a family of parameterised operators. The characteristics of this multiway synchronisation are important in the biological context as they allow one copy of a process (molecule) within a set of identical processes to undertake a reaction individually, something that would not be possible in CSP[1].

2.2 π-Calculus

The π-calculus [Mil99] (and its stochastic form [Pri95]) was designed to express mobility, represented by the passing of channel names. It evolved from CCS [Mil80] and includes the operations of a constant, action prefix, choice, parallel composition, communication and scope restriction. There are variants of the syntax, here we use the following form with events π and processes P:

$$\pi ::= \tau \mid x \mid \overline{x} \mid x(y) \mid \overline{x}\langle y\rangle$$

$$P ::= \mathbf{0} \mid \pi.P \mid P|P \mid P + P \mid \nu x P$$

Following CCS [Mil80], τ is the unobservable event. All other events are observable and paired, e.g. $x(y)$ with $\overline{x}\langle y\rangle$, with $x(y)$ denoting input y on channel x, and $\overline{x}\langle y\rangle$ denoting output y on channel x. $\mathbf{0}$ is the inactive process and $\nu x P$ restricts the scope of the name x to P. In the stochastic form, rates are bound to channels, but as with the other process algebras, we will omit rates here.

A structural congruence, denoted $\equiv$, determines when two syntactic expressions are equivalent, and an operational semantics is given by a set of reaction rules that define how a system evolves following communication. We do not give the full definitions of the congruence and reaction rules, but note two distinguishing features. First, the constant, $\mathbf{0}$, is an identity for parallel composition, i.e. there is a *syntactic* equality $P \mid \mathbf{0} \equiv P$. Second, interaction only occurs when there is a complementary pair of input and output events. The relevant reduction rule is $(\ldots + \overline{x}\langle y\rangle.Q) \mid (\ldots + x(z).P) \rightarrow Q \mid P\{y/z\}$.

There have been numerous applications of the π-calculus to biomolecular processes, starting with the work of Regev *et al.* [RSS01]. An interesting aspect of the application of π-calculus is that it was designed to facilitate modelling mobility and name passing, thus in the original π-calculus events are parameterised, e.g. $x(y)$. Yet, most biological applications do not exploit mobility — the parameter is not relevant, except when modelling compartments, or internal communications. So, in many models unparameterised events are also permitted, e.g. x and $\overline{x}$, and we have also included them here. We note the recent work of Cardelli [Car08] on translations between process algebra and chemical reactions that introduces a subset of the π-calculus and CCS suitable for modelling chemical reactions. It is similar to the syntax above, but excludes event parameters and the ν operator. Additionally, it includes an expression of initial components.

---

[1] This might explain why, to the best of our knowledge, there has been no work applying CSP to biomolecular modelling.

A further distinctive aspect of the π-calculus/CCS paradigm for biomolecular modelling is the underlying assumption of two-way synchronous communication. This means that a a binary chemical reaction, e.g. of the form $A + B \rightarrow^r C$, is modelled by processes A and B offering events r and $\bar{r}$, whereas a unary chemical reaction, e.g. of the form $A \rightarrow^r B$, must be modelled by an unobservable τ event.

2.3   Beta-Binders

Beta-binders [DPPQ06] is a process algebra based on the π-calculus, designed for modelling and simulation of biological processes. A biological process is modelled by a *bio-process*, which is a π-calculus process encapsulated in a box with interaction capabilities expressed as beta-binders. Each communication channel has a set of associated types and there are three kinds of binder: visible, hidden, and complexed. Additionally, there are rates, but these are omitted here. A bio-process is either a constant or pair of encapsulated π-calculus processes composed with a synchronous parallel operator.

The language has evolved over a number of years, here we use the following syntax for boxes B and beta-binders **B**, assuming π-calculus processes P:

$$B ::= Nil \mid \mathbf{B}[P] \mid B \parallel B$$

$$\mathbf{B} ::= \beta(x, \Gamma) \mid \beta^h(x, \Gamma) \mid \beta^c(x, \Gamma)$$

Further, there is a additional syntactic category for *events*, which include functions on boxes to join, split, create and destroy boxes; these are called *join*, *split*, *new* and *delete*, respectively. These functions are only applied when a condition, defined over binders and π processes, is fulfilled.

Interaction is two-way and is either *intra-box*, in which case it is standard π-calculus interaction, or it is *inter-box* in which case it is specified by the beta-binders and it is between (visible) input/output pairs, but now the types have only to be compatible (rather than identical). There are additional actions (within boxes) that include changing the status of binders (e.g. unhide or change type). There are three structural congruences: $\equiv_p$, the standard congruence on π processes, $\equiv_b$, a congruence on boxes (e.g. $\parallel$ is associative, commutative), and $\equiv_e$, a congruence on events (e.g. join, split have substitution property).

2.4   PEPA

Performance Evaluation Process Algebra (PEPA) was introduced in the early 1990s as a formalism for building Markovian-based performance models of computer and communication systems [Hil96]. All actions in PEPA consist of an action type and a *rate*, which specifies the average duration of the action as an exponentially distributed random variable. The language has a small set of combinators (prefix, choice, parallel composition/cooperation, hiding and constant).

Recursive behaviour is specified by mutually recursive definitions. As PEPA was designed for specifying ergodic continuous time Markov chains (CTMC), a restriction is often placed on model construction via a two level syntax, meaning that models consist of parallel compositions of sequential components (constructed using only prefix and choice):

$$S := \alpha.S \mid S + S \mid C$$

$$P := P \bowtie_L P \mid P/L \mid S$$

where S denotes a *sequential component*, P a *model component* and C is a constant defined by a declaration such as

$$C \stackrel{def}{=} S$$

$\alpha.S$ carries out activity α (with an exponentially distributed duration, but omitted here), and it subsequently behaves as S. As discussed above, PEPA supports multi-way cooperations between components: the result of synchronising on an activity α is thus another α, available for further synchronisation. We write $P \bowtie_L Q$ to denote cooperation between P and Q over L. The set which is used as the subscript to the cooperation symbol, the *cooperation set L*, determines those activities on which the *cooperands* are forced to synchronise. For action types not in L, the components proceed independently and concurrently with their enabled activities. We write $P \parallel Q$ as an abbreviation for $P \bowtie_L Q$ when L is empty. P/L denotes the component P in which all actions with types in L are *hidden* meaning that their type is no longer visible but is replaced by the distinguished type τ. We do not consider hiding in the remainder of this paper.

The stochastic nature of the actions means that the choice becomes a probabilistic choice governed by a *race condition* between the involved actions. Similarly actions of parallel components that are not forced to cooperate are also subject to a race condition. When components cooperate on actions but have different definitions of the rate of the action, the rate of the synchronised action is defined to be that of the slowest of the components. While these dynamic considerations do not concern us in this paper, and PEPA has been used for modelling a number of biological examples, we note that the form of the dynamics of synchronisation (the rate of the slowest component) is not always appropriate in this context.

2.5   Bio-PEPA

Bio-PEPA [CH08] is a newly defined modification of the PEPA formalism that has been specifically designed for modelling biochemical networks. It shares many features with PEPA but also has some characteristics to tailor it to the biological application.

Functional rates: In contrast to PEPA, individual processes are not able to define their own rates for actions. Instead the rate associated with an action is specified once, independently of the processes in which the action occurs.

The value of this rate can be specified to be a function that depends on the current state of the system.

Stoichiometry: For each action, as well as its type, the stoichiometry or degree of involvement is also specified.

Parameterised processes: Bio-PEPA has been designed to support the population-based reagent-centric style of modelling and so a model consists of a number of sequential components each representing a distinct species which evolve quantitatively (increasing or decreasing amounts). Thus in order to capture the state of a system each component is parameterised recording its current level.

Differentiated prefix: For each action (reaction) that a component is involved in it records its *role* within that reaction, e.g. reactant, product, inhibitor etc. This enables the appropriate values to be used in the functional rate associated with this reaction.

As with PEPA, Bio-PEPA has a two level grammar. The syntax of the sequential (species) components is defined as:

$$S ::= (\alpha, \kappa) \text{ op } S \mid S + S \mid C \qquad \text{op} ::= \downarrow \mid \uparrow \mid \oplus \mid \ominus \mid \odot.$$

In the prefix term (α, κ) op S, α is an action name and can be viewed as the name or label of a reaction, κ is the stoichiometry coefficient of the species and the prefix combinator op represents the role of the element in the reaction. Specifically, $\downarrow$ denotes the role of reactant, $\uparrow$ product, $\oplus$ activator, $\ominus$ inhibitor and $\odot$ generic modifier. The operator $+$ expresses the choice between possible actions and the constant C is defined by an equation $C \stackrel{def}{=} S$.

The syntax of model components is defined as:

$$P ::= P \bowtie_{\mathcal{L}} P \mid S(x)$$

The process $P \bowtie_{\mathcal{L}} Q$ denotes the synchronisation between components P and Q and the set $\mathcal{L}$ specifies those activities on which the components must synchronise. In the model component $S(x)$, the parameter $x \in \mathbb{R}$ represents the initial concentration by default, although according to the analysis to be carried out the parameter may also be interpreted as number of molecules or molecular level after appropriate conversion.

2.6 sCCP

In the Concurrent Constraint Programming (CCP) process algebra, rather than components and actions, there are components and constraints [BJG96]; there are also variables. The components evolve by adding constraints to a constraint store (tell) or checking the current state of the constraint store (ask). This leads to an asynchronous form of communication between components (via global variables in the constraint store) and there is no direct synchronisation. In addition to tell and ask components may also have choice, parallel composition, procedure call and local variables. In the stochastic form of CCP, sCCP [Bor06], a stochastic duration is associated with the ask and tell operators in a manner analogous to the durations of actions in other stochastic process algebras.

sCCP has been proposed as a modelling formalism for biological networks, and stochastic, deterministic and hybrid semantics have been associated with models in this context [BP08]. The style of modelling is similar to that of Bio-PEPA in that a population-based view is taken, although here explicit variables record the quantitative state of species, rather than parameterised components. At a high level the abstraction is that measurable entities (molecules etc.) are associated with stream variables, logical entities are associated with processes or control variables and reactions are associated with processes. In general a reaction is modelled as a sequence of interactions with the constraint store: first checking that there is sufficient amount of the substrates and then updating the amounts of the products. For mass action reactions the ask step of this sequence will be given a rate equal to the product of the kinetic constant and the amounts of the substrates; the tell step is assumed to be instantaneous. Thus an arbitrary mass action reaction

$$R_1 + \ldots + R_n \longrightarrow_k P_1 + \ldots + P_m$$

will be represented as

$$\mathsf{reaction}(\mathsf{k}, [\mathsf{R}_1, \ldots, \mathsf{R}_n], [\mathsf{P}_1, \ldots, \mathsf{P}_m]) : -$$

$$\mathsf{ask}_{r_{MA}(\mathsf{k},\mathsf{R}_1,\ldots,\mathsf{R}_n)} \left(\bigwedge_{i=1}^{n} (\mathsf{R}_i > 0) \right).$$

$$\left(\|_{i=1}^{n} \mathsf{tell}_\infty (\mathsf{R}_i \; \$= \mathsf{R}_i - 1) \; \|_{j=1}^{m} \mathsf{tell}_\infty (\mathsf{P}_j \; \$= \mathsf{P}_j + 1) \right)$$

Here R_i and P_j are stream variables and r_{MA} is a predefined function with the obvious definition.

3 Example Pathway

We refer to a small synthetic pathway when exploring how design decisions with respect to the the process algebra determine the mappings available to the modeller. The pathway consists of five representative reactions. The reactions are given in chemical notation in Figure 1, and presented graphically in Figure 2. While the pathway is a synthetic example, it is based on behaviour we have observed in various pathways, including the ubiquitous Raf/MEK/ERK signalling pathway.

The equations exhibit various combinations of increasing/decreasing/preserved reagents between the left and right hand sides. Specifically, r_1 and r_4 have a decreasing number of reagents, r_2 and r_5 have an increasing number of reagents, and r_3 has the same number of reagents on the left and right hand sides. Note that r_5 has no reagent on the left hand side; we might use a reaction like this

$$A + B \to^{r_1} C$$
$$C \quad \to^{r_2} A + B$$
$$B \quad \to^{r_3} D$$
$$D + E \to^{r_4} B$$
$$\to^{r_5} E$$

Fig. 1. Example pathway in chemical notation

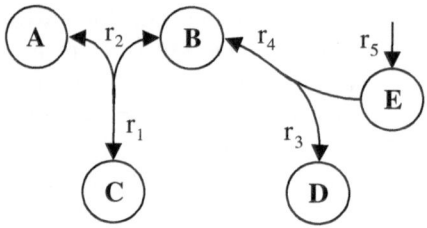

Fig. 2. Example pathway

to indicate that E is plentiful, or that it is produced by another pathway that is irrelevant to this abstraction. We will find it useful to refer to the *degree* of a chemical reaction, meaning the number of reactants that it has i.e. the number of reagents on the left hand side.

We have not included a homeo-reaction [Car08], where the components on the left hand side are identical, as it is only relevant to distinguish this case when rates are determined. In the example pathway, we assume initial concentrations of A, B and E, unless stated otherwise.

4   PEPA Models

4.1   Reagent-Centric Style

In the reagent-centric view, first proposed in [CGH06], species concentrations are discretised into *levels*; the granularity of the system is determined by the number of levels n and the concentration *step size* h, where there is a given maximum concentration max, $h = max/n$. As the number of levels increases/step size decreases, the granularity of the model increases.

For each species, there is a family of processes, each defining the behaviour for that (abstraction of) concentration. The system is defined by the parallel composition of a number of initial components.

The simplest abstraction is obtained when the number of levels is two, so that for each species there are two processes, denoting behaviour in the *presence* and *absence* of that species, respectively. We often refer to this kind of model as the *high/low model*. For example, for species A, A_H denotes presence and A_L denotes absence (alternatively A_1 and A_0, respectively). Figure 3 gives the PEPA high/low model for the example pathway, consisting of a set of equations and a system definition. Figure 4 illustrates the state space for this model.

$$A_H \stackrel{def}{=} r1.A_L$$
$$A_L \stackrel{def}{=} r2.A_H$$
$$B_H \stackrel{def}{=} r1.B_L + r3.B_L$$
$$B_L \stackrel{def}{=} r2.B_H + r4.B_H$$
$$C_H \stackrel{def}{=} r2.C_L$$
$$C_L \stackrel{def}{=} r1.C_H$$

$$D_H \stackrel{def}{=} r4.D_L$$
$$D_L \stackrel{def}{=} r3.D_H$$
$$E_H \stackrel{def}{=} r4.E_L$$
$$E_L \stackrel{def}{=} r5.E_H$$

$$System \stackrel{def}{=} A_H \bowtie_* B_H \bowtie_* C_L \bowtie_* D_L \bowtie_* E_H$$

Fig. 3. Example pathway: PEPA reagent-centric high/low model

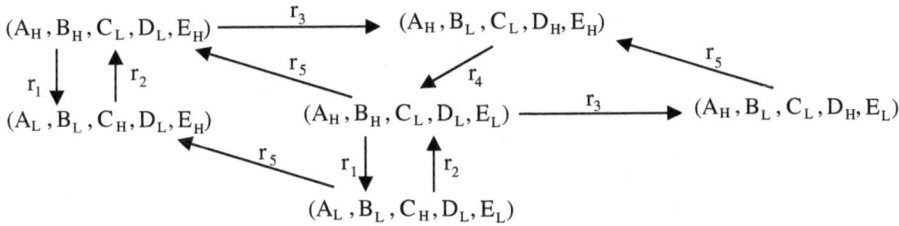

Fig. 4. State space of the PEPA high/low model. Note that we use $(A_X, B_X, C_X, D_X, E_X)$ to denote the state since the number of components is fixed and the synchronisation structure does not change.

$$A_0 \stackrel{def}{=} r2.A_1$$
$$A_1 \stackrel{def}{=} r1.A_0 + r2.A_2$$
$$A_2 \stackrel{def}{=} r1.A_1$$
$$B_0 \stackrel{def}{=} r2.B_1 + r4.B_1$$
$$B_1 \stackrel{def}{=} r1.B_0 + r3.B_0 + r2.B_2 + r4.B_2$$
$$B_2 \stackrel{def}{=} r1.B_1 + r3.B_1$$
$$C_0 \stackrel{def}{=} r1.C_1$$
$$C_1 \stackrel{def}{=} r2.C_0 + r1.C_2$$
$$C_2 \stackrel{def}{=} r2.C_1$$

$$D_0 \stackrel{def}{=} r3.D_1$$
$$D_1 \stackrel{def}{=} r4.D_0 + r3.D_2$$
$$D_2 \stackrel{def}{=} r4.D_1$$
$$E_0 \stackrel{def}{=} r5.E_1$$
$$E_1 \stackrel{def}{=} r4.E_0 + r5.E_2$$
$$E_2 \stackrel{def}{=} r4.E_1$$

$$System \stackrel{def}{=} A_2 \bowtie_* B_2 \bowtie_* C_0 \bowtie_* D_0 \bowtie_* E_2$$

Fig. 5. Example pathway: PEPA reagent-centric model with $n = 3$

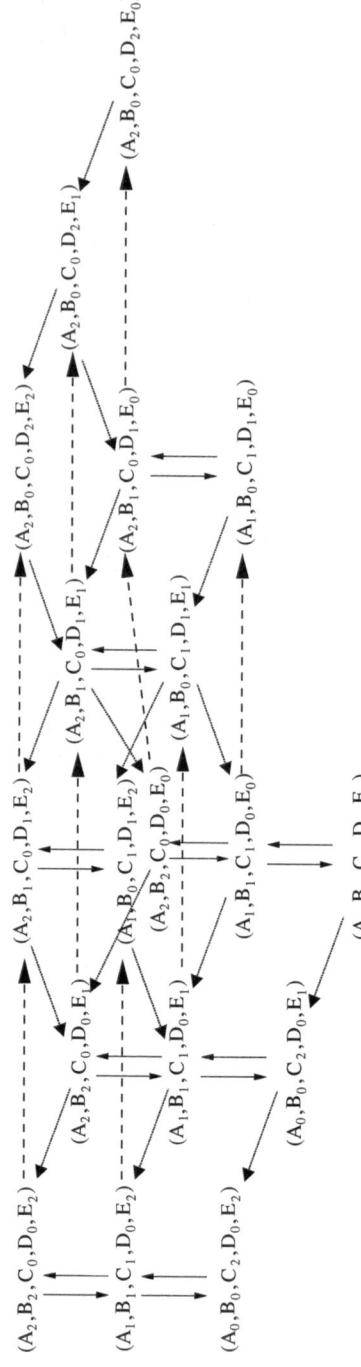

Fig. 6. State space of the PEPA reagent-centric model with $n = 3$. To avoid clutter in the diagram reaction labels are omitted, but r_1 and r_2 are shown in solid lines, r_3 in dashed lines and r_4 and r_5 in dotted lines.

As an example of a model with a different granularity, Figure 5 contains a reagent-centric model with $n = 3$ (i.e. levels 0, 1, and 2). The state space is in Figure 6. Note that regardless of the number of levels, the number of (system) components is constant during system evolution, i.e. there are always five components (the number of species).

Process as molecule in reagent-centric style. The granularity of the reagent-centric style depends on the step size h. In the limit, the finest grained model has a step size of one molecule. In general, it is impractical to increase n to its corresponding limit, but one alternative is to take a reagent-centric model with $n = 1$ and interpret each process as denoting the presence or absence of a *molecule*. An approach based on this abstraction has been used for studying the FGF pathway using stochastic model checking in [HKNT06]. For our example, for species A, A_H denotes presence of a molecule and A_L denotes absence. So, the population based high/low model model in Figure 3 can also be interpreted as an individuals model, with at most one molecule for each species. Similarly, a model consisting of (at most) two molecules for each species, is given by replacing the system definition of Figure 3 by the system definition:

$$(A_H \parallel A_H) \underset{*}{\bowtie} (B_H \parallel B_H) \underset{*}{\bowtie} (C_L \parallel C_L) \underset{*}{\bowtie} (D_L \parallel D_L) \underset{*}{\bowtie} (E_H \parallel E_H).$$

Figure 7 illustrates a small portion of the corresponding state space (one transition step). Notice that this system describes the possible evolution of *every* molecule: it is very fine grained. For example, from the initial state there are 8 possible transitions for reaction r_1, because there are two possible molecules of A that can be consumed, two possible molecules of B that can be consumed, and two possible molecules of C that can be produced (2^3 combinations). Similarly, there are 4 possibilities for reaction r_3.

In many cases this degree of granularity is inappropriate. By appealing to symmetry (i.e. composition is commutative), we can use a form of counter abstraction to represent the molecules $\underbrace{A_H \parallel \ldots \parallel A_H}_{n}$ by A_n, $\underbrace{A_H \parallel \ldots \parallel A_H}_{n-1} \parallel A_L$

by A_{n-1}, and so on. This counter abstraction involves identifying an *equivalence class* of states in a high/low model of m molecules, with a state in a model n levels, where $n = m$. In other words, we define the processes as in the high/low

$$(A_H, A_H, B_H, B_H, C_L, C_L, D_L, D_L, E_H, E_H) \xrightarrow{r_3} (A_H, A_H, B_L, B_H, C_L, C_L, D_H, D_L, E_H, E_H)$$

$$r_1 \downarrow \quad \uparrow r_2 \qquad \vdots$$

$$(A_L, A_H, B_L, B_H, C_H, C_L, D_L, D_L, E_H, E_H)$$

$$\vdots$$

Fig. 7. One transition step in PEPA reagent-centric process-as-molecule model with two molecules

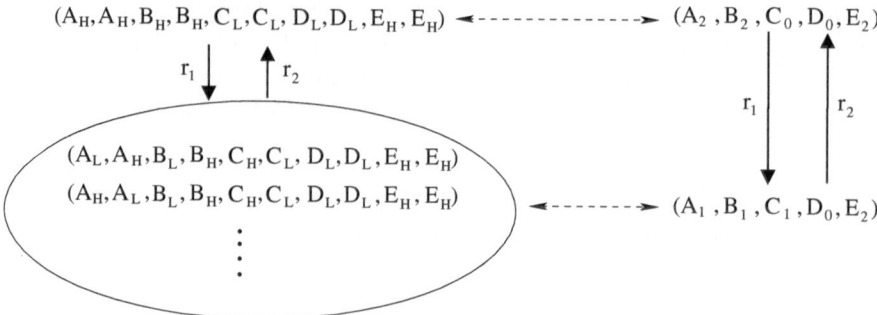

Fig. 8. One transition step in the state space of the PEPA counter abstraction model

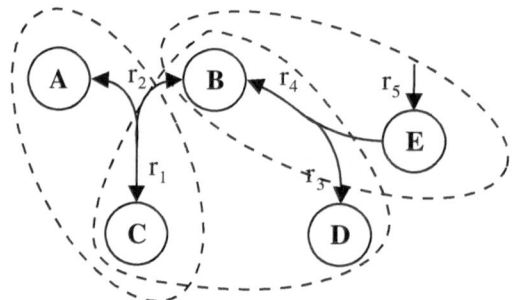

Fig. 9. Example set of reactions with pathways indicated

model of Figure 3, then compose multiple copies of each process and interpret A_n as representing n molecules. This is illustrated in Figure 8, for the example pathway with two molecules for each species. States in the fine-grained individuals model are quotiented and dashed lines indicate how the quotient class relates to a state in the counter abstraction model.

4.2 Pathway-Centric Style

An alternative style of modelling that has been proposed in PEPA is the pathway-centric style. In this style, we specify the sub-pathways that consume and replenish the *initial species*, which are the species with significant initial concentrations. In the example pathway, this involves defining the sub-pathways starting from A, B, and E. Call these $Path_1$, $Path_2$ and $Path_3$, respectively. The example pathway is given in Figure 10, with corresponding state space in Figure 11.

Notice that although the system definition has only 3 components, this space is isomorphic to the high/low reagent-centric model (Figure 4). Notice also implicitly, the model has two levels. For example, $Path_1$ denotes high concentration of both A and B. We could make levels explicit in this style, by composing

$$Path_1 \stackrel{def}{=} r_1.r_2.Path_1$$
$$Path_2 \stackrel{def}{=} r_1.r_2.Path_2 + r_3.r_4.Path_2$$
$$Path_3 \stackrel{def}{=} r_4.r_5.Path_3$$

$$System \stackrel{def}{=} Path_1 \underset{*}{\bowtie} Path_2 \underset{*}{\bowtie} Path_3$$

Fig. 10. Example pathway: PEPA pathway-centric model

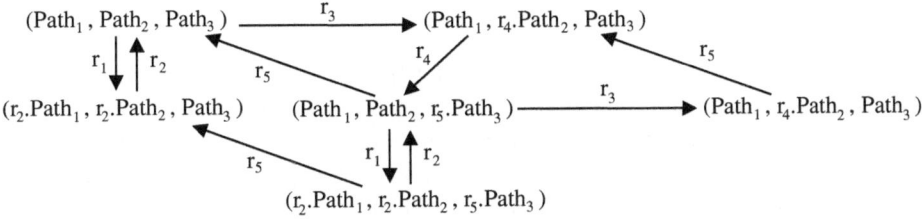

Fig. 11. Pathway-centric model state space

multiple copies of each pathway (with parallel composition, no synchronisation). For example the three level model would be:

$$(Path_1 \parallel Path_1) \underset{*}{\bowtie} (Path_2 \parallel Path_2) \underset{*}{\bowtie} (Path_3 \parallel Path_3)$$

In this case, similarly to the individuals reagent-centric style, there are more potential interleavings than in the reagent-centric population-based representation, and so the explicit state space here will be larger. However, again, by appealing to symmetry, we can work at the aggregate level. Thus for a given number of levels, the state space size and structure of both the pathway-centric and the reagent-centric models should be the same, as established in [CGH06]. Note that tools like the PEPA workbench [TDG09] can automatically detect such symmetries. We observe that assuming chemical reactions of at most degree two, we only require binary synchronisation, for this style of model.

5   Bio-PEPA

The Bio-PEPA formulation [CH08] of the reagent-centric style for the example pathway is given in Figure 12. This example does not fully exploit the power of Bio-PEPA, since the stochiometric coefficients are all simple (1) and the functional rates are omitted. However, it does illustrate how the language focuses on the role of each species, in each reaction. Initial concentrations are denoted A_0 for species A, etc. An integral part of a Bio-PEPA specification (omitted here) is a definition of h and n, for every species, as well as initial concentrations (expressed as levels).

The state space of this model depends upon the levels, for example, if the number of levels is uniformly 2, then the state space is the same as Figure 6.

$$A \stackrel{def}{=} (r1,1){\downarrow}A + (r2,1){\uparrow}A$$
$$B \stackrel{def}{=} (r1,1){\downarrow}B + (r2,1){\uparrow}B + (r3,1){\downarrow}B$$
$$C \stackrel{def}{=} (r2,1){\uparrow}C$$
$$D \stackrel{def}{=} (r4,1){\downarrow}D + (r3,1){\uparrow}D$$
$$E \stackrel{def}{=} (r4,1){\downarrow}E + (r5,1){\uparrow}E$$

$$System \stackrel{def}{=} A(A_0) \underset{*}{\bowtie} B(B_0) \underset{*}{\bowtie} C(C_0) \underset{*}{\bowtie} D(D_0) \underset{*}{\bowtie} E(E_0)$$

Fig. 12. Example pathway: Bio-PEPA model

Note that in the corresponding PEPA model (i.e. Figure 5), the number of levels is "hardwired" into the equations, whereas in the Bio-PEPA model, it is given as a parameter offering more flexibility to the modeller. If the number of levels is set sufficiently high the model has a state space corresponding to an individuals model (i.e. if n is chosen to be the number of molecules).

6 π-Calculus

Models in the π-calculus and its stochastic variants predominantly follow the reactant style (e.g. [TK08]), based on the molecules-as-processes abstraction. Thus these are individuals based models. Figure 13 gives the π-calculus model in this style for the example pathway; since each reagent in the example also occurs on the left hand side of a chemical equation, there are processes for $A \dots E$.

The example pathway highlights an interesting aspect of this style because in the biochemistry there are

1. equations with a decreasing number of components, and
2. an equation with no left hand side.

Consider the former case. Since synchronisations are between reagents on the left hand side of an equation only, there is an arbitrary (and inconsequential) choice between which component is output and which is input. Further, the components on the left hand side, when translated into processes, evolve into components on the right hand side. If the number of components decreases, then we have to nominate one or more to evolve to 0, the null process. For example, $A + B \rightarrow^r C$ could map to $A = r.C$ and $B = \bar{r}.0$; equally, it could map to $A = \bar{r}.0$ and $B = r.C$, or $A = r.0$ and $B = \bar{r}.C$, etc. Taking the first choice, $A \mid B$ evolves to $C \mid 0$. This is an example of a "trailing 0", which is removed through application of the *syntactic* equality $P \mid 0 \equiv P$, i.e. $A \mid B$ evolves to C.

Now consider the second case. We cannot model an equation without a left hand side explicitly, e.g. r_5, but since E is present initially, we could represent the infinite supply of E by a τ event, after offering the output event $\bar{r_4}$. However, this would constrain the creation of E to occur only after a molecule has been

$$A = r_1.C$$
$$B = \overline{r_1}.0 + \tau.D$$
$$C = \tau.A \mid B$$
$$D = r_4.\, B$$
$$E = \overline{r_4}.0$$
$$Env = \tau.Env \mid E$$

$$System = A \mid B \mid E \mid Env$$

Fig. 13. Example pathway: π-calculus model

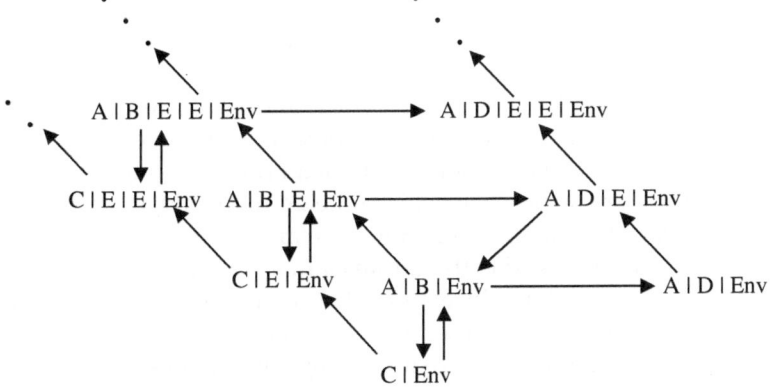

Fig. 14. π-calculus model state space

consumed in the reaction r_4. An alternative, which we use, is to introduce a representation of the environment Env and define it as follows:

$$Env = \tau.Env \mid E$$

This presents the possibility that an unbounded number of E molecules may be introduced into the system, which is true when we represent the system only qualitatively. In the biological reality and when quantitative information is included in the model in the form of rates the system will become *pragmatically bounded* meaning that the probability for E to grow unboundedly is extremely small.

Figure 14 illustrates possible evolutions for the system with one molecule of A, B and E initially, i.e. the evolution of $A \mid B \mid E \mid Env$. We have not labelled the transitions since events are either unobservable or become so after synchronisation. Notice that in this state space the number of system components fluctuates, it both increases and decreases. Moreover the state space is infinite due to the potentially unbounded number of E, although a graph isomorphic to the state space of the pathway-centric model is embedded within it. An alternative interpretation of this model is therefore a fine-grained pathway-centric view based on molecules. Or rather, it is a mixture of two styles: equations are defined for each reagent, but the system definition has the form of a pathway-centric model.

While this approach provides a faithful overall system model, it is not compositional. Specifically, one equation incorporates aspects of the initial system and it would be misleading to a reader who inspected the behaviour only of a process that arbitrarily terminates, e.g. B, which can evolve into 0. Moreover, some reactions are represented explicitly by named events, i.e. r_1 and r_4, whereas the unary or nullary reactions r_2, r_3 and r_5 are represented by the τ event. Thus, there are no occurrences of the reaction names r_2, r_3 and r_5 in the model.

7 Beta-Binders

There are several ways to map a chemical reaction in this formalism. For example, we could define a mapping very similar to the π-calculus mapping, with boxes for the processes that are initial, i.e. the system is given by $[A] \parallel [B] \parallel [E]$, with suitable beta-binders defined for each box, and each encapsulated process is defined as in Figure 13. The authors recommend this mapping when the reaction denotes a collision of entities, the collision being mapped to (inter-box) communication. However, if we use this mapping, we are left with boxes containing the π-calculus constant process (i.e. 0) and we cannot remove them by the structural congruences: we need to introduce an explicit *delete* event to remove them.

Alternatively, instead of representing reactions by inter-box communication, we could represent reactions by events, i.e. by the box operations. In this case, a reaction such as $A + B \to^r C$ maps to (A, B) *join* C, where A, B, and C are constant bio-processes. Figure 15 gives a Beta-binders model of the example pathway using events. Notice that there are four events and no communication: the encapsulated processes are constants, except for process B, which changes its interaction type (to that of D). The state space is given in Figure 16; the space is isomorphic to the π-calculus model, though we could bound the occurrences of *new E* with a condition. The model is also a mixture of styles: equations are defined for each reagent, but it is not reagent-centric: there is no communication and the system definition has the form of a reaction-centric model.

There is a third possible mapping when the reaction denotes a binding (e.g. ligand to receptor); this is usually written in chemical notation as: $A + B \to^r [A + B]$. In this case we could we use the complex/decomplex beta-binder operations to create and delete dedicated communication channels between boxes $[A]$ and

$$
\begin{array}{ll}
(A, B) \; join \; C & \textbf{where} \quad A = \beta(x, \Gamma_A) \, [nil] \\
C \; split \; (A, B) & \qquad\qquad B = \beta(x, \Gamma_B) \, [chtype(x, \Gamma_D).\, nil] \\
(D, E) \; join \; B & \qquad\qquad C = \beta(x, \Gamma_C) \, [nil] \\
new \; E & \qquad\qquad D = \beta(x, \Gamma_D) \, [nil] \\
& \qquad\qquad E = \beta(x, \Gamma_E) \, [nil]
\end{array}
$$

Fig. 15. Example pathway in Beta-binders

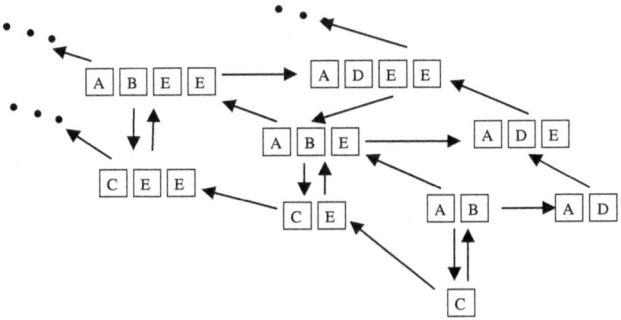

Fig. 16. State space of Beta-binders model. Note that following the graphical notation for Beta-binders, we omit the parallel composition operation ‖ on bio-processes.

$[B]$. That is, the two boxes $[A]$ and $[B]$ would evolve into a complex of two boxes, instead of into two separate boxes.

8   sCCP

Our last example is a model in sCCP. This is shown in Figure 17. There are five processes: one for each reaction, with stream variables representing the species. Each process has the form **ask** (check that there is sufficient of a species) followed by the parallel composition of all the possible effects of the reaction (i.e. production or consumption) expressed by **tell**. The state space of the model is shown in Figure 18. Unsurprisingly this includes the state space which has been retrieved from the other models, such as the reagent-centric PEPA models (shown in the shaded area in the diagram). However note that this model also permits the unbounded growth of the population of E (as in the π-calculus model), leading to an infinite state space unless an explicit guard is inserted which disables reaction r_5 when the population of E reaches a given size.

This model bears some similarity to the state based PRISM model given in [CVOG06], where species are represented by state variables. This is not surprising, since the PRISM modelling language is essentially the language of reactive modules [AH90]. However, in [CVOG06], there is still explicit synchronisation and commands are grouped by species, rather than by reaction. The reactions-as-processes models of sCCP can therefore be considered to be *reaction-centric* and in that they are similar to other rule-based formalisms such as the κ-calculus [VFF+07] and BIOCHAM [CRCD+04].

9   Discussion

The three main abstractions for mapping chemical equations to process algebras are *molecule-as-process*, *species-as-process*, and *reaction-as-process*. We have

reaction(r_1, [A, B], [C]) : −
    ask(A > 0 ∧ B > 0). (tell(A $= A − 1) ∥ tell(B $= B − 1) ∥ tell(C $= C + 1))

reaction(r_2, [C], [A, B]) : −
    ask(C > 0). (tell(C $= C − 1) ∥ tell(A $= A + 1) ∥ tell(B $= B + 1))

reaction(r_3, [B], [D]) : −
    ask(B > 0). (tell(B $= B − 1) ∥ tell(D $= D + 1))

reaction(r_4, [D, E], [B]) : −
    ask(D > 0 ∧ E > 0). (tell(D $= D − 1) ∥ tell(E $= E − 1) ∥ tell(B $= B + 1))

reaction(r_5, [], [E]) : −
    (tell(E $= E + 1))

5_reaction_system : −
    reaction(r_1, [A, B], [C]) ∥ reaction(r_2, [C], [A, B]) ∥ reaction(r_3, [B], [D])
    ∥ reaction(r_4, [D, E], [B]) ∥ reaction(r_5, [], [E])

Fig. 17. Example pathway: sCCP model

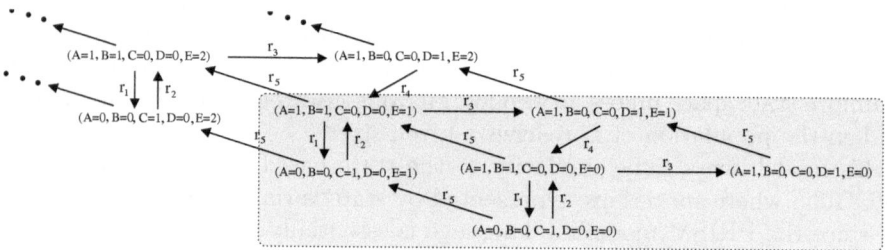

Fig. 18. State space of the sCCP model of the example

further defined four styles: *reagent-centric, pathway-centric, reactant-centric*, and *reaction-centric*. We have presented reactant-centric π-calculus and Beta-binders models, and (individuals-based) reagent-centric PEPA models as examples of the the molecule-as-process abstraction, (population-based) pathway-centric PEPA and (population-based) reagent-centric Bio-PEPA models as examples of the species-as-process abstraction, and a reaction-centric sCCP model as an example of the reaction-as-process abstraction.

The styles of modelling supported by a process algebra is strongly influenced by the form of synchronisation available. Whilst languages with multiway synchronisation are capable of representing models in reagent-centric, reactant-centric or pathway-centric style, the same is not true for languages with conjugate actions and binary synchronisation. These languages cannot generally represent reactions in the reagent-centric style. Only first degree, or unary, reactions could be modelled in this style in these languages.

In the example considered we have only considered reactions with degree one and two — indeed there are thermodynamic arguments for restricting consideration to such reactions if we wish to be faithful to biochemistry. However, abstractions which lead to higher degree reactions are often applied by biologists for a variety of reasons. For example, consider the enzyme-enabled association of two smaller molecules (A and B) into a complex C. In terms of elementary reactions this might proceed as follows:

$$A + B + E \;_{r_2}\!\longleftrightarrow^{r_1} A + B{:}E \longrightarrow^{r_3} C + E$$

where E is the enzyme and $B{:}E$ is a complex formed from B and the enzyme. This could abstracted as $A + B \overset{E}{\longrightarrow}^{r} C$. The abstraction has the advantage that the number of reagents considered in the transformation is reduced, and that the number of reaction rates which have to be measured, estimated or fitted is cut from three to one. Moreover this is typically more consistent with what can be observed in the lab as $r_1, r_2 \gg r_3$. It may not even be known whether the enzyme binds with A or B, leaving uncertainty about how to model the reaction without the abstraction. However, representing this in even the reactant-centric style requires three-way synchronisation, and four-way synchronisation in the reagent-centric style, assuming that the enzyme is modelled as both a reactant and a product in the abstracted reaction. In other words, it is not possible to support modelling such biological abstractions using strictly binary synchronisations.

As with reagent-centric style, reaction-centric style seems to implicitly assume a multi-way synchronisation. However note that in the way that this style is captured in sCCP, the only process algebra that currently supports reaction-centric modelling to the best of our knowledge, the requirement is not so strong. What is needed is atomic multi-way composition of updates to the constraint store, but this is not necessarily a synchronisation. Whilst sCCP is the only process algebra supporting reaction-centric, or reaction-as-process, modelling, conversely it is difficult to see sCCP being used to construct models in any of the other styles or abstractions.

In process algebras with conjugate actions, each partner in an action/reaction must be assigned an input/output role. In general this will be rather arbitrary and somewhat artificial from the perspective of the biochemistry. Consider the reaction r_1 in our example. When A and B form the complex C there does not appear to be a natural way to choose which of A and B should receive input and which provide output. Furthermore, reactions of degree one, such as r_5 in the example, must be represented as a τ action. This means that the textual representation of the model does not clearly articulate the biologists' notion of the system. This problem becomes even worse at the level of the state space where all transitions are labelled τ and information about the reactions that gave rise to them is lost.

If we consider the contrast between population-based and individuals-based modelling we can observe that population-based modelling is more compact both from the point of view of the textual model expression and the underlying state space. This means that for such models it can be feasible to use explicit state space representations and the analysis techniques associated with them such as model checking, equivalence checking and numerical analysis of the continuous time Markov chain. Of course, such techniques reply on the state space being finite. In contrast individuals-based modelling has a clear association with stochastic simulation as proposed by Gillespie [GP06]. These models can be used in association with explicit state space techniques, such as those listed above, but only for very small systems or in combination with abstractions such as the assumption of single molecules, as discussed in Section 4.1.

In the PEPA and Bio-PEPA models, as a consequence of the two level grammar used to define these languages as compositions of sequential components, the number of system components is constant, regardless of whether individuals-based or population-based. This matches the species-as-process abstraction since the *possible* species of the pathway will be known and fixed and is particularly natural in the population-based modelling where the state of the system is a count for each species. In contrast, in the π-calculus and Beta-binder models, which are without the syntactic restriction, the number of system components fluctuates throughout system evolution. This is in keeping with the molecules-as-processes abstraction since we would expect the visible molecules within a system to change as complexes are formed and dissociated etc. In sCCP, based on the reaction-as-process abstraction, the number of species is fixed as the variables in the variable store remain fixed. Here as in the PEPA/Bio-PEPA population-based modelling the state of the system is captured in terms of the number of each species so each species must always be present, even if to record that its current count is zero.

The conservative nature of the PEPA/Bio-PEPA models (in terms of number of components, and fixed number of levels) also means that the state space underlying such models is necessarily finite. This is not the case in the other process algebras as we have seen. It can be argued that if we consider the example as presented there is the potential for unbounded numbers of E via reaction r_5 and π-calculus, beta binder and sCCP correctly capture this. But on the other hand, in a biological system unbounded growth like this will lead to cell death, and when we introduced the example we explained that this reaction would be used as an abstraction of some more complex, but bounded, situation. The beta binders and sCCP formalisms do offer language mechanisms which allow the number of E to remain bounded by introducing guards on the reaction, but there is no such possibility in the π-calculus.

In this paper we have focussed on the standard discrete state spaces. However analysis based on these state spaces is rarely feasible. Therefore for all the languages there are alternative semantics given by ordinary differential equations (population-based) and/or Gillespie simulations (individuals-based). The discrete state space does of course form the basis of the Gillespie simulation but it is never considered explicitly and the offered semantics avoid the construction. Additionally, PEPA and Bio-PEPA support an alternative representation, which is based on an explicit discrete state space but seeks to avoid the state space explosion. Rather than states representing the count of molecules of each species, the states represent the current *level* of concentration for each species. In other words, the range of possible concentration values is discretised into intervals, and these intervals constitute the states of the CTMC. In such models the stochastic element of Gillespie's approach is retained but the resulting CTMCs can be considerably smaller. Keeping the state space manageable means that the CTMCs can be solved explicitly and the repeated runs necessitated by stochastic simulation are avoided. Further, in addition to quantitative analysis on the CTMC, analysis by model checking of stochastic properties is possible, as illustrated in [CVOG06] or [HKNT06].

10   Conclusions

As highlighted by Regev and Shapiro computational abstractions have already brought considerable benefit to the study of biological phenomena [RS01]. For example the *DNA-as-string* abstraction has been hugely successful and allowed significant leaps forward. In the context of biomolecular processes the potential benefit seems equally large. However further work is needed to assess the abstractions that are on offer, and their suitability to the systems under study. Research in this direction has been enthusiastically taken up by theoretical computer scientists as witnessed by the plethora of formal languages currently proposed for modelling such systems. In this paper we have aimed to extract the general paradigms of expression which underlie process algebras which aim to model biomolecular processes. We have discovered that there are genuine differences in the form of expression used, and this can impact on the form of analysis that is readily applied.

In the long term all research on formal description techniques for biomolecular systems has the objective of attracting biological users, and contributing to the growing body of knowledge on how cells function. However in the medium term we need to develop closer links with biologists, not only as users of our formal description techniques, but also in the important work of evaluating them.

References

[AH90]   Alur, R., Henzinger, T.A.: Reactive modules. Formal methods in System Design 15(1), 7–48 (1990)

[BMMT06]   Barbuti, R., Maggiolo-Schettini, A., Milazzo, P., Troina, A.: A Calculus of Looping Sequences for Modelling Microbiological Systems. Fundamenta Informaticae 72(1-3), 21–35 (2006)

[BJG96]   Brim, L., Jacquet, J.-M., Gilbert, D.: A process algebra for synchronous concurrent programming. In: Hanus, M., Rodríguez-Artalejo, M. (eds.) ALP 1996. LNCS, vol. 1139, pp. 165–178. Springer, Heidelberg (1996)

[Bor06]   Bortolussi, L.: Stochastic concurrent constraint programming. In: Proceedings of QAPL 2006: 4th International workshop on quantitative aspects of programming languages, vol. 164, pp. 65–80 (2006)

[BP08]   Bortolussi, L., Policriti, A.: Modelling biological systems in stochastic constraint programming. Constraints 13, 66–90 (2008)

[CGH06]   Calder, M., Gilmore, S., Hillston, J.: Modelling the influence of RKIP on the ERK signalling pathway using the stochastic process algebra PEPA. In: Priami, C., Ingólfsdóttir, A., Mishra, B., Riis Nielson, H. (eds.) Transactions on Computational Systems Biology VII. LNCS (LNBI), vol. 4230, pp. 1–23. Springer, Heidelberg (2006)

[Car04]   Cardelli, L.: Brane Calculus. In: Danos, V., Schachter, V. (eds.) CMSB 2004. LNCS (LNBI), vol. 3082, pp. 257–278. Springer, Heidelberg (2005)

[Car08]   Cardelli, L.: On process rate semantics. Theoretical Computer Science 391(1), 190–215 (2008)

[CPR+04]   Cardelli, L., Panina, E.M., Regev, A., Shapiro, E., Silverman, W.: BioAmbients: An Abstraction for Biological Compartments. Theoretical Computer Science 325(1), 141–167 (2004)

[CG09]   Ciocchetta, F., Guerriero, M.L.: Modelling Biological Compartments in Bio-PEPA. ENTCS 227, 77–95 (2009)

[CH08]   Ciochetta, F., Hillston, J.: Bio-PEPA: a framework for modelling and analysis of biological systems. Theoretical Computer Science (to appear)

[CRCD+04]   Chabrier-Rivier, N., Chiaverini, M., Danos, V., Fages, F., Schächter, V.: Modeling and querying biomolecular interaction networks. Theoretical Computer Science 325(1), 25–44 (2004)

[CVOG06]   Calder, M., Vyshemirsky, V., Orton, R., Gilbert, D.: Analysis of signalling pathways using Continuous Time Markov Chains. In: Priami, C., Plotkin, G. (eds.) Transactions on Computational Systems Biology VI. LNCS (LNBI), vol. 4220, pp. 44–67. Springer, Heidelberg (2006)

[VFF+07]   Danos, V., Feret, J., Fontana, W., Harmer, R., Krivine, J.: Rule-based modelling of cellular signalling. In: Caires, L., Vasconcelos, V.T. (eds.) CONCUR 2007. LNCS, vol. 4703, pp. 17–41. Springer, Heidelberg (2007)

[DPPQ06]   Degano, P., Prandi, D., Priami, C., Quaglia, P.: Beta-binders for biological quantitative experiments. Electronic Notes in Computer Science 164, 101–117 (2006)

[GP06]   Gillespie, D., Petzold, L.: Numerical Simulation for Biochemical Kinetics. In: System Modelling in Cellular Biology. MIT Press, Cambridge (2006)

[HKNT06]   Heath, J., Kwiatkowska, M., Norman, G., Parker, D., Tymchyshyn, O.: Probabilistic model checking of complex biological pathways. In: The Proceedings of 4th International Workshop on Computational Methods in Systems Biology 2006, Trento, Italy, October 18-19 (2006)

[Hil96]   Hillston, J.: A Compositional Approach to Performance Modelling. Cambridge University Press, Cambridge (1996)

[Hoa85]   Hoare, C.A.R.: Communicating Sequential Processes. Prentice-Hall, Englewood Cliffs (1985)

[Mil80]   Milner, R.: A Calculus for Communicating Systems. LNCS, vol. 92. Springer, Heidelberg (1980)

[Mil99]   Milner, R.: Communicating and Mobile Systems: the π-Calculus. Cambridge University Press, Cambridge (1999)

[Pri95]   Priami, C.: Stochastic π-calculus. The Computer Journal 38, 578–589 (1995)

[RS01]   Regev, A., Shapiro, E.: Cellular abstractions: cells as computation. Nature 419, 343 (2001)

[RSS01]   Regev, A., Silverman, W., Shapiro, E.: Representation and simulation of biochemical processes using π-calculus process algebra. In: Pacific Symposium on Biocomputing 2001 (PSB 2001), pp. 459–470 (2001)

[TK08]   Tymchyshyn, O., Kwiatkowska, M.: Combining intra- and inter-cellular dynamics to investigate intestinal homeostasis. In: Fisher, J. (ed.) FMSB 2008. LNCS (LNBI), vol. 5054, pp. 63–76. Springer, Heidelberg (2008)

[TDG09]   Tribastone, M., Duguid, A., Gilmore, S.: The PEPA Eclipse Plug-in. Performance Evaluation Review 36(4), 28–33 (2009)

[V07]   Versari, C.: A Core Calculus for a Comparative Analysis of Bio-inspired Calculi. In: De Nicola, R. (ed.) ESOP 2007. LNCS, vol. 4421, pp. 411–425. Springer, Heidelberg (2007)

Simple, Enhanced and Mutual Mobile Membranes

Bogdan Aman and Gabriel Ciobanu

Romanian Academy, Institute of Computer Science, Iaşi, Romania
A.I.Cuza University, 700506 Iaşi, Romania
baman@iit.tuiasi.ro, gabriel@info.uaic.ro

Abstract. The operations governing the movement of biological membranes are endocytosis and exocytosis. New models of computation are inspired by these biological operations. In this paper we present the models defined by simple, enhanced and mutual mobile membranes, together with their biological motivations. Some results concerning their computational power are presented, including the first universality result for mutual mobile membranes. In the case of simple and enhanced mobile membranes, we improve the existing results by reducing the number of membranes needed to get computational universality.

1 Introduction

Simple, enhanced and mutual mobile membranes represent new variants of membrane systems. Membrane systems (also called P systems) were introduced in [16]; standard P systems and several variations are presented in the monograph [17]. Membrane systems were introduced as distributed, parallel and nondeterministic computing models inspired by the compartments of eukaryotic cells and by their biochemical reactions. The cellular components are formally represented in the definition of membrane systems. The structure of the cell is represented by a set of hierarchically embedded regions, each one delimited by a surrounding boundary (called membrane), and all of them contained inside an external special region called the skin membrane. The molecular species (ions, proteins, etc.) floating inside cellular compartments are represented by multisets of objects described by means of symbols or strings over a given alphabet, objects which can be modified or communicated between adjacent compartments. Chemical reactions are represented by evolution rules given in the form of rewriting rules which operate on the objects, as well as on the compartmentalized structure (by dissolving, dividing, creating, or moving membranes).

A membrane system can perform computations in the following way: starting from an initial configuration which is defined by the multiset of objects initially placed inside the compartmentalized structure, the system evolves by applying the evolution rules of each membrane in a nondeterministic and maximally parallel manner. A rule is applicable when all the objects that appear in its left hand side are available in the region where the rule is placed. The maximal

C. Priami et al. (Eds.): Trans. on Comput. Syst. Biol. XI, LNBI 5750, pp. 26–44, 2009.
© Springer-Verlag Berlin Heidelberg 2009

parallelism of rule application means that every rule that is applicable inside a region *has to* be applied in that region. A halting configuration is reached when no rule is applicable. The result is represented by the number of objects from a specified region.

Several variants of membrane systems are inspired by different aspects of living cells (symport and antiport-based communication through membranes, catalytic objects, membrane charge, etc.). Their computing power and efficiency have been investigated using the approaches of formal languages and grammars, register machines and complexity theory. An updated bibliography can be found at the webpage http://ppage.psystems.eu

A first definition of mobile P systems is given in [21] with rules coming from mobile ambients [5]. Inspired by the operations of endocytosis and exocytosis, namely moving a membrane inside a neighbouring membrane (endocytosis) and moving a membrane outside the membrane where it is placed (exocytosis), the P systems with mobile membranes are introduced in [14] as a variant of P systems with active membranes [17]. We use *simple mobile membranes* instead of P systems with mobile membranes. The computational power of simple mobile membranes is treated in [12,14]: Turing completeness is obtained by using nine membranes together with the operations of endocytosis and exocytosis [14], while only four mobile membranes are enough using additional contextual evolution rules [12]. In this paper we look at certain biological phenomena which motivate and inspire new specific rules in simple mobile membranes.

Endocytosis is a general term for a group of processes that bring macro-molecules, large particles, small molecules, and even small cells into another cell. There are three types of endocytosis: *phagocytosis* ("cellular eating"), *pinocytosis* ("cellular drinking"), and *receptor-mediated endocytosis* in which the membrane infolds around materials from the environment, forming a small pocket. The pocket deepens, forming a vesicle which separates from the membrane and migrates with its contents to the cell's interior.

While *pinocytosis* can be modelled using communication rules of usual P systems, there is no rule capable to model the process of engulfing a cell by another one in phagocytosis. This is the reason why we define the enhanced mobile membranes in Subsection 2.2; an example on how the new rules work is also presented.

The *enhanced mobile membranes* represent a variant of simple mobile membranes; they have been proposed in [3] for describing some biological mechanisms of the immune system. The operations governing the mobility of the enhanced mobile membrane systems are endocytosis (endo), exocytosis (exo), enhanced endocytosis (fendo) and enhanced exocytosis (fexo). The computational power of the enhanced mobile membranes using these four operations was studied in [13] where it is proved that twelve membranes can provide the computational universality. It is worth noting that unlike the results for simple mobile membranes, the context-free evolution of objects is not used in proving any of these results.

Receptor-mediated endocytosis is used by animal cells to capture specific macromolecules from the cell's environment. This process depends on receptor proteins, i.e., integral membrane proteins that can bind to a specific molecule in the cell's environment. The uptake process is similar to nonspecific endocytosis. However, in receptor-mediated endocytosis, the receptor proteins at particular sites on the extracellular surface of the plasma membrane bind to specific substances. These sites are called coated pits because they form a slight depression in the plasma membrane. The cytoplasmic surface of a coated pit is coated by proteins, such as clathrin. Strengthened and stabilized by clathrin molecules, this vesicle carries the macromolecule into the cell [23].

SNARE-mediated exocytosis is the movement of materials out of a cell via vesicles. SNARES (Soluble NSF Attachment Protein Receptor)) located on the vesicles (v-SNARES) and on the target membranes (t-SNARES) interact to form a stable complex that holds the vesicle very close to the target membrane.

There is no rule capable to model the mutual agreement between membranes for the receptor-mediated endocytosis and SNARE-mediated exocytosis. This is the reason why we define the mutual mobile membranes in Subsection 2.3; an example on how the new rules work is also presented.

The *mutual mobile membranes* represent a variant of simple mobile membranes in which the endocytosis and exocytosis work whenever the involved membranes "agree" on the movement; this agreement is described by using dual objects a and $\bar{a}$ in the involved membranes. The operations governing the mobility of the mutual mobile membranes are mutual endocytosis (mutual endo), and mutual exocytosis (mutual exo).

In this paper we study the computational power of simple, enhanced and mutual mobile membranes. For simple mobile membranes we obtain the computational universality by using three membranes, and in this way improving the result presented in [12] where four membranes are used. For enhanced mobile membranes we obtain the computational universality by using nine membranes, thus improving the result from [13] where twelve membranes are used. For mutual mobile membranes we show that by using dual objects a and $\bar{a}$ in the involved membranes, only seven membranes are enough to obtain the computational universality.

The structure of the paper is as follows. In Section 2 we formally define the simple, enhanced and mutual mobile membranes, and give biological motivations for the enhanced and mutual mobile membranes. Section 3 contains a first universality result for mutual mobile membranes, and improvements of the existing results for simple and enhanced mobile membranes. Section 4 presents related results in P systems with active membrane from which the simple mobile membranes originate. Conclusions and references end the paper.

2 Mobile Membranes

In this section we define the simple, enhanced and mutual mobile membranes, describing some biological phenomena inspiring their rules.

2.1   Simple Mobile Membranes

Definition 1 ([14]). *A simple mobile membrane is a construct*
$$\Pi = (V, H, \mu, w_1, \ldots, w_n, R)$$
where:

1. $n \geq 1$ *(the initial* degree *of the system);*
2. V *is an alphabet (its elements are called* objects*);*
3. H *is a finite set of* labels *for membranes;*
4. $\mu \subset H \times H$ *describes the* membrane structure, *such that* $(i, j) \in \mu$ *denotes that the membrane labelled by j is contained in the membrane labelled by i; we distinguish the external membrane (usually called the "skin" membrane) and several internal membranes; a membrane without any other membrane inside it is said to be elementary;*
5. $w_1, \ldots, w_n$ *are strings over V, describing the* multisets of objects *placed in the n regions of μ;*
6. R *is a finite set of* developmental rules, *of the following forms:*

object evolution

(a) $[a \to v]_m$*, for $m \in H$, $a \in V$, $v \in V^*$;*
   An object a placed inside a membrane labelled m evolves into a multiset of objects v.

endocytosis

(b) $[a]_h[\ ]_m \to [[b]_h]_m$*, for $h, m \in H$, $a, b \in V$;*
   An elementary membrane labelled h enters the adjacent membrane labelled m, under the control of object a; the labels h and m remain unchanged during the process; however the object a may be modified to b during the operation; m is not necessarily an elementary membrane.

exocytosis

(c) $[[a]_h]_m \to [b]_h[\ ]_m$*, for $h, m \in H$, $a, b \in V$;*
   An elementary membrane labelled h is sent out of a membrane labelled m, under the control of object a; the labels of the two membranes remain unchanged, but the object a of membrane h may be modified during this operation; membrane m is not necessarily elementary.

The rules are applied according to the following principles:

1. Rules are applied in parallel, non-deterministically choosing the rules, the membranes, and the objects in such a way that the parallelism is maximal; this means that in each step we apply a certain set of rules such that no further rule can be added to the set.
2. The membrane m from the rules of type $(a) - (c)$ is said to be *passive*, while the membrane h is said to be *active*. In any step of a computation, any object and any active membrane can be involved in at most one rule. However, the passive membranes can be used by several rules at the same time. In a rule $[a \to v]_m$ of type (a), object a is active, while membrane m is passive.

3. When a membrane is moved across another membrane, by endocytosis or exocytosis, its whole contents (its objects) are moved; the inner objects evolve first (if rules are applicable for them), and then any membrane is moved with the contents as obtained after its internal evolution.
4. If a membrane exits the system (by exocytosis), then its internal evolution stops, even if there are rules of type (a) which could be applied.
5. The objects and membranes which do not evolve at a given step are passed unchanged to the next configuration of the system.

2.2   Enhanced Mobile Membranes

The enhanced mobile membranes have been introduced in [3] for describing some biological mechanisms of the immune system. The presentation of the immune system is taken from [10], a book which is revised every few years to keep the pace with the new discoveries in this field. The cells of the immune system work together with different proteins to seek out and destroy anything foreign or dangerous which enters our body. It takes some time for the immune cell to be activated, but once this happens there are very few hostile organisms having a chance. There are several types of immune cells, each of them with its own strength and weakness. Some seek out and engulf the invaders, while other destroy the infected or mutated body cells. A type of immune cells are the B cells which have the ability to release special proteins called antibodies which mark intruders in order to be destroyed by macrophages. The immune system has also the ability to produce some cells able to remember enemies which it fought in the past. In this way, once the immune system recognizes an invader it attacks more quickly and strongly against it.

Dendritic cells can engulf bacteria, viruses, and other cells. Once a dendritic cells engulfs a bacterium, it dissolves this bacterium and places portions of

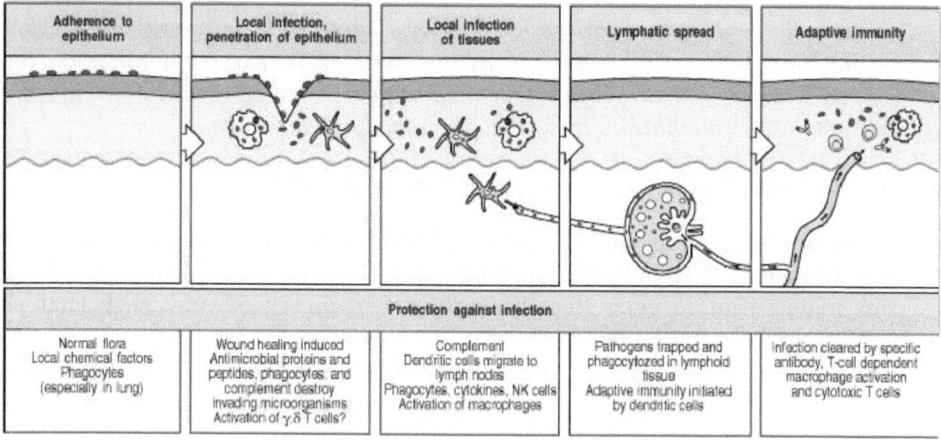

Fig. 1. Immune System Mechanisms [10]

bacterium proteins on its surface (see Figure 1). These surface markers serve as an alarm to other immune cells, namely helper T cells, which then infer the form of the invader. This mechanism makes sensitive the T cells to recognize the antigens or other foreign agents which triggers a reaction of the immune system. Antigens are often found on the surface of bacterium and viruses.

New rules are introduced according to this biological example. We define a new variant of mobile membranes, namely the enhanced mobile membranes, originally introduced in [3]. The multiset u is the one indicating the membrane which initializes the move in the rules of type $(b) - (e)$.

Definition 2 ([3]). *An enhanced mobile membrane is a construct*
$$\prod = (V, H, \mu, w_1, \ldots, w_n, R), \quad where:$$

1. *n, V, H, μ, $w_1, \ldots, w_n$ are as in Definition 1;*
2. *R is a finite set of developmental rules of the following forms:*

local evolution

(a) $[\, [u \rightarrow v]_m]_h$ *for $h, m \in H, u \in V^+, v \in V^*$;*
These rules are called local because the evolution of a multiset of objects u of membrane m is possible only when membrane m is inside membrane h. If the restriction of nested membranes is not imposed, that is, the evolution of the multiset of objects u in membrane m is allowed wherever membrane m is placed, then we say that we have a global evolution rule, and write it simply as $[u \rightarrow v]_m$.

endocytosis

(b) $[uv]_h[v']_m \rightarrow [[w]_h w']_m$ *for $h, m \in H; u \in V^+, v, v', w, w' \in V^*$;*
An elementary membrane labelled h enters the adjacent membrane labelled m, under the control of the multisets of objects uv and v'. The labels h and m remain unchanged during this process; however the multisets of objects uv and v' are replaced with the multisets of objects w and w', respectively.

exocytosis

(c) $[[uv]_h v']_m \rightarrow [w]_h[w']_m$, *for $h, m \in H; u \in V^+, v, v', w, w' \in V^*$;*
An elementary membrane labelled h is sent out of a membrane labelled m, under the control of the multisets of objects uv and v'. The labels of the two membranes remain unchanged, but the multisets of objects uv and v' are replaced with the multisets of objects w and w', respectively.

enhanced endocytosis

(d) $[v]_h[uv']_m \rightarrow [[w]_h w']_m$ *for $h, m \in H, u \in V^+, v, v', w, w' \in V^*$;*
An elementary membrane labelled h is engulfed into the adjacent membrane labelled m, under the control of the multisets of objects uv' and v. The labels h and m remain unchanged during the process; however, the multisets of objects uv' and v are transformed into the multisets of objects w' and w, respectively. The effect of this rule is similar to the effect of rule (b); the difference is that the movement is not controlled by a multiset of objects inside the moving membrane h, but by a multiset

of objects uv′ placed inside the membrane m which engulfs membrane h. This means that the membrane which initiates the movement is membrane m, and not the membrane h as in rule (b).

enhanced exocytosis

(e) $[[v]_h uv']_m \rightarrow [w]_h [w']_m$ *for* $h, m \in H, u \in V^+, v, v', w, w' \in V^*$;

An elementary membrane labelled h is pushed out of a membrane labelled m under the control of the multisets of objects uv′ and v. The labels of the two membranes remain unchanged; however, the multisets of objects uv′ and v evolve into the multisets of objects w′ and w, respectively. The effect of this rule is similar to the one of rule (c); the difference is that the movement is not controlled by an object inside the moving membrane h, but by a multiset of objects uv′ placed inside the membrane m which expels membrane h. This means that the membrane which initiates the movement is membrane m, and not the membrane h as in rule (c).

The rules of enhanced mobile membranes are applied according to the principles of simple mobile membranes.

Using the rules of the enhanced mobile membranes we can describe the immune system mechanisms of Figure 1. We associate a membrane to each cell, and objects to the signals, states and parts of molecules. For the steps done by the dendritic cells presented in Figure 1, we use the following encodings:

- dendritic cell: $[eat]_{DC}$

  An immature dendritic cell is willing to eat any bacterium it encounters, so we translate it into a membrane labelled by DC which has inside an object *eat* used to engulf the bacterium.
- bacterium cell: $[antigen]_{bacterium}$

  A bacterium cell contains antigen so we simply represent it as a membrane labelled by *bacterium* containing a single object *antigen* that encodes the information of the bacterium.
- lymph node: $[\ ]_{lymph\ node}$

  The lymph node is the place where the mature dendritic cells migrate in order to start the immune response, so we translate it into a membrane labelled by *lymph node*.

Using these membranes, we describe the system as follows (here *body* stands for the body skin):

$$[[eat]_{DC}[\ ]_{lymph\ node}]_{body}[antigen]_{bacterium}$$

The evolution is described by following rules:

* $[antigen]_{bacterium}[\ ]_{body} \rightarrow [[antigen]_{bacterium}]_{body}$

A bacterium enters through the body skin by performing an endocytosis rule in order to infect the body. The bacterium contains an object *antigen* which represent its signature.

* $[eat]_{DC}[\ ]_{bacterium} \rightarrow [eat[\ ]_{bacterium}]_{DC}$

  Once an immature dendritic cell becomes sibling to a bacterium, it "eats" the bacterium by performing an enhanced endocytosis rule. Until now the bacterium has controlled its own movement; in this step its movement becomes controlled by the dendritic cell which engulfs it.

* $[[antigen]_{bacterium}]_{DC} \rightarrow [antigen]_{DC}$

  Once the bacterium is engulfed into the dendritic cell, it is dissolved and its content is released into the dendritic cell.

* $[antigen]_{DC}[\ ]_{lymph\ node} \rightarrow [[antigen]_{DC}]_{lymph\ node}$

  Once the dendritic cell contains parts of the antigen, it enters the lymph node in order to activate a special class of T cells, namely the helper T cells.

* $[[eat]_{DC}]_{lymph\ node} \rightarrow [[\ ]_{DC}]_{lymph\ node}$

  Once the dendritic cell enters the lymph node, it matures and the capacity to engulf bacteria disappears; the *eat* object is consumed.

2.3   Mutual Mobile Membranes

In a receptor-mediated endocytosis a cell engulfs a particle of low-density lipoprotein (LDL) from the outside [23]. To do this, the cell uses receptors that specifically recognize and bind to the LDL particle. The receptors are clustered together. An LDL particle contains one thousand or more cholesterol molecules. A monolayer of phospholipid surrounds the cholesterol and its embedded with proteins called apoB. This apoB proteins are specifically recognized by receptors on the cell membrane. The receptors of the coated pit bind to the apoB proteins of the LDL particle. The pit is reenforced by a lattice-like network of proteins called clathrin. Additional clathrin molecules are added to the lattice which eventually pinches off apart from the membranes.

SNARE-mediated exocytosis is the movement of materials out of a cell via vesicles. SNARES located on the vesicles (v-SNARES) and SNARES located on the target membranes (t-SNARES) interact to form a stable complex that holds the vesicle very close to the target membrane as in Figure 3.

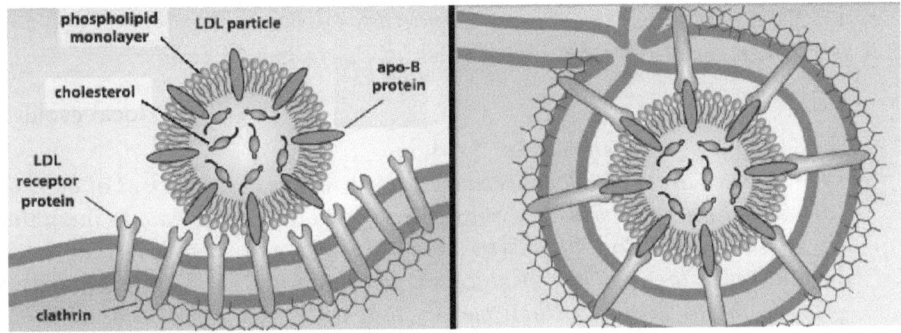

Fig. 2. Receptor-Mediated Endocytosis [23]

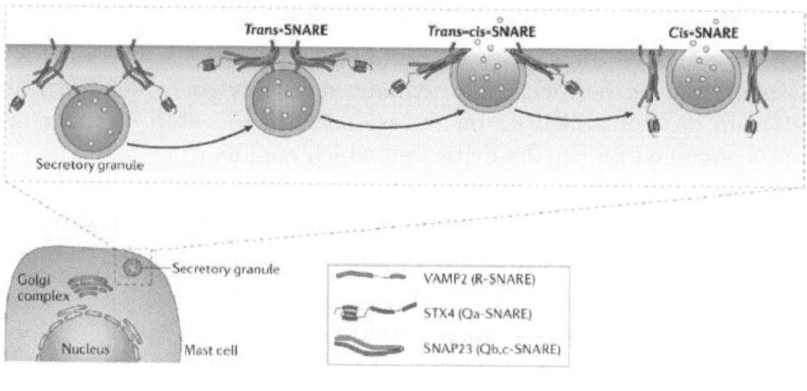

Fig. 3. SNARE-Mediated Exocytosis

Starting from these biological examples we see the necessity to introduce new rules. The rules of enhanced mobile membranes allow a membrane to enter, exit, to engulf or to push out another membrane. The second membrane just undergoes the movement; no permission is required from the second membrane which may not even be aware that a movement involving it has taken place. Following an approach described initially in [4], we introduce a new variant of mobile membranes called mutual mobile membranes.

In mutual mobile membranes, a movement takes place only if the involved membranes *agree* on the movement. This can be described by means of objects a and co-objects $\bar{a}$ present in the membranes involved in such a movement. Since we have the equality $\bar{\bar{a}} = a$, we have that mutual endocytosis is the same as mutual enhanced endocytosis and mutual exocytosis is the same as mutual enhanced exocytosis. The mutual mobile membranes are defined as follows:

Definition 3 ([4]). *A* mutual mobile membrane *is a construct*
$$\textstyle\prod = (V, H, \mu, w_1, \ldots, w_n, R), \quad where:$$

1. n, V, H, μ, $w_1, \ldots, w_n$ *are as in Definition 1;*
2. *R is a finite set of* developmental rules *of the following forms:*

local evolution

(a) $[\,[u \to v]_m]_h$ *for* $h, m \in H, u, v \in V^*$;
   These rules are called local because the evolution of a multiset of objects u of membrane m is possible only when membrane m is inside membrane h. If the restriction of nested membranes is not imposed, that is, the evolution of the multiset of objects u in membrane m is allowed wherever membrane m is placed, then we say that we have a global evolution rule, and write it simply as $[u \to v]_m$.

mutual endocytosis

(b) $[uv]_h[\overline{u}v']_m \rightarrow [\,[w]_hw']_m$ for $h, m \in H, u, \overline{u} \in V^+, v, v', w, w' \in V^*$;

An elementary membrane labelled h enters the adjacent membrane labelled m under the control of the multisets of objects uv and $\overline{u}v'$. The labels h and m remain unchanged during this process; however the multisets of objects uv and $\overline{u}v'$ are replaced with the multisets of objects w and w', respectively.

mutual exocytosis

(c) $[\overline{u}v'[uv]_h]_m \rightarrow [w]_h[w']_m$ for $h, m \in H, u, \overline{u} \in V^+, v, v', w, w' \in V^*$;

An elementary membrane labelled h exits a membrane labelled m, under the control of the multisets of objects uv and $\overline{u}v'$. The labels of the two membranes remain unchanged, but the multisets of objects uv and $\overline{u}v'$ are replaced with the multisets of objects w and w', respectively.

The rules of the mutual mobile membranes are applied according to principles of simple mobile membranes. An object u indicates the membrane which initializes the move in the rules of type $(b) - (c)$, while an object $\overline{u}$ indicates the membrane which accepts the movement.

Using the rules of the mutual mobile membranes we can describe the receptor-mediated endocytosis of Figure 2. We associate a membrane to each cell, and objects to the signals, states and parts of molecules. For the steps done by the cells presented in Figure 2, we use the following encodings:

- LDL particle: $[cholesterol \dots cholesterol\ apoB \dots apoB]_{LDL}$

  An LDL particle contains one thousand or more *cholesterol* molecules and some *apoB* proteins.
- cell membrane: $[receptor \dots receptor\ clarithin \dots clarithin]_{cell}$

  The cell contains *receptors* which are able to recognize apoB proteins and also some proteins *clathrin* which enforce the pit containing the receptors.

Using the above membranes, and the equality $\overline{apoB} = receptor$, we can describe the membrane system as follows:

$$[cholesterol \dots\ apoB \dots]_{LDL}\ [\overline{apoB} \dots\ clarithin \dots]_{cell}$$

The evolution is described by applying a rule of type (b):

$$[cholesterol \dots\ apoB \dots]_{LDL}\ [\overline{apoB} \dots\ clarithin \dots]_{cell} \rightarrow$$
$$\rightarrow [\,[cholesterol \dots\ apoB \dots]_{LDL}\ \overline{apoB} \dots\ clarithin \dots]_{cell}$$

3 Computability Power of Mobile Membranes

In this section we present some existing results and also new results related to the computational power of simple, enhanced and mutual mobile membranes. First we present some notations from the field of formal languages which are used throughout this section. More notions from formal languages can be found in [7] and [22].

For an alphabet $V = \{a_1, \dots, a_n\}$, we denote by V^* the set of all strings over V; λ denotes the empty string. V^* is a monoid with λ as its unit element. For a string $x \in V^*$, $|x|_a$ denotes the number of occurrences of symbol a in x. A multiset over V is represented by a string over V (together with all its permutations), and each string precisely identifies a multiset. For an alphabet V, the *Parikh vector* is $\psi_V : V^* \to \mathbf{N}^n$ with $\psi_V(x) = (|x|_{a_1}, \dots, |x|_{a_n})$, for all $x \in V^*$. For a language L, the Parikh vector is $\psi_V(L) = \{\psi_V(x) \mid x \in L\}$, while for a family FL of languages, it is $PsFL = \{\psi_V(L) \mid L \in FL\}$.

A *matrix grammars with appearance checking* is a construct $G = (N, T, S, M, F)$ where N, T are disjoint alphabets of non-terminals and terminals, $S \in N$ is the axiom, M is a finite set of matrices of the form $(A_1 \to x_1, \dots, A_n \to x_n)$ of context-free rules, and F is a set of occurrences of rules in M. For $w, z \in (N \cup T)^*$, we write $w \Rightarrow_m z$ if there is a matrix $(A_1 \to x_1, \dots, A_n \to x_n)$ in M and the strings $w_i \in (N \cup T)^*$, $1 \le i \le n+1$, such that $w = w_1$, $z = w_{n+1}$, and for all i, $1 \le i \le n$, either (1) $w_i = w_i' A_i w_i''$, $w_{i+1} = w_i' x_i w_i''$, for some w_i', $w_i'' \in (N \cup T)^*$, or (2) $w_i = w_{i+1}$, A_i does not appear in w_i, and the rule $A_i \to x_i$ appears in F. The language generated by G is $L(G) = \{x \in T^* \mid S \Rightarrow^* x\}$. A *matrix grammar in the strong binary normal form* is a construct $G = (N, T, S, M, F)$, where $N = N_1 \cup N_2 \cup \{S, \#\}$, with these three sets mutually disjoint, two distinguished symbols $B^{(1)}$, $B^{(2)} \in N_2$, and the matrices in M of one of the following forms:

(1) $(S \to XA)$, with $X \in N_1, A \in N_2$,
(2) $(X \to Y, A \to x)$, with $X, Y \in N_1, A \in N_2, x \in (N_2 \cup T)^*$,
(3) $(X \to Y, B^{(j)} \to \#)$, with $X, Y \in N_1, j = 1, 2$,
(4) $(X \to \lambda, A \to x)$, with $X \in N_1, A \in N_2, x \in T^*$.

If we ignore the empty string when comparing languages, then the rules of type (4) are of the form $(X \to a, A \to x)$, with $X \in N_1$, $a \in T$, $A \in N_2$, $x \in T^*$.

3.1   Simple Mobile Membranes

The computational power of simple mobile membranes is treated in [14].

$PsMM_n(levol, endo, exo)$ denotes the family of all sets $Ps(\Pi)$ generated by systems using local evolution rules, together with endocytosis and exocytosis rules and at most n membranes. If the number of membrane is not bounded, this is denoted by $PsMM_*(levol, endo, exo)$. When global evolution rules are used, *levol* is replaced by *gevol*. If a type of rules is not used, then its name is omitted from the list of parameters. The number of membranes does not increase during the computation, but it can decrease by sending membranes out of the skin.

The following result establishes an universality result using nine membranes and the operations of endocytosis and exocytosis:

Theorem 1 ([14]). $PsMM_9(endo, exo) = PsRE$.

A strengthening of the previous universality result is:

Corollary 1 ([14]). $PsMM_*(endo, exo) = PsMM_n(endo, exo) = PsMM_n(gevol, endo, exo) = PsMM_n(levol, endo, exo) = PsRE$, *for all* $n \ge 9$.

An improvement of the result presented in Theorem 1 is:

Theorem 2 ([12]). $PsMM_4(gevol, endo, exo) = PsRE$.

We improve the previous result by decreasing the number of membranes to three.

Theorem 3. $PsMM_3(levol, endo, exo) = PsRE$.

Proof. Consider a matrix grammar $G = (N, T, S, M, F)$ in the improved strong binary normal form (hence with $N = N_1 \cup N_2 \cup \{S; \#\}$), having n_1 matrices of types (2) and (4) (that is, not used in the appearance checking mode), and n_2 matrices of type (3) (with appearance checking rules). Let $B^{(1)}$ and $B^{(2)}$ be the two objects in N_2 for which we have rules $B^{(j)} \to \#$ in matrices of M. The matrices of the form $(X \to Y, B^{(j)} \to \#)$ are labelled by m_i', with $i \in lab_j$, for $j \in \{12\}$, such that lab_1, lab_2, and $lab_0 = \{1, \ldots, n_1\}$ are mutually disjoint sets.

We construct a mobile membrane system $\Pi = (V, H, \mu, w_1, w_2, w_3, R, 2)$ of degree three, where:

$$V = N \cup \{X, X_{i,j} \mid X \in N_1, 1 \leq i \leq n_1, 0 \leq j \leq n_1\}$$
$$\cup \{a, a' \mid a \in T\} \cup \{x \mid x \in (N_2 \cup T)^*\}$$
$$\cup \{A, A_{i,j} \mid A \in N_2, 1 \leq i \leq n_1, 0 \leq j \leq n_1\}$$
$$H = \{1, 2, 3\}$$
$$\mu = [[\ ]_2[\ ]_3]_1$$
$$w_2 = XA, \text{ where } (S \to XA) \text{ is the initial matrix of } G$$
$$w_h = \lambda, \text{ for all } h \in \{1, 3\}$$

The set R of rules is constructed as follows:

(i) For each (nonterminal) matrix $m_i : (X \to Y, A \to x)$, $X, Y \in N_1$, $A \in N_2$, $x \in (N_2 \cup T)^*$, with $1 \leq i \leq n_1$, we consider the rules:
 1. $[X]_2[\ ]_3 \to [[X_{i,0}]_2]_3$ (endo)
 2. $[[A]_2]_3 \to [A_{i,0}]_2[\ ]_3$ (exo)
 3. $[[X_{i,j} \to X_{i,j+1}]_2]_1$, $j < i$ (levol)
 4. $[[A_{i,j} \to A_{i,j+1}]_2]_1$, $j < i$ (levol)
 5. $[[A_{i,i}X_{i,i} \to xY]_2]_1$ (levol)
 6. $[[A_{i,j}X_{j,j} \to \#]_2]_1$, $j < i$ (levol)
 7. $[[A_{j,j}X_{i,j} \to \#]_2]_1$, $j < i$ (levol)

In the initial configuration, we have the objects X and A corresponding to the initial matrix in membrane 2. To simulate a matrix of the above type we start by applying the endocytosis rule 1, thus replacing X with $X_{i,0}$, followed by the exocytosis rule 2, thus replacing a single $A \in N_2$ with $A_{i,0}$. No other $A \in N_2$ can be replaced until membrane 2 enters membrane 3. Rule 3 (for X) and rule 4 (for A) are used to increment the second indices of X and A. This is done to check if the indices of X and A are the same, and in this case to rewrite A according to the matrix m_i. Once the indices are equal, rule 5 is applied to complete the simulation of matrix m_i. If the indices of X and A are not the same, rule 6 (if the indices of X is lower than the indices of A) or rule 7 (if the indices of X is bigger than the indices of A) is applied, the computation is blocked without producing any output.

(ii) For a terminal matrix $m_i : (X \to a, A \to x)$, $X \in N_1$, $a \in T$, $A \in N_2$, $x \in T^*$, where $1 \le i \le n_1$, we use the rules 1-7, where the rule 5 is replaced by the rules:

8. $[a_{i,i}X_{i,i} \to a'Y]_1$ (levol)
9. $[[a']_2]_1 \to [a]_2[\ ]_1$ (exo)

Observe that simulation of a type (4) matrix is along similar steps, except that we have an a in place of Y. During the finishing stages of a type (4) simulation, we use rule 8 to replace $a_{i,i}$ by a', and then to rewrite it to a when sending the membrane 2 out of the skin membrane, namely membrane 1.

(iii) For each matrix $m'_i : (X \to Y, B^{(k)} \to \#)$, $X, Y \in N_1$, $A \in N_2$, where $n_1 + 1 \le j \le n_1 + n_2$, $j \in lab_k$, $k = 1, 2$, we consider the rules:

10. $[X]_2[\ ]_3 \twoheadrightarrow [[X_k]_2]_3$, for $i \in lab_k$ (endo)
11. $[[X_k B^{(k)} \to \#]_2]_3$, $k = 1, 2$ (levol)
12. $[[X_k]_2]_3 \to [Y]_2[\ ]_3$, $k = 1, 2$ (exo)

The simulation of matrices of type (3) begins by a rule of type 10. This is followed by a rule 11 in case $B^{(k)}$ exists, blocking membrane 2 inside membrane 3 and the computation stops without producing any output. If no $B^{(k)}$ exists, then rule 12 can be used to send out membrane 2, successfully completing the simulation.

3.2   Enhanced Mobile Membranes

The operations governing the mobility of the enhanced mobile membranes are endocytosis (endo), exocytosis (exo), enhanced endocytosis (fendo) and enhanced exocytosis (fexo). The interplay between these four operations is quite powerful, and the computational power of a Turing machine is obtained using twelve membranes without using the context-free evolution of objects [13].

The family of all sets $Ps(\Pi)$ generated by systems of degree at most n using rules $\alpha \subseteq \{exo, endo, fendo, fexo, cevol\}$ is denoted by $PsEMM_n(\alpha)$. Here endo and exo represent endocytosis and exocytosis, fendo and fexo represent enhanced endocytosis and enhanced exocytosis, and cevol represents contextual evolution. The main results are the following.

Theorem 4 ([13]). $PsEMM_{12}(endo, exo, fendo, fexo) = PsRE$.

Theorem 5 ([13]). $PsEMM_3(cevol) = PsRE$.

Theorem 6 ([13]). $PsEMM_3(endo, exo) = PsEMM_3(fendo, fexo)$.

We improve the result of Theorem 4 as follows:

Theorem 7. $PsEMM_9(endo, exo, fendo, fexo) = PsRE$.

Proof. Consider a matrix grammar $G = (N, T, S, M, F)$ in the improved strong binary normal form (hence with $N = N_1 \cup N_2 \cup \{S; \#\}$), having n_1 matrices $m_1, \ldots, m_{n_1}$ of types (2) and (4) (that is, not used in the appearance checking mode), and n_2 matrices of type (3) (with appearance checking rules). The initial matrix is $m_0 : (S \to XA)$. Let $B^{(1)}$ and $B^{(2)}$ be the two objects in N_2 for which we have rules $B^{(j)} \to \#$ in matrices of M. The matrices of the form $(X \to Y, B^{(j)} \to \#)$ are labelled by m_i', $1 \leq i \leq n_2$ with $i \in lab_j$, for $j \in \{12\}$, such that lab_1, lab_2, and $lab_0 = \{1, 2, \ldots, n_1\}$ are mutually disjoint sets.

We construct a mobile membrane system $\Pi = (V, H, \mu, w_1, \ldots, w_9, R, 7)$ of degree nine, where:

$$V = N \cup T \cup \{X_{0i}', A_{0i}' \mid X \in N_1, A \in N_2, 1 \leq i \leq n_1\}$$
$$\cup \{X_{ji}, A_{ji} \mid 0 \leq i, j \leq n_1\} \cup \{X_i^j, X_j \mid X \in N_1, j \in \{1, 2\}, 1 \leq i \leq n_2\}$$
$$H = \{1, \ldots, 9\}$$
$$\mu = [[\ ]_7[\ ]_8[\ ]_9[[\ ]_3[\ ]_4[\ ]_5[\ ]_6]_2]_1$$
$$w_7 = XA, \text{ where } (S \to XA) \text{ is the initial matrix of } G$$
$$w_h = \lambda, \text{ for all } h \in \{1, \ldots, 9\} \backslash \{7\}$$

The set R of rules is constructed as follows:

(i) For each (nonterminal) matrix $m_i : (X \to Y, A \to x)$, $X, Y \in N_1$, $A \in N_2$, $x \in (N_2 \cup T)^*$, with $1 \leq i \leq n_1$, we consider the rules:

1. $[X]_7[\ ]_8 \to [[X_{i,i}]_7]_8$ (endo)
2. $[[A]_7]_8 \to [A_{i,i}]_7[\ ]_8$ (exo)
3. $[X_{j,i}]_7[\ ]_9 \to [[X_{j-1,i}]_7]_9$ (endo)
4. $[[A_{j,i}]_7]_9 \to [A_{j-1,i}]_7[\ ]_9$ (exo)
5. $[\ ]_8[X_{0,i}]_7 \to [X_{0,i}'[\ ]_8]_7$ (fendo)
6. $[\ ]_9[A_{0,i}]_7 \to [A_{0,i}'[\ ]_9]_7$ (fendo)
7. $[\ ]_8[X_{0,i}]_7 \to [\#[\ ]_8]_7$ (fendo)
8. $[[A_{0,i}]_7]_9 \to [\#]_7[\ ]_9$ (exo)
9. $[X_{0,i}'[\ ]_8]_7 \to [\ ]_8[Y]_7$ (fexo)
10. $[A_{0,i}'[\ ]_9]_7 \to [\ ]_9[x]_7$ (fexo)

In the initial configuration, we have the objects X, A corresponding to the initial matrix in membrane 7. To simulate a matrix of type (2), we start by applying the endocytosis rule 1, thus replacing X with $X_{i,i}$, followed by the exocytosis rule 2, thus replacing a single $A \in N_2$ with $A_{i,i}$. Rule 3 (for X) and rule 4 (for A) are used to decrement the first indices of X and A. This is done to check if the indices of X and A are the same, and in this case to rewrite A according to the matrix m_i. By using fendo rules 5 and 6, membranes 8 and 9 enter membrane 7 replacing $X_{0,i}$ and $A_{0,i}$ with $X_{0,i}'$ and $A_{0,i}'$, respectively. This is then followed by rules 9 and 10, when membranes 8 and 9 exit membrane 7 by fexo rules replacing $X_{0,i}'$ and $A_{0,i}'$ with Y and x, respectively. If $i > j$, then we obtain $A_{0,j}$ before $X_{0,i}$. In this case, we have a configuration where membrane 7 is inside membrane 9 containing $A_{0,j}$. Then rule 8 is used, replacing $A_{0,j}$ with $\#$, and an infinite computation is obtained (rule 17). If $j > i$, then we obtain $X_{0,i}$ before $A_{0,j}$.

In this case, we reach a configuration with $X_{0,i}A_{k,j}$, $k > 0$ in membrane 7, and membrane 7 is in the skin membrane. Rule 3 cannot be used now, and the only possibility is to use rule 7, which leads to an infinite computation. Thus, if $i = j$, then we can correctly simulate a matrix of type (2).

(ii) For each matrix $m'_i : (X \rightarrow Y, B^{(k)} \rightarrow \#)$, $X, Y \in N_1$, $A \in N_2$, where $n_1 + 1 \leq j \leq n_1 + n_2$, $j \in lab_k$, $k = 1, 2$, we consider the rules:

11. $[X]_7[\ ]_2 \rightarrow [[X_i^{(j)}]_7]_2$, $j = 1, 2$ (endo)
12. $[\ ]_{j+2}[X_i^{(j)}]_7 \rightarrow [X_i^{(j)}[\ ]_{j+2}]_7$, $j = 1, 2$ (fendo)
13. $[\ ]_{j+4}[B^{(j)}]_7 \rightarrow [\#[\ ]_{j+4}]_7$, $j = 1, 2$ (fendo)
14. $[X_i^{(j)}[\ ]_{j+2}]_7 \rightarrow [\ ]_{j+2}[Y_j]_7$, $j = 1, 2$ (fexo)
15. $[[Y_j]_7]_2 \rightarrow [Y]_7[\ ]_2$, $j = 1, 2$ (exo)

The simulation of matrices of type (3) begins by a rule of type 11. Inside membrane 2, rules 12 and 13 are used, and so membrane $(j + 2)$ enters membrane 7, and membrane $(j + 4)$ enters membrane 7 if the symbol $B^{(j)}$ is present. In this case, $B^{(j)}$ is replaced with $\#$. Otherwise, membrane $(j+2)$ comes out of the membrane 7 replacing $X_i^{(j)}$ with Y_j. Then membrane 7 exits membrane 2, by replacing Y_j with Y thus successfully simulating a matrix of type (3).

(iii) For a terminal matrix $m_i : (X \rightarrow a, A \rightarrow x)$, $X \in N_1$, $a \in T$, $A \in N_2$, $x \in T^*$, where $1 \leq i \leq n_1$:

16. $[[a']_7]_1 \rightarrow [a]_7[\ ]_1$ (exo)
17. $[\ ]_8[\#]_7 \rightarrow [\#[\ ]_8]_7$ (fendo)
    $[\#[\ ]_8]_7 \rightarrow [\ ]_8[\#]_7$ (fexo)

Observe that simulation of a matrix of type (4) matrix is similar to that of a matrix of type (2), except that we have an a' in place of Y in rule 9. During the finishing stages of a matrix of type (4) simulation, we use rule 16 to replace a' with a when sending the membrane 7 out of the skin membrane.

3.3   Mutual Mobile Membranes

Similar to other classes of mobile membranes, we try to establish the number of membranes in mutual mobile membranes in order to obtain a system which is equivalent to Turing machines. The following result offers an answer. The minimum number of membranes needed remains an open problem.

The family of all sets $Ps(\Pi)$ generated by systems of degree at most n using rules $\alpha \subseteq \{mutual\ exo, mutual\ endo\}$ is denoted by $PsMMM_n(\alpha)$. Here *mutual endo* and *mutual exo* represent mutual endocytosis and mutual exocytosis rules. By using objects and co-objects, the computational power is obtained using a lower number of membranes than for enhanced mobile membranes, namely:

Theorem 8. $PsMMM_7(mutual\ endo, mutual\ exo) = PsRE.$

Proof (Sketch). Consider a matrix grammar $G = (N, T, S, M, F)$ in the improved strong binary normal form (hence with $N = N_1 \cup N_2 \cup \{S; \#\}$), having n_1 matrices $m_1, \ldots, m_{n_1}$ of types (2) and (4) (that is, not used in the appearance checking mode), and n_2 matrices of type (3) (with appearance checking rules). The initial matrix is $m_0 : (S \to XA)$. Let $B^{(1)}$ and $B^{(2)}$ be the two objects in N_2 for which we have rules $B^{(j)} \to \#$ in matrices of M. The matrices of the form $(X \to Y, B^{(j)} \to \#)$ are labelled by m'_i, $1 \leq i \leq n_2$ with $i \in lab_j$, for $j \in \{1, 2\}$, such that lab_1, lab_2, and $lab_0 = \{1, 2, \ldots, n_1\}$ are mutually disjoint sets.

We construct a mobile membrane system $\Pi = (V, H, \mu, w_1, \ldots, w_7, R, 7)$ of degree seven, where:

$$V = N \cup T \cup \{X'_{0i}, A'_{0i} \mid X \in N_1, A \in N_2, 1 \leq i \leq n_1\}$$
$$\cup \{X_{ji}, A_{ji} \mid 0 \leq i, j \leq n_1\} \cup \{X_i^j, X_j \mid X \in N_1, j \in \{1, 2\}, 1 \leq i \leq n_2\}$$
$$\cup \{\beta, \overline{\beta}, \gamma, \overline{\gamma}\}$$
$$H = \{1, \ldots, 7\}$$
$$\mu = [[\,]_7[\,]_5[\,]_6[[\,]_3[\,]_4]_2]_1$$
$$w_1 = \overline{\beta}, \ w_2 = \overline{\beta}, \ w_3 = \beta, \overline{\beta}, \gamma, \overline{\gamma}, \ w_4 = w_5 = w_6 = \overline{\beta}, \overline{\gamma}$$
$$w_7 = XA\beta\gamma, \text{ where } (S \to XA) \text{ is the initial matrix of } G.$$

In what follows we present only a part of the set of rules R constructed, namely the ones used to simulate m'_i matrices.

(i) For each matrix $m'_i : (X \to Y, B^{(k)} \to \#)$, $X, Y \in N_1$, $A \in N_2$, where $n_1 + 1 \leq j \leq n_1 + n_2$, $j \in lab_k$, $k = 1, 2$, we consider the rules:

1. $[\beta X]_7[\overline{\beta}]_2 \to [\overline{\beta}[\beta X_i^{(j)}]_7]_2$, $j = 1, 2$ (mutual endo)
2. $[\beta X_i^{(j)}]_7[\overline{\beta}]_{j+2} \to [[\beta X_i^{(j)}]_7\overline{\beta}]_{j+2}$, $j = 1, 2$ (mutual endo)
3. $[\gamma]_3[\overline{\gamma}]_4 \to [\gamma[\overline{\gamma}]_4]_3$ (mutual endo)
4. $[\gamma B^{(1)}]_7[\overline{\gamma}]_4 \to [\gamma\#[\overline{\gamma}]_4]_7$ (mutual endo)
5. $[\gamma[\overline{\gamma}]_4]_3 \to [\gamma]_3[\overline{\gamma}]_4$ (mutual exo)
6. $[\beta]_3[\overline{\beta}]_4 \to [\beta[\beta]_3]_4$ (mutual endo)
7. $[\gamma B^{(2)}]_7[\overline{\gamma}]_3 \to [\gamma\#[\overline{\gamma}]_3]_7$ (mutual endo)
8. $[\overline{\beta}[\beta]_3]_4 \to [\beta]_3[\overline{\beta}]_4$ (mutual exo)
9. $[\overline{\beta}[\beta X_i^{(j)}]_7]_{j+2} \to [\overline{\beta}]_{j+2}[\beta Y_j]_7$, $j = 1, 2$ (mutual exo)
10. $[\overline{\beta}[\beta Y_j]_7]_2 \to [\beta Y]_7[\overline{\beta}]_2$, $j = 1, 2$ (mutual exo)

The simulation of matrices of type (3) begins by a rule of type 1. Inside membrane 2, rules 2 is used, by which membrane 7 enters membrane $(j+2)$. Rules 3 and 6 are used to introduce the remaining membrane (3 or 4) near membrane 7. In this case, if $B^{(j)}$ exists this is replaced with $\#$ and the computation is stopped. Otherwise, membrane 7 comes out of membrane $(j+2)$ replacing $X_i^{(j)}$ with Y_j. After the other membrane is removed from $(j+2)$, membrane 7 exits membrane 2, successfully simulating a matrix of type (3).

4   Related Work: P Systems with Active Membranes

The mobile membranes derive from the P systems with active membranes introduced in [18]. P systems with active membranes are a variant of P systems in which each membrane is supposed to have an "electrical polarization" (also called *charge*): *positive*, *negative*, or *neutral*.

A P system with active membranes has a finite set of developmental rules, of the following forms:

object evolution

(a) $[a \rightarrow v]_h^\alpha$, for $h \in H$, $\alpha \in \{+, -, 0\}$, $a \in V$, $v \in V^*$

communication

(b) $a[\ ]_h^{\alpha_1} \rightarrow [b]_h^{\alpha_2}$, for $h \in H$, $\alpha_1, \alpha_2 \in \{+, -, 0\}$, $a, b \in V$

communication

(c) $[a]_h^{\alpha_1} \rightarrow [\ ]_h^{\alpha_2} b$, for $h \in H$, $\alpha_1, \alpha_2 \in \{+, -, 0\}$, $a, b \in V$

dissolving

(d) $[a]_h^\alpha \rightarrow b$, for $h \in H$, $\alpha \in \{+, -, 0\}$, $a, b \in V$

division of elementary membranes

(e) $[a]_h^{\alpha_1} \rightarrow [b]_h^{\alpha_2} [c]_h^{\alpha_3}$; for $h \in H$, $\alpha_1, \alpha_2, \alpha_3 \in \{+, -, 0\}$, $a, b, c \in V$

division of non-elementary membranes

(f) $[\ [\ ]_{h_1}^{\alpha_1} \cdots [\ ]_{h_k}^{\alpha_1} [\ ]_{h_{k+1}}^{\alpha_2} \cdots [\ ]_{h_n}^{\alpha_2}]_{h_0}^{\alpha_0} \rightarrow [\ [\ ]_{h_1}^{\alpha_3} \cdots [\ ]_{h_k}^{\alpha_3}]_{h_0}^{\alpha_5} [\ [\ ]_{h_{k+1}}^{\alpha_4} \cdots [\ ]_{h_n}^{\alpha_4}]_{h_0}^{\alpha_6}$
for $k \geq 1$, $n > k$, $h_i \in H$, $0 \leq i \leq n$,
and $\alpha_0, \ldots, \alpha_6 \in \{+, -, 0\}$ with $\{\alpha_1, \alpha_2\} = \{+, -\}$

More details about these rules and how they are applied may be found in [18].

By denoting with LPA the family of languages $L(\Pi)$ generated by P systems with active membranes, we have the following result:

Theorem 9 ([18]). $PsRE = PsLPA$.

NPA_{rd} denotes the family of vectors of natural numbers $N(\Pi)$ computed by non-cooperative systems Π which do not use division rules of type (f). The subscript rd stands for "restricted division". This restriction does not decrease the power of P systems with active membranes:

Theorem 10 ([20]). $PsRE = NPA_{rd}$.

NPA_r denotes the family of natural numbers $N(\Pi)$ computed by systems Π which use only rules of types (a), (b) and (c).

Theorem 11 ([15]). $PsRE = NPA_r$.

The set of numbers generated in the minimally parallel way by a system Π is denoted by $N_{min}(\Pi)$. The family of sets $N_{min}(\Pi)$ generated by systems with rules of the non-restricted form, having initially at most n_1 membranes and using configurations with at most n_2 membranes during any computation is denoted by $N_{min}OP_{n_1,n_2}$. When a type of rule is not used, it is not mentioned in the notation. If any of the parameters n_1, n_2 is not bounded, then it is replaced by $\star$. If the system do not use polarizations for membranes, then we write (a_0), (b_0), (c_0), (d_0), (e_0) instead of (a), (b), (c), (d), (e). When using the maximal parallel way we replace the subscript min by max.

Theorem 12 ([1,6,17])

1. $N_{max}OP_{3,3}((a),(b),(c)) = NRE.$ [17]
2. $N_{max}OP_{\star,\star}((a_0),(b_0),(c_0),(d_0),(e_0)) = NRE.$ [1]
3. $N_{min}OP_{3,3}((a),(b),(c)) = NRE.$ [6]

Theorem 13 ([8]). $N_{max}OP_{n_1,\star}((a_1),(b_1),(c_1),(d_1),(e_1))=NRE,$ for all $n_1 \geq 5$.

Theorem 14 ([8]). $N_{min}OP_{n_1,\star}((a_1),(b_1),(c_1),(d_1),(e_1))=NRE,$ for all $n_1 \geq 7$.

When the rules of a given type (α_0) are able to change the labels of the involved membranes, then we denote that type of rules by (α_0').

Using the power of label changing, the following results are obtained:

Theorem 15 ([2]). $PsOP(a_0, b_0, c_0, e_0') = PsOP(a_0, b_0, c_0') =$
$$= PsOP(a_0, b_0', c_0) = PsOP(a_0, c_0, e_0') = PsRE.$$

By introducing replicative-distribution rules for nested membranes:

(l_0) $[a[\,]_{h_1}]_{h_2} \to [[u]_{h_1}]_{h_2} v$, for $h_1, h_2 \in H$, $a \in V$, $u, v \in V^*$;

the following result is obtained:

Theorem 16 ([9]). $PsOP_4(l_0') = PsRE.$

5 Conclusions

Simple, enhanced and mutual mobile membranes are new models of computation inspired from the biological operations governing the movement of biological membranes: endocytosis and exocytosis. After defining these classes of mobile membranes according to their biological motivations, some results concerning their computational power are presented. For mutual mobile membranes this is the first universality result, while for simple and enhanced mobile membranes the results are improvements for existing ones by reducing the number of membranes needed.

Acknowledgements

Many thanks to the referees for their helpful remarks and comments. This work has been partially supported by research grants CNCSIS IDEI 402/2007 and CNCSIS TD 345/2008.

References

1. Alhazov, A.: P Systems without Multiplicities of Symbol-Objects. Information Processing Letters 100, 124–129 (2006)
2. Alhazov, A., Pan, L., Păun, G.: Trading Polarizations for Labels in P Systems with Active Membranes. Acta Informatica 41(2-3), 111–144 (2004)

3. Aman, B., Ciobanu, G.: Describing the Immune System Using Enhanced Mobile Membranes. Electronic Notes in Theoretical Computer Science, vol. 194, pp. 5–18 (2008)
4. Aman, B., Ciobanu, G.: Resource Competition and Synchronization in Membranes. In: Proceedings SYNASC 2008. IEEE Computing Society, Los Alamitos (2009)
5. Cardelli, L., Gordon, A.: Mobile Ambients. In: Nivat, M. (ed.) FOSSACS 1998. LNCS, vol. 1378, pp. 140–155. Springer, Heidelberg (1998)
6. Ciobanu, G., Pan, L., Păun, G., Pérez-Jiménez, M.J.: P Systems with Minimal Parallelism. Theoretical Computer Science 378, 117–130 (2007)
7. Dassow, J., Păun, G.: Regulated Rewriting in Formal Language Theory. Springer, Heidelberg (1990)
8. Freund, R., Păun, G., Pérez-Jiménez, M.J.: Polarizationless P Systems with Active Membranes Working in the Minimally Parallel Mode. In: Akl, S.G., Calude, C.S., Dinneen, M.J., Rozenberg, G., Wareham, H.T. (eds.) UC 2007. LNCS, vol. 4618, pp. 62–76. Springer, Heidelberg (2007)
9. Ishdorj, T.-O., Ionescu, M.: Replicative-Distribution Rules in P Systems with Active Membranes. In: Liu, Z., Araki, K. (eds.) ICTAC 2004. LNCS, vol. 3407, pp. 68–83. Springer, Heidelberg (2005)
10. Janeway, C.A., Travers, P., Walport, M., Shlomchik, M.J.: Immunobiology - The Immune System in Health and Disease, 5th edn. Garland Publishing, New York (2001)
11. Krishna, S.N.: On the Efficiency of a Variant of P Systems with Mobile Membranes. In: Cellular Computing: Complexity Aspects, Fenix Editora, Sevilla, pp. 237–246 (2005)
12. Krishna, S.N.: The Power of Mobility: Four Membranes Suffice. In: Cooper, S.B., Löwe, B., Torenvliet, L. (eds.) CiE 2005. LNCS, vol. 3526, pp. 242–251. Springer, Heidelberg (2005)
13. Krishna, S.N., Ciobanu, G.: On the Computational Power of Enhanced Mobile Membranes. In: Beckmann, A., Dimitracopoulos, C., Löwe, B. (eds.) CiE 2008. LNCS, vol. 5028, pp. 326–335. Springer, Heidelberg (2008)
14. Krishna, S.N., Păun, G.: P Systems with Mobile Membranes. Natural Computing 4, 255–274 (2005)
15. Păun, A.: On P Systems with Membrane Division. In: Unconventional Models of Computation, pp. 187–201 (2000)
16. Păun, G.: Computing with Membranes. Journal of Computer and System Sciences 61, 108–143 (2000)
17. Păun, G.: Membrane Computing. An Introduction. Springer, Berlin (2002)
18. Păun, G.: P Systems with Active Membranes: Attacking NP-Complete Problems. Journal of Automata, Languages and Combinatorics 6, 75–90 (2001)
19. Păun, G., Rozenberg, G., Salomaa, A.: Membrane Computing with External Output. Fundamenta Informaticae 41, 259–266 (2000)
20. Păun, G., Suzuki, Y., Tanaka, H., Yokomori, T.: On the Power of Membrane Division in P Systems. Theoretical Computer Science 324, 61–85 (2004)
21. Petre, I., Petre, L.: Mobile Ambients and P Systems. Journal of Universal Computer Science 5, 588–598 (1999)
22. Salomaa, A.: Formal Languages. Academic Press, London (1973)
23. http://bcs.whfreeman.com/thelifewire

Bio-PEPA with Events

Federica Ciocchetta

Laboratory for Foundations of Computer Science,
The University of Edinburgh, Edinburgh EH8 9AB, Scotland

Abstract. In this work we present an extension of Bio-PEPA, a language recently defined for the modelling and analysis of biochemical systems, to handle *events*. Events are constructs that represent changes in the system due to some trigger conditions. The events considered here are simple, but nevertheless able to describe most of the discontinuous changes in models and experiments.

Events are added to our language without any modification to the rest of the syntax in order to keep the specification of the model as straightforward as possible. Some maps are defined from Bio-PEPA with events to analysis tools. Specifically, we map our language to *Hybrid Automata (HA)* and we consider a modification of Gillespie's algorithm for stochastic simulation. In order to test our approach, we present the translation in Bio-PEPA of a biochemical network describing the functional properties of the Acetylcholine receptor with the addition of an event that causes the inactivation of some reactions at a given time.

1   Introduction

Computational models play an important role in systems biology. Indeed they help to study, analyze and predict the behaviour of biological systems. In recent years there have been some applications of process algebras for the analysis of biological systems (e.g. [27,25,8,9]). In most cases the analysis is performed using Gillespie's stochastic simulation algorithm [18]. Other possibilities exist, such as the mapping to differential equations [7].

Many biological models need to capture both discrete and continuous phenomena [1,4,23]. These models are called *hybrid systems*. A first example of a hybrid system describes the activation of a certain activity when the concentration of enabling quantities is above the desired threshold. A second example considers a signal or stimuli that becomes null after some time leading to some changes in the interactions of the system. Other examples describe some experiments, where it may be necessary to render the possible change to the system, due, for instance, to the introduction or the removal of some reagents.

In this work we present an extension of Bio-PEPA [9,10], a language recently defined for the modelling and analysis of biological systems, to handle *events*. Broadly speaking, events are constructs that represent changes in the system due to some trigger conditions.

Here we are interested in simple forms of events. Specifically, we refer to the definition of events reported in the SBML specification [22]. These kinds of

C. Priami et al. (Eds.): Trans. on Comput. Syst. Biol. XI, LNBI 5750, pp. 45–68, 2009.

events can be found in biochemical networks, such as the ones in the BioModels database [24] or defined in some experimental settings.

The idea underlying our work is the following:

Biochemical networks with events $\Longrightarrow$ *Bio-PEPA with events* $\Longrightarrow$ *Analysis*

Starting from a biochemical network with one or more events, we want to map it into a Bio-PEPA system. From that, we can then consider different kinds of analysis. In this, view Bio-PEPA is a formal, intermediate, compositional representation of the biochemical network. This idea is the one proposed for (the standard) Bio-PEPA.

A first challenge concerns the *modelling*: we need to add events to the Bio-PEPA system. Events are added to our language as a set of elements and the rest of the syntax is unchanged. There are two motivations for this choice. First, we keep the specification of the model as simple as possible. Second, this approach is appropriate when we study the same biochemical system but with different experimental regimes as we can modify the list of events without any changes to the rest of the system.

A second aspect is the *analysis*. Some maps must be defined from Bio-PEPA to analysis tools. Specifically, we map our language to *Hybrid Automata (HA)* [19]. HA are a formalism that consider both continuous and discrete changes. The continuous part is expressed by a set of variables evolving in each state according to a set of differential equations and the discrete dynamics is given by transitions between states, triggered by some conditions on variables. Furthermore, we can consider a modification of Gillespie's algorithm [18] in order to tackle events.

A preliminary version of this work has been presented in [11]. Here we add some definitions concerning the kind of events and further details concerning the mappings from Bio-PEPA with events to the Hybrid Automata and Gillespie's algorithm. Furthermore, we consider more general kinds of events, such as simultaneous events or events with a delay different from zero.

The rest of the paper is organised as follows. Section 2 reports a description of Bio-PEPA. In Section 3 we define the events we are considering in this work and then we extend Bio-PEPA in order to handle them. Section 4 describes the mapping from our language to Hybrid Automata. The mapping to stochastic simulation is reported in Section 5. After that, Section 6 illustrates the modelling in Bio-PEPA of a biochemical network describing the functional properties of the Acetylcholine receptor with an event that is triggered at a given time and causes the inactivation of some reactions. In Section 7 we overview some related work. Finally, in Section 8, some conclusions are reported.

2   Bio-PEPA

Bio-PEPA [9,10] is a language for the modelling and analysis of biochemical networks. The syntax of Bio-PEPA is defined as:

$$S ::= (\alpha, \kappa) \text{ op } S \mid S + S \mid C \qquad P ::= P \bowtie_{\mathcal{I}} P \mid S(x)$$

where $\text{op} = \downarrow \mid \uparrow \mid \oplus \mid \ominus \mid \odot$.

The component S (*species component*) abstracts a biological species and the component P (*model component*) describes the system and the interactions among components. The prefix term (α, κ) **op** S contains information about the role of the species in the reaction associated with the action type α: κ is the *stoichiometry coefficient* of the species and the *prefix combinator* "op" represents the role of the element in the reaction. Specifically, $\downarrow$ indicates a *reactant*, $\uparrow$ a *product*, $\oplus$ an *activator*, $\ominus$ an *inhibitor* and $\odot$ a generic *modifier*. The operator "+" expresses choice between possible actions and the constant C is defined by an equation $C \stackrel{def}{=} S$. The parameter $x \in \mathbb{R}^+$ in $S(x)$ represents the initial quantity (for instance the concentration) of the species. Finally, the process $P \bowtie_{\mathcal{I}} Q$ denotes the cooperation between components: the set $\mathcal{I}$ determines those activities on which the operands are forced to synchronize. In Bio-PEPA the rates are not expressed in the syntax of components but are defined as functional rates. These allow us to express any kind of kinetic law. Each action is associated with a specific functional rate.

A possible modelling style supported by Bio-PEPA is in terms of concentration levels. This is the style considered in the derivation of the transition system for Bio-PEPA. The species concentrations can be discretized into a number of levels. The granularity of the system is expressed in terms of the *step size h*, i.e. the length of the concentration interval representing a level. The information about the step sizes and the number of levels for each species is collected in a set $\mathcal{N}$. Specifically, the elements of the set $\mathcal{N}$ have the form: "$C : h = value\_h, N = value\_N, M = value\_M, V = value\_V, unit = value\_u$", where C is the species component name, h is the step size, N is the maximum level, M is the maximum concentration, V is the name of the enclosing compartment and unit is the unit for concentration.

In order to fully describe a biochemical network in Bio-PEPA we need to define structures that collect information about the compartments, the maximum concentrations, number of levels for all the species, the constant parameters and the functional rates. The Bio-PEPA system is defined in the following way:

Definition 1. *A Bio-PEPA system $\mathcal{P}$ is a 6-tuple $\langle \mathcal{V}, \mathcal{N}, \mathcal{K}, \mathcal{F}_R, Compon - ents, P \rangle$, where: $\mathcal{V}$ is the set of compartments, $\mathcal{N}$ is the set of quantities describing species, $\mathcal{K}$ is the set of parameter definitions, $\mathcal{F}_R$ is the set of functional rates, $Components$ is the set of definitions of sequential components, P is the model component describing the system.*

For details see [9,10].

The behaviour of the system is defined in terms of an operational semantics. This refers to the level-based modelling style and in this context the parameter in the species components stands for the concentration level. We define two relations. The former, called *capability relation*, is indicated by $\stackrel{\theta}{\rightarrow}_c$. The label θ is of the form (α, w), where $w := [S : op(l, \kappa)] \mid w :: w$, with S a species component, op a symbol representing the role of the species in the reaction, l the level and κ the stoichiometry coefficient. This relation is defined as the minimum relation satisfying the rules reported in Table 1.

Table 1. Axioms and rules for Bio-PEPA

prefixReac $((\alpha,\kappa)\downarrow S)(l) \xrightarrow{(\alpha,[S:\downarrow(l,\kappa)])}_c S(l-\kappa)\quad \kappa \le l \le N$

prefixProd $((\alpha,\kappa)\uparrow S)(l) \xrightarrow{(\alpha,[S:\uparrow(l,\kappa)])}_c S(l+\kappa)\quad 0 \le l \le (N-\kappa)$

prefixMod $((\alpha,\kappa)\,op\,S)(l) \xrightarrow{(\alpha,[S:op(l,\kappa)])}_c S(l)\quad$ with $op = \odot, \oplus, \ominus$ and

$$0 < l \le N \text{ if } op = \oplus,\ 0 \le l \le N \text{ otherwise}$$

choice1

$$\frac{S_1(l) \xrightarrow{(\alpha,w)}_c S_1'(l')}{(S_1 + S_2)(l) \xrightarrow{(\alpha,w)}_c S_1'(l')}$$

choice2

$$\frac{S_2(l) \xrightarrow{(\alpha,w)}_c S_2'(l')}{(S_1 + S_2)(l) \xrightarrow{(\alpha,w)}_c S_2'(l')}$$

constant

$$\frac{S(l) \xrightarrow{(\alpha,S:[op(l,\kappa)])}_c S'(l')}{C(l) \xrightarrow{(\alpha,C:[op(l,\kappa)])}_c S'(l')}\quad \text{with } C \stackrel{def}{=} S$$

coop1

$$\frac{P_1 \xrightarrow{(\alpha,w)}_c P_1'}{P_1 \bowtie_{\mathcal{L}} P_2 \xrightarrow{(\alpha,w)}_c P_1' \bowtie_{\mathcal{L}} P_2}\quad \text{with } \alpha \notin \mathcal{L}$$

coop2

$$\frac{P_2 \xrightarrow{(\alpha,w)}_c P_2'}{P_1 \bowtie_{\mathcal{L}} P_2 \xrightarrow{(\alpha,w)}_c P_1 \bowtie_{\mathcal{L}} P_2'}\quad \text{with } \alpha \notin \mathcal{L}$$

coop3

$$\frac{P_1 \xrightarrow{(\alpha,w_1)}_c P_1' \quad P_2 \xrightarrow{(\alpha,w_2)}_c P_2'}{P_1 \bowtie_{\mathcal{L}} P_2 \xrightarrow{(\alpha,w_1::w_2)}_c P_1' \bowtie_{\mathcal{L}} P_2'}\quad \text{with } \alpha \in \mathcal{L}$$

The latter relation, called *stochastic relation*, is $\rightarrow_s \subseteq \tilde{\mathcal{P}} \times \Gamma \times \tilde{\mathcal{P}}$, where $\tilde{\mathcal{P}}$ is the set of well-defined Bio-PEPA systems[1] and Γ is the set of labels $\gamma = (\alpha, r)$, with α the action type and r the associated rate. This relation is defined as the minimal relation satisfying the rule:

---

[1] In a *well-defined* Bio-PEPA system each element has to satisfy some conditions. For instance, we have that each species component $C \in Comp$ must have subterms of the form "$(\alpha, \kappa)\,op\,C$" and the action types in each single component must be all distinct. Furthermore, the model component P must be defined in terms of the species components defined in $Comp$ and, for each cooperation set $\mathcal{L}_j$ in P, $\mathcal{L}_j \subseteq \mathcal{A}(P)$. For details see [12].

Final
$$\frac{P\xrightarrow{(\alpha_j,w)}_c P'}{\langle \mathcal{V},\mathcal{N},\mathcal{K},\mathcal{F},Comp,P\rangle\xrightarrow{(\alpha_j,r_\alpha[w,\mathcal{N},\mathcal{K}])}_s\langle \mathcal{V},\mathcal{N},\mathcal{K},\mathcal{F},Comp,P'\rangle}$$

The element $r_\alpha[w,\mathcal{N},\mathcal{K}]$ is the rate associated with the action α and is defined as:
$$r_\alpha[w,\mathcal{N},\mathcal{K}] = \frac{f_\alpha[w,\mathcal{N},\mathcal{K}]}{h}$$

where h is the step size for the species involved in the reaction and the notation $f_\alpha[w,\mathcal{N},\mathcal{K}]$ means that the function f_α is evaluated over w and the information about parameters and species components contained in the sets $\mathcal{N}$ and $\mathcal{K}$.

In this definition r_α represents the parameter of a negative exponential distribution. The dynamic behaviour of processes is determined by a *race condition*: all activities enabled attempt to proceed but only the fastest succeeds.

A *Stochastic Labelled Transition System (SLTS)* is defined for a Bio-PEPA system. From this we can obtain a *continuous time Markov Chain (CTMC)*. Both the SLTS and the CTMC derived from Bio-PEPA are defined in terms of levels of concentration. We call this Markov chain the *CTMC with levels*.

Bio-PEPA can be seen as an *intermediate, formal, compositional* representation of biological systems, from which different kinds of analysis can be performed. We have defined some mappings from Bio-PEPA to ODEs, CTMC with levels, stochastic simulation and PRISM [26]. Some tools for the analysis of Bio-PEPA system have been implemented [3]. In the following we report a brief description of the mapping from Bio-PEPA to ODE, as it is used later in the paper. For further details and the other mappings see [10].

2.1   From Bio-PEPA to ODE System (π_{ODE})

Let π_{ODE} be the mapping from Bio-PEPA system to the associated ODE system.
The mapping π_{ODE} entails three steps:

1. definition of the stoichiometry ($n \times m$) matrix $\mathbf{D}$, where n is the number of species and m is the number of reactions;
2. definition of the *kinetic law vector* ($m \times 1$) $\mathbf{v_{KL}}$ containing the kinetic laws of each reaction;
3. definition of the vector ($n \times 1$) $\mathbf{x}$, with $\mathbf{x}^T = (x_1, x_2, ..., x_n)$.

A crucial part is the derivation of the stoichiometry matrix $\mathbf{D} = \{d_{ij}\}$. The entries of the matrix are obtained as follows: for each sequential component C_i consider the prefix subterms C_{ij} representing the contribution of the species i to the reaction j. If the term represents a reactant we write the corresponding stoichiometry κ_{ij} as $-\kappa_{ij}$ in the entry d_{ij}. In the case of a product we write $+\kappa_{ij}$. All other cases are null. The kinetic law vector is derived from the functional rates and its definition is straightforward.

The ODE system thus obtained has the form:

$$\frac{d\mathbf{x}}{dt} = \mathbf{D} \times \mathbf{v_{KL}}$$

where the vector of initial concentrations is $\mathbf{x_0}$, with $x_{i,0}$ the initial concentration of the species i, as given in the specification of the system.

2.2   Example

In order to show how to model biochemical systems in Bio-PEPA we consider the network presented in Fig. 1 and we translate it into Bio-PEPA. This network is then used as a running example in the rest of the paper.

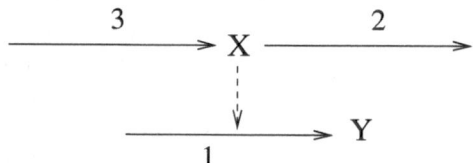

Fig. 1. Biochemical network composed of two proteins X and Y. The numbers indicate the reactions. Reaction 1 is the translation of Y enhanced by X, reaction 2 is the degradation of X and reaction 3 the translation of X.

The network is composed of two proteins, X and Y. These are involved in the following interactions:

- Translation of Y enhanced by X (reaction 1): $X \xrightarrow{r_1} X + Y$.
  The kinetic law is mass-action with constant parameter $r_1 = 0.01$;
- Degradation of the protein X (reaction 2): $X \xrightarrow{r_2} \emptyset$.
  The kinetic law is mass-action with constant parameter $r_2 = 0.02$;
- Translation of the protein X (reaction 3): $\emptyset \xrightarrow{r_3} X$.
  The kinetic law is mass-action with constant parameter $r_3 = 0.01$.

Each reaction i is represented by an action type α_i. The kinetic laws are represented by the following functional rates:

$$f_{\alpha_1} = fMA(0.01); \quad f_{\alpha_2} = fMA(0.02); \quad f_{\alpha_3} = 0.01;$$

where $fMA(r)$ stands for mass-action kinetic law with rate r.

The Bio-PEPA species components[2] corresponding to the two proteins are:

$$X \stackrel{def}{=} (\alpha_1, 1) \oplus X + (\alpha_2, 1) \downarrow X + (\alpha_3, 1) \uparrow X \qquad Y \stackrel{def}{=} (\alpha_1, 1) \uparrow Y$$

---

[2] Note that we use X and Y (capital letters) to indicate the names of the species and the name of the Bio-PEPA components, whereas x and y indicate the associated species concentrations.

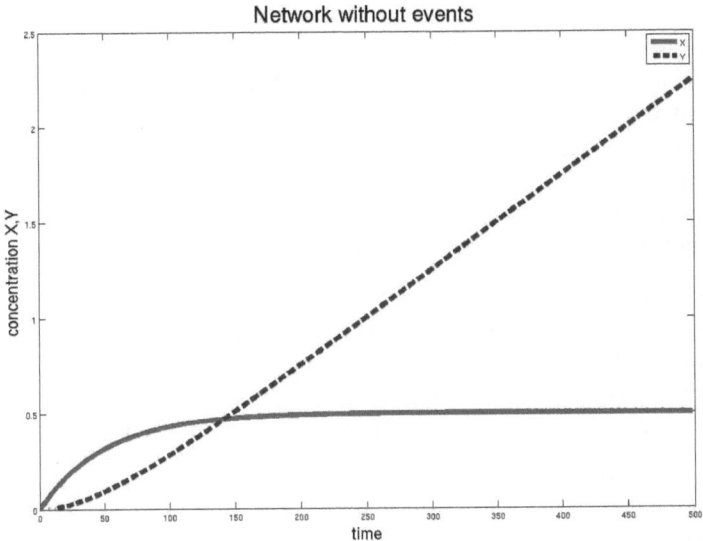

Fig. 2. ODE integration results for the network

whereas the model component is:

$$X(0) \underset{\{\alpha_1\}}{\bowtie} Y(0)$$

where the initial values are zero for both the proteins.

The set of compartments and the set $\mathcal{N}$ are not reported.

Applying the mapping π_{ODE} we obtain the ODE system:

$$\begin{array}{l} \frac{dx}{dt} = -0.02 \cdot x + 0.01 \\ \frac{dy}{dt} = 0.01 \cdot x \end{array}$$

where x and y are the two variables describing X and Y. The result of ODE integration is reported in Fig. 2. The protein X reaches a steady-state whereas Y increases infinitely.

3   Bio-PEPA with Events

3.1   SBML-Like Events: Some Definitions

In this work we consider events as defined in the SBML specification [22]. SBML events describe explicit discontinuous state changes in the model. Specifically, an SBML event has the following structure:

"event_id, if *trigger* then *event_assignment_list* with *delay"*

where

- *event_id* is the event identifier,
- *trigger* is a mathematical expression that, when it is evaluated to true, makes the event fire. It can be composed of one or more conditions;
- *event_assignment_list* is a list of assignments that are made when the event is executed;
- *delay* is the length of time between the time when the event fires and the time when the event assignments are executed.

The trigger and the list of assignments are both mandatory and can involve parameters, species concentrations and compartment sizes. All the triggers are initially evaluated to false. An SBML-like event is *immediate* if *delay* is equal to zero. Otherwise, the event is called *delayed*.

The definition of sequential and simultaneous events is reported below.

Definition 2. *Two or more SBML-like events are sequential if they are fired one after the other in a given order. They are said to be simultaneous if they happen at the same point in time.*

In most biochemical systems which we are interested in we have sequential events. In the general situation of simultaneous events, sometimes some *tie-breaking* rules are necessary to decide which of any set of events is simulated first. The most common way to do this is to assign a *priority* to each event [13]: when there are two or more simultaneous events, the event with the highest priority is defined to be the next event to fire. However, the order in which a set of simultaneous events is fired is not always important, for instance when the assignments of the events influence different variables. We have the following definition:

Definition 3. *Two simultaneous events are independent if their event assignments do not effect each other. Otherwise, they are called dependent.*

If we have simultaneous independent events we may abstract them as a single event and the system is reset according to the assignments of all the set of simultaneous events. Simultaneous independent events are dealt with similarly to sequential ones.

3.2   Assumptions

We make the following assumptions for the events considered in this work.

1. Triggers can involve time and species components' names, while assignments can involve species components (concentrations), compartments (size), parameters (values) and functional rates (function definitions).
2. Triggers are deterministic, i.e. when they become true they are fired.
3. Triggers are only unidirectional, i.e. describing the change from one mode to another, but not vice versa. Bidirectional triggers can be decomposed into two unidirectional triggers.
4. Events are either sequential or simultaneous and independent.

These assumptions are not restrictive. Indeed the events satisfying these assumptions allow us to represent a large number of discontinuous changes that we can find in biological systems.

3.3   The Definition of the Language

We can add events to a Bio-PEPA system by introducing a *set* of elements that have the form $(id, trigger, event\_assignment, delay)$, where *id* is the name of the event, *trigger* is a mathematical expression involving the components of the Bio-PEPA model and time, *event_assignment* is a list of assignments, *delay* is 0 (*immediate events*) or positive real value (*delayed events*). Formally, we have the following definitions:

$$
\begin{aligned}
trigger \quad &::= \quad \text{cond} \mid \text{cond } \textbf{or } \text{cond} \mid \text{cond } \textbf{and } \text{cond} \mid \textbf{not } \text{cond} \\
cond \quad &::= \quad t \text{ eq value} \mid expression(\bar{C}, \bar{k}) \text{ eq value} \mid \\
&\qquad expression(\bar{C}, \bar{k}) \text{ eq } expression(\bar{C}, \bar{k}) \\
eq \quad &::= \quad = \mid \neq \mid > \mid < \mid \leq \mid \geq \\
delay \quad &::= \quad value \\
event\_assignment \quad &::= \quad assignment \text{ ; } event\_assignment \\
assignment \quad &::= \quad \text{k} \leftarrow value \mid C \leftarrow value \mid f_\alpha \leftarrow expression(\bar{C}, \bar{k}) \\
&\qquad V \leftarrow value \mid t \leftarrow value \\
event \quad &::= \quad (\text{id, trigger, event\_assignment, delay})
\end{aligned}
$$

where C stands for any species component, k for any parameter and V for any compartment, f_α is the functional rate associated with the action type α, the variable $t \in \mathbb{R}^+$ represents the global time of the system, $expression(\bar{C}, \bar{k})$ is an arithmetic expression involving a set of components (denoted $\bar{C}$) and a set of parameters (denoted $\bar{k}$), $value \in \mathbb{R}^+$ and *id* is a string indicating the event name. Note that in the assignment definition, C indicates the concentration of the associated species component and V the size of the associate compartment. The assignment involving time is just auxiliary to express delayed events (see Sec. 4.2).

The set of events is then defined as:

$$Events \quad ::= \quad [] \mid event :: Events$$

Definition 4. *A Bio-PEPA system with events $\mathcal{P}$ is an 8-tuple $\langle \mathcal{V}, \mathcal{N}, \mathcal{K}, \mathcal{F}, Comp, P, Events, t \rangle$, where Events is the set of events, $t \in \mathbb{R}^+$ is the variable expressing time and the other elements are as in standard Bio-PEPA.*

A Bio-PEPA system is well-defined if all the elements are well-defined. The definition of well-definedness for all the elements, with the exception of events, is reported in [10].

Definition 5. *The set Events is well-defined if and only if the following conditions hold:*

- *triggers involve time or the name of species components, assignments involve species components, compartments, parameters and functional rates;*

- *all the elements used in the events are defined in the Bio-PEPA system;*
- *all the triggers are different and do not overlap in their values;*
- *given an event, the different assignments are independent (i.e. involve different elements).*

In the following we refer to Bio-PEPA with events simply as Bio-PEPA. Only well-defined Bio-PEPA systems are considered.

3.4   Example (Continued)

Consider the simple network described in Sect. 2.2. We can assume that the translation of the protein X is possible only when the concentration of Y is less than 0.8. This is expressed in Bio-PEPA by the following immediate event involving concentrations:

$$(event_1,\ Y = 0.8,\ r_3 \leftarrow 0,\ 0)$$

where r_3 is the constant rate for the translation of protein X. When the concentration of Y reaches the value 0.8, the value of r_3 becomes 0 and therefore the creation of X is not possible anymore.

In addition to this events, we can assume that when the concentration of X is less or equal than a given value (0.2, for instance) and the concentration of Y greater than 0.8 the creation of X is enabled again but with a smaller rate than before ($r_3 = 0.005$).

This is expressed in Bio-PEPA by the following immediate event:

$$(event_2,\ Y > 0.8\ \textbf{and}\ X = 0.2,\ r_3 \leftarrow 0.005,\ 0)$$

These two events are sequential and clearly satisfy the assumptions discussed above.

4   Mapping to Hybrid Automata

4.1   Hybrid Automata

Hybrid automata (HA) [19] combine discrete transition graphs with continuous dynamical systems. They are used to formally model hybrid systems, dynamical systems with both discrete and continuous components. An hybrid automaton consists of a finite set of *real-valued variables* $\{X_1, X_2, ..., X_n\}$ and a finite labelled graph, whose vertices correspond to *control modes* (states), described by differential equations, and whose edges are *control switches*, corresponding to discrete events. In addition, we have some labels for the edges, specifying the *jump conditions* (activation conditions) and labels for the vertices, containing information about initial and invariant conditions. The variables evolve continuously in time, apart from some changes induced by events. When an event happens there is a change in the mode. The dynamic behaviour of each mode is described by a set of differential equations, generally different from mode to mode. We can use HA both for simulation (see for instance *the SHIFT language* [15]) and model checking (see *HyTech* [20]). In this work we limit our attention to simulation. For a formal definition and details of the formalism see [19].

4.2   Definition of the Mapping

Here we present the map from Bio-PEPA to HA. First, we limit our attention to the case of immediate events and then we show a way to represent delayed events. Indeed, the translation of the delay associated with an event is not straightforward in the usual definition of HA.

Let $\mathcal{P}_0 = \langle \mathcal{V}_0, \mathcal{N}_0, \mathcal{K}_0, \mathcal{F}_0, Comp_0, P_0, Events, t_0 \rangle$ be the initial Bio-PEPA system and let N_{events} be the number of events. We have the following correspondences:

1. Each species component C_i in $Comp$ is associated with a variable X_i. The set of variables is then given by $\{X_1, X_2, \cdots, X_{N_{Comp}}, t\}$, where t is the variable expressing the time and N_{Comp} is the number of species components. The evolution of the variable t is described by the trivial differential equation $dt/dt = 1$.
2. The initial conditions of the variables are derived from the initial model component P_0. The variable t is initially set to 0.
3. For each event i in $Events$, we can consider the trigger tr_i. We use these triggers to define the jump conditions. In the case we have only sequential events, the number of possible jump conditions N_{jump} is just N_{Events}. If simultaneous independent events are possible, we may combine them together in order to define a new jump condition representing the union of the triggers of the simultaneous events. In this case, the system is reset according to the union of the assignment lists of the events involved.
4. Each mode is described by a specific instance of the Bio-PEPA system. Indeed modes are defined according to either the initial system or the system modified with the event assignments relative to a trigger. The number of modes is $N_{jump} + 1$. σ is used to indicate a mode and the Σ the set of all modes. In each mode some invariant conditions are added in order to force the change of mode when the trigger becomes true. We have that:
   - The initial mode σ_0 is defined from the initial system $\mathcal{P}_0$. It is described in terms of an ODE system and this is derived from the Bio-PEPA model by considering the map π_{ODE}. Therefore, we have $\sigma_0 = \pi_{ODE}(\mathcal{P}_0)$.
   - Given a mode $\sigma_i = \pi_{ODE}(\mathcal{P}_i)$, let tr_{ij} be one possible jump condition that can be satisfied from it. We define the Bio-PEPA system $\mathcal{P}_j = \mathcal{P}_i[event\_assignment_{ij}]$ as the modification of the previous system $\mathcal{P}_i$ according to the event assignments associated with the trigger. The mode σ_j is then defined as $\sigma_j = \pi_{ODE}(\mathcal{P}_j)$.

Case of delayed events. The delay associated with an event represents the time interval between when the event is fired and when its assignments are executed. This information cannot be directly translated in any of the components of standard HA. In the following we report as we handle the delay in HA.

First, we introduce a new variable t_{mode} representing the time when the system enters in a specific mode. It is initially set to zero. The differential equation associated with this new variable is $dt_{mode}/dt = 0$, i.e. this variable is constant in each mode.

Second, given an event $(id, trigger, event\_assignment, delay)$, we split it into two immediate events, defined as:

1. $(id_1, trigger, t_{mode} \leftarrow t, 0)$;
2. $(id_2, t = t_{mode} + delay, event\_assignment, 0)$.

The role of the former event is to introduce the delay whereas the role of the second is to guarantee that the assignments of the initial event are executed after the given delay.

4.3   Example (Continued 2)

Consider the network presented in Sect. 2.2 with the addition of the set of events (see Sect. 3.4):
$$[(event_1, \; Y = 0.8, \; r_3 \leftarrow 0, \; 0)].$$
A schema of the HA associated with this network is reported in Fig. 3. The set of variables is $\{x, y, t\}$, where x and y are the two variables representing the two proteins. The initial concentrations, derived from the initial condition in the Bio-PEPA model, are $x = 0$, $y = 0$ and $t = 0$. We have just one event so we have two modes and the jump condition (guard) is $y = 0.8$. The former mode is described by the invariant condition $y < 0.8$ and the latter by $y \geq 0.8$.

The ODE system corresponding to the initial mode ($S1$) is derived by applying the mapping π_{ODE} to the initial Bio-PEPA system ($\mathcal{P}_0$) and is:

$$\begin{aligned}\frac{dx}{dt} &= -0.02 \cdot x + 0.01 \\ \frac{dy}{dt} &= 0.01 \cdot x\end{aligned}$$

For the second mode, the ODE system ($S2$) is obtained as $\pi_{ODE}(\mathcal{P}_0[r_3 \leftarrow 0])$ and is:

$$\begin{aligned}\frac{dx}{dt} &= -0.02 \cdot x \\ \frac{dy}{dt} &= 0.01 \cdot x\end{aligned}$$

If we consider both $event_1$ and $event_2$, we have the HA represented in Fig. 4. There are three modes, representing the network at the initial state, when $y < 0.8$ and when $y \geq 0.8$ and $x \geq 0.2$. Two jumps conditions are defined in terms of the trigger conditions.

The ODE systems describing the first and second modes are as above, whereas the ODE system for the third mode ($S3$) is obtained from $\pi_{ODE}(\mathcal{P}_1[r_3 \leftarrow 0.005])$ (where $\mathcal{P}_1$ is the Bio-PEPA system corresponding to the second mode) and is:

Fig. 3. HA representation for the network composed of the two proteins X and Y and with an event involving concentrations

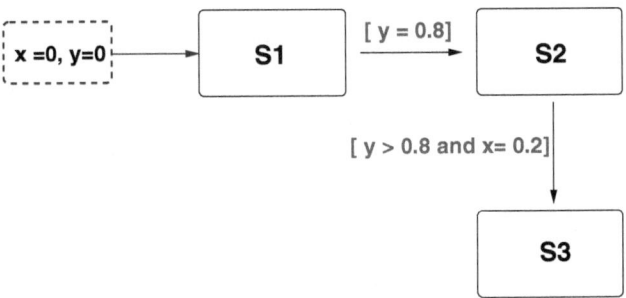

Fig. 4. HA representation for the network composed of the two proteins X and Y and with two sequential events

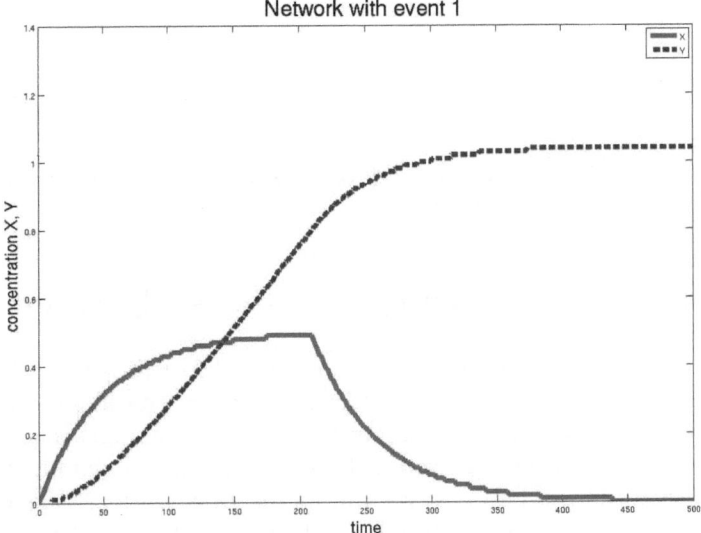

Fig. 5. Simulation results for the network composed of the two proteins X and Y and with the addition of $event_1$

$$\frac{dx}{dt} = -0.02 \cdot x + 0.005$$
$$\frac{dy}{dt} = 0.01 \cdot x$$

Some results for the network with just $event_1$ are reported in Fig. 5. The protein X increases until time 200 s when Y reaches the value 0.8 and then decreases to 0. The protein Y increases, but after the event, its rate of increase is much lower than the case without the event.

The results for the network with both $event_1$ and $event_2$ is reported in Fig. 6. In this case, when the second event is fired, the protein X starts to increase again and this has effect on the production of Y as well.

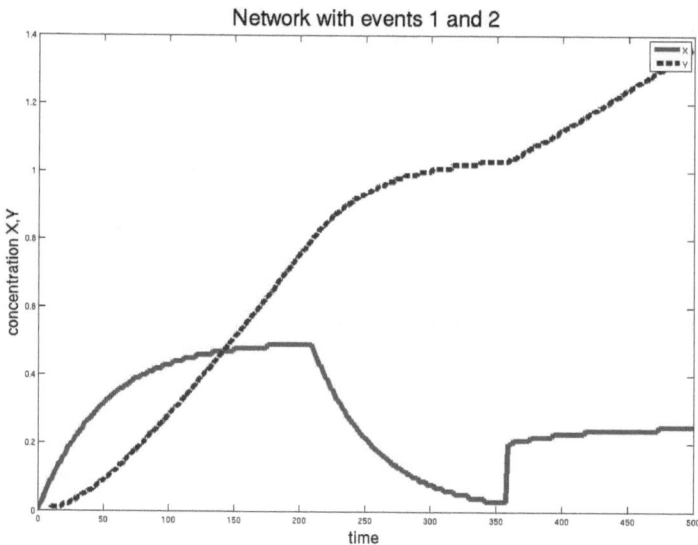

Fig. 6. Simulation results for the network composed of the two proteins X and Y and with the addition of $event_1$ and $event_2$

5   Stochastic Simulation by Gillespie's Algorithm

One of the possible kinds of analysis supported by Bio-PEPA is stochastic simulation using Gillespie's algorithm [10]. When events are considered the algorithm has to be modified in order to handle them. Broadly speaking, events are tackled by adding some conditions and some checks along the simulation. We start at time $t = 0$, with the Bio-PEPA system in its initial conditions. We assume that initially all the triggers evaluate to false. When one of the conditions is satisfied, the simulation stops and the system is modified according to the event assignments associated with the trigger. After that, the simulation can start again until another condition becomes true or the simulation time is reached. The use of triggers involving time can be challenging since it can happen that the time of the event does not coincide with any of the simulation time points. Our approach to deal with this case is discussed below.

Note that if the events involve species concentrations, we have to change concentrations into number of molecules for stochastic simulation. Specifically, we have to multiply each concentration by $Na V$, where Na is the Avogadro number[3] and V is the volume of the compartment. In the the rest of this section we assume that the events are in terms of number of molecules.

---

[3] This is the number of "entities" (atoms or molecules) in one mole of substance. Its value is $6.022 \times e{+}23 \ (mol)^{-1}$.

We propose the following procedure for each simulation run.

1. Let $\mathcal{P}_0$ be the initial Bio-PEPA system and t_s the maximum simulation time.
2. While $t < t_s$ and $trigger_i = false$ for $i = 1, 2, ..., N_{Events}$, simulate.
3. If $t \geq t_s$ then stop.
4. If $t < t_s$ and there exists a $trigger_i$ such that it is true, we have that:
   (a) if $delay = 0$ modify the Bio-PEPA system according to the event assignments associated with that trigger: $\mathcal{P}'(t) = \mathcal{P}(t)[event\_assignment_i]$. Go to (2).
   (b) if $delay > 0$ go on with the simulation until time $t + delay$ and then proceed as in (a).

Some final observations concern how to use the algorithm in two particular situations.

- In the case of two or more independent simultaneous events we proceed as observed in Section 3.1: we can abstract these events as a single event, whose trigger is defined in terms of the triggers of the two events and the event assignments are the union of the assignments. Therefore, we modify the system according to the assignments associated with all the events involved.
- When we have an event with a trigger involving time $t = \tilde{t}$, the time value $\tilde{t}$ may not correspond to any of the simulation time point obtained by using Gillespie's simulation algorithm. Specifically, there exist two consecutive simulation time points t_j and t_{j+1} such that $t_j < \tilde{t} < t_{j+1}$. If this happens, we have to decide when the system has to be modified. In order to handle this situation we consider the following approach:

  1. if $t_j < \tilde{t} + delay < t_{j+1}$ with $delay \geq 0$ consider the system at time $\tilde{t} + delay$ and modify it at that time point. The simulation restarts from $\tilde{t} + delay$.
  2. If $delay > 0$ and $\tilde{t} + delay \geq t_{j+1}$ consider the last simulation time point $t_h \leq \tilde{t} + delay$ and run the simulation until t_h. Then, modify the system at time $\tilde{t} + delay$ and restart the simulation from that time point.

6   The Acetylcholine Receptor Model

This example concerns the functional properties of the *nicotin Acetylcholine Receptors (nAChR)*. These are transmembrane proteins that mediate interconversions between open and closed channel states under the control of neurotransmitters. The detailed description of the model is reported in [16].

A schema of the model is shown in Figure 7. B (*Basal state*), A (*Active state*), D (*Desensitized state*) and I (*Inactivable state*) represent the different states of the Acetylcholine receptors. The numbers 0, 1, 2 associated with the state are the number of ligands (denoted X) bound to a receptor. In the model the ligands are not modelled explicitly. Each column corresponds to a series of ligand binding actions at two identical sites per receptor whereas each row corresponds to a series of transactions between conformational states. All the

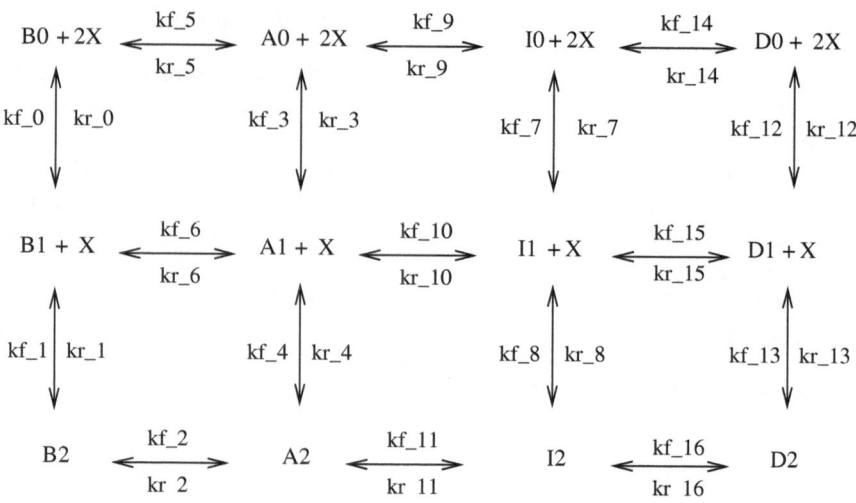

Fig. 7. Schema of the Acetylcholine receptor model

Table 2. The Acetylcholine receptor model. The list of parameters. The unit is s^{-1}.

parameter	value	parameter	value	parameter	value	parameter	value
kf_0	3000	kr_0	8000	kf_1	1500	kr_1	16000
kf_2	30000	kr_2	700	kf_3	3000	kr_3	8.64
kf_4	1500	kr_4	17.28	kf_5	0.54	kr_5	10800
kf_6	130	kr_6	2740	kf_7	3000	kr_7	4
kf_8	1500	kr_8	8	kf_9	19.7	kr_9	3.74
kf_{10}	19.85	kr_{10}	1.74	kf_{11}	20	kr_{11}	0.81
kf_{12}	3000	kr_{12}	4	kf_{13}	1500	kr_{13}	8
kf_{14}	0.05	kr_{14}	0.0012	kf_{15}	0.05	kr_{15}	0.0012
kf_{16}	0.05	kr_{16}	0.0012				

reactions are reversible and the dynamics are described by mass-action laws. For each reaction i, with $i = 1, 2, ...16$, the rate of the forward direction is kf_i and the rate of the reverse direction kr_i.

In addition to these elements, there is an event to describe the recovery upon removal of free agonist at a given time. This is expressed by constraining the reaction rates of each second-order ligand-receptor reaction to zero. These constraints prevent ligand binding reactions from happening after that time, hence the states evolve as if the free ligands were completely removed from the system. The event is immediate, the trigger is "$t = t_2$", where $t_2 = 20$ s, and the event assignments are $kf_0 \leftarrow 0$, $kf_1 \leftarrow 0$, $kf_3 \leftarrow 0$, $kf_4 \leftarrow 0$, $kf_7 \leftarrow 0$, $kf_8 \leftarrow 0$, $kf_{12} \leftarrow 0$, $kf_{13} \leftarrow 0$.

The Bio-PEPA system associated with the Acetylcholine receptor model. In the following we report briefly the definition of the Bio-PEPA system representing the Acetylcholine receptor model. The complete system is reported in the Appendix A.

- *Definition of the compartment list* $\mathcal{V}$. In the model we have a single three-dimensional compartment, defined as "$compl : 1e\text{-}16, l;$", where l is litre.
- *Definition of the set* $\mathcal{N}$. Each species is associated with a species component. For each species component we have to declare the step size, the number of levels, the initial and maximum concentrations and the compartment where the species is. The ligand is not represented explicitly. For instance, in the case of $B0$, $B1$ and $B2$ we have:

$$B0 : H = h, \ N = N_{B0}, \ M = M_{B0}, \ V = compl, \ unit = \mu M;$$
$$B1 : H = h, \ N = N_{B1}, \ M = M_{B1}, \ V = compl, \ unit = \mu M;$$
$$B2 : H = h, \ N = N_{B2}, \ M = M_{B2}, \ V = compl, \ unit = \mu M;$$

where the step size is $1.66e\text{-}5$, the number of levels $N_{B0} = N_{B1} = N_{B2}$ is 1 (i.e. the species can be present, 1, or absent, 0), the maximum concentration $M_{B0} = M_{B1} = M_{B2}$ is $1.66e\text{-}5$ and coincides with the initial concentration of channels at the basal state. Note that the information about the step size and the number of levels is not used in this work, as we do not consider CTMC with levels, however we define them for completeness.

- *Definition of functional rates* ($\mathcal{F}_R$) *and parameters* ($\mathcal{K}$). Each reversible reaction i, $i = 0, 1, 2, \cdots, 16$, is decomposed in two irreversible reactions, f_i and r_i, representing the forward and inverse directions respectively. The associated kinetic laws are $f_{\alpha\_f_i} = fMA(kf\_i)$; and $f_{\alpha\_r_i} = fMA(kr\_i)$, where fMA denotes mass-action. All the parameters are defined in the set $\mathcal{K}$. The values are the ones reported in the paper [16].
- *Definition of species components* ($Comp$) *and of the model component* (P). In the following we report the definition for $B0$, $B1$ and $B2$; the other species are dealt with similarly.

$$B0 \stackrel{def}{=} (\alpha\_f_0, 1)\downarrow B0 + (\alpha\_r_0, 1)\uparrow B0 + (\alpha\_f_5, 1)\downarrow B0 + (\alpha\_r_5, 1)\uparrow B0$$
$$B1 \stackrel{def}{=} (\alpha\_f_0, 1)\uparrow B1 + (\alpha\_r_0, 1)\downarrow B1 + (\alpha\_f_6, 1)\downarrow B1 + (\alpha\_r_6, 1)\uparrow B1 +$$
$$(\alpha\_f_1, 1)\uparrow B1 + (\alpha\_r_1, 1)\downarrow B1$$
$$B2 \stackrel{def}{=} (\alpha\_f_2, 1)\downarrow B2 + (\alpha\_r_2, 1)\uparrow B2 + (\alpha\_f_1, 1)\uparrow B2 + (\alpha\_r_1, 1)\downarrow B2$$

The system is described as:

$$B0(1.66e\text{-}5) \underset{L1}{\bowtie} B1(0) \underset{L2}{\bowtie} B2(0) \underset{L3}{\bowtie} A0(0) \underset{L3}{\bowtie} A1(0) \underset{L4}{\bowtie} A2(0) \underset{L5}{\bowtie}$$
$$I0(0) \underset{L6}{\bowtie} I1(0) \underset{L7}{\bowtie} I2(0) \underset{L8}{\bowtie} D0(0) \underset{L9}{\bowtie} D1(0) \underset{L10}{\bowtie} D2(0)$$

where L_i, $i = 1, ..., 10$ are the cooperation sets and the initial values for the species are 0 with the exception of the species $B0$.

- *Definition of events*. We have only one event, describing a change in the system at time 20 s:

$$[(event, t = 20, \ kf_0 \leftarrow 0; \ kf_1 \leftarrow 0; \ kf_3 \leftarrow 0 \ kf_4 \leftarrow 0; \ kf_7 \leftarrow 0; \ kf_8 \leftarrow 0;$$
$$kf_{12} \leftarrow 0; \ kf_{13} \leftarrow 0, \ 0)]$$

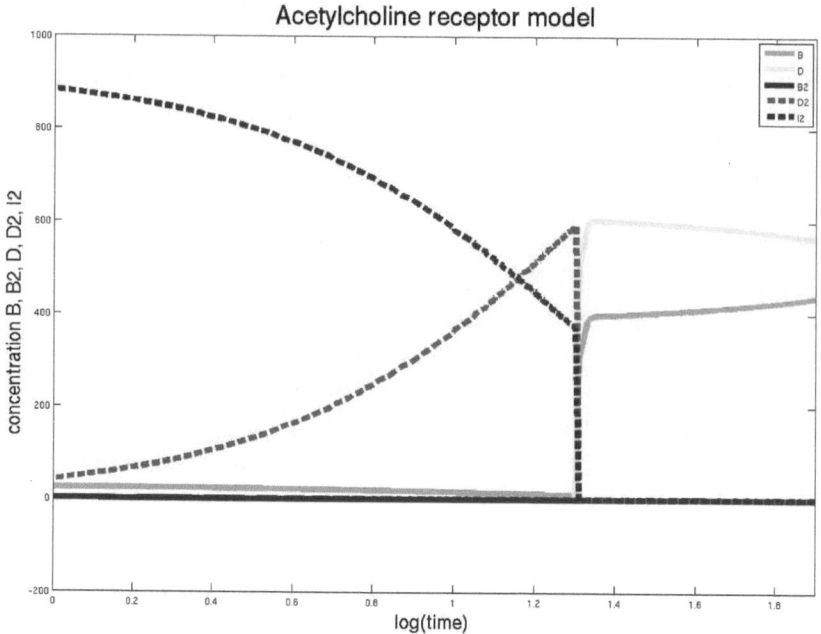

Fig. 8. Stochastic simulation results for the Acetylcholine receptor model (average over 100 runs)

Analysis results. The HA associated with the Acetylcholine receptor model is similar to the one for the network presented in Sect.2.2 with the addition of the set of events. We have two modes, described by two different sets of differential equations. The trigger condition involves time and is "$t = 20\ s$". The details of the two systems describing each mode are not reported.

Simulation results made using Gillespie's algorithm are reported in Fig. 8. The initial number of molecules for $B0$ is given $M_0 \times V \times Na = (1.66e\text{-}5\ \mu M) \times (1.e\text{-}16\ l) \times (6.022 \times e+23\ (mol)^{-1}) = 1000$, where Na is the Avogadro number. All the other species are initially null. The number of runs is 100. The graph reproduces results in agreement with the ones reported in the paper [16]. Following the ligand removal, the state $I2$ loses agonist molecules and is transformed to the state $B0$ very rapidly, while $D2$ loses ligand molecules to form $D0$. Since the data occur on a wide range of times we represent the time on a logarithmic scale.

7   Related Works

The use of mathematical formalisms in order to represent discrete changes in biological systems is not new [1,4,23,17,5,6]. In [1] the authors proposed a hybrid system approach to modelling an intra-cellular network using continuous differential equations to model some part of the system and mode-switching to describe the changes in the underlying dynamics. Some models with hybrid behaviour are presented and described using *CHARON* [2], a language that allows formal description of hybrid systems. The authors of [23] discussed the use of discrete changes in biological systems and presented some examples using the formalism *HybridSAL* [21]. *Hybrid Concurrent Constraint Programming* is used to model some biological systems with both discrete and continuous changes in [4]. In [5] the authors presented a map from *stochastic Concurrent Constraint Programming (sCCP)* to HA. The HA generated in this way are said to be able to capture some aspects of the dynamics which are lost if standard differential equations are used. A discussion of hybrid systems and biology is reported in [6]. Finally, in [17] the authors presented *HYPE*, a process algebra for the modelling of hybrid systems and used it to represent the *repressilator*, an artificial genetic network composed of three genes and their respective proteins with oscillatory behaviour. In none of these works are SBML-like events considered explicitly, but the focus is on general hybrid systems.

Events have been proposed in the *Beta Workbench (BetaWB)* [14] and in the associated programming language *BlenX* [28]. In BlenX events can be considered as global rules of the environment, triggered only when the conditions associated with them are satisfied. Each event is the composition of a condition (*cond*) and an action (*verb*) and is associated with a rate. Conditions can involve number of entities, the simulation time or the simulation step. The possible actions are the join of two entities, the split of one entity into two, the update of a variable of the system and the deletion or the creation of a new entity.

The concept of events proposed in BlenX is quite similar to the one considered for Bio-PEPA. The BlenX condition and action correspond to the trigger and event assignment in Bio-PEPA events. However, rates in BlenX have a different meaning from the delay in Bio-PEPA. Indeed, in Bio-PEPA an events occurs when the trigger is satisfied and the role of the delay is to postpone when the event is executed. BlenX events with a finite rate can happen only when the trigger is satisfied but it is in competition with other actions that are enabled contemporaneously (race condition). BlenX events with infinite rate correspond to immediate events in Bio-PEPA.

In order to compare the definition of events in the two languages, we show how the events proposed in this paper can be described in BlenX. The event $event_1$ defined in Sec. 3.4 is represented in BlenX as:

$$\text{when } (Y \rightarrow value) \text{ update } (r_3, change\_par)$$

where *value* is $0.8 \cdot NaV$ molecules and the function *change_par* is defined as *change_par : function* $= 0$. The operator "$\rightarrow$" recognizes when the quantity bound to Y becomes greater than the specified value, whereas the action " update $(r_3, change\_par)$" means that the parameter r_3 is updated according to the function *change_par* (in our case it assigns the value 0). The rate associated with the update action is always infinite and not reported. Concerning the event in the Acetylcholine receptor model, it is not possible to represent this event in BlenX as conditions involving time are not allowed with the action update.

Note that BlenX events represent more general kinds of interactions than Bio-PEPA events. For instance, they are used to model the formation of a complex (by using the action join) or the split of a complex into two parts (by using the split action). These reactions (as all the other kinds) are represented in Bio-PEPA by synchronization of the species components over the action types abstracting the reactions. Bio-PEPA events have been introduced specifically to represent experimental situations when there is change in the system due to some conditions.

8   Conclusions

In this work we have presented an extension of Bio-PEPA to handle *SBML-like events*. Events are constructs that represent changes in the system due to some trigger conditions. The events considered here are simple, but nevertheless able to describe most of the discontinuous changes in models and experiments. Events are added to our language without any modification to the rest of the syntax. The main motivation of this choice is that we want to keep the specification of the model as simple as possible. Furthermore, this approach is appropriate when we study the same biochemical system but with different experimental regimes.

A topic for the future concerns the study of more general events and the possible extension to other kinds of hybrid systems in biology. Furthermore, we plan to exploit the possible kinds of analysis involving hybrid systems in the context of systems biology. In this paper we focus on the mapping to Hybrid Automata and stochastic simulation by (a modification of) Gillespie's algorithm. Further investigation will concern the application of model checking for the study of the properties of biological systems.

Acknowledgements

The author thanks Jane Hillston, Vashti Galpin and Adam Duguid for their helpful comments. The author is supported by the EPSRC under the CODA project "Process Algebra Approaches for Collective Dynamics" (EP/c54370x/01).

References

1. Alur, R., Belta, C., Ivancic, F., Kumar, V., Mintz, M., Pappa, G., Rubin, H., Schug, J.: Hybrid modeling and simulation of biomolecular networks. In: Di Benedetto, M.D., Sangiovanni-Vincentelli, A.L. (eds.) HSCC 2001. LNCS, vol. 2034, pp. 19–32. Springer, Heidelberg (2001)
2. Alur, R., Grosu, R., Hur, Y., Kumar, V., Lee, I.: Modular Specification of Hybrid Systems in CHARON. In: Lynch, N.A., Krogh, B.H. (eds.) HSCC 2000. LNCS, vol. 1790, p. 6. Springer, Heidelberg (2000)
3. Bio-PEPA Workbench Home Page,
   http://www.dcs.ed.ac.uk/home/stg/software/biopepa/

4. Bockmayr, A., Courtois, A.: Using hybrid concurrent constraint programming to model dynamic biological systems. In: Stuckey, P.J. (ed.) ICLP 2002. LNCS, vol. 2401, p. 85. Springer, Heidelberg (2002)

5. Bortolussi, L., Policriti, A.: Hybrid Approximation of Stochastic Concurrent Constraint Programming. Constraints 13(1-2), 66–90 (2008)

6. Bortolussi, L., Policriti, A.: Hybrid Systems and Biology. Continuous and Discrete Modeling for Systems Biology. In: Bernardo, M., Degano, P., Zavattaro, G. (eds.) SFM 2008. LNCS, vol. 5016, pp. 424–448. Springer, Heidelberg (2008)

7. Calder, M., Gilmore, S., Hillston, J.: Automatically deriving ODEs from process algebra models of signalling pathways. In: Proc. of CMSB 2005, pp. 204–215 (2005)

8. Calder, M., Gilmore, S., Hillston, J.: Modelling the influence of RKIP on the ERK signalling pathway using the stochastic process algebra PEPA. In: Priami, C., Ingólfsdóttir, A., Mishra, B., Riis Nielson, H. (eds.) Transactions on Computational Systems Biology VII. LNCS (LNBI), vol. 4230, pp. 1–23. Springer, Heidelberg (2006)

9. Ciocchetta, F., Hillston, J.: Bio-PEPA: an extension of the process algebra PEPA for biochemical networks. In: Proc. of FBTC 2007. ENTCS, vol. 194(3), pp. 103–117 (2008)

10. Ciocchetta, F., Hillston, J.: Bio-PEPA: a framework for the modelling and analysis of biological systems. Theoretical Computer Science (to appear)

11. Ciocchetta, F.: Bio-PEPA with SBML-like Events. In: Proc. of the Workshop Computational Models for Cell Processes, TUCS general publication, vol. 47 (2008)

12. Ciocchetta, F., Hillston, J.: Bio-PEPA: a framework for the modelling and analysis of biological systems. School of Informatics University of Edinburgh Technical Report, EDI-INF-RR-1231 (2008)

13. Cota, B.A., Sargent, R.B.: Simultaneous events and distributed simulation. In: Proc. of the Winter Simulation Conference (1990)

14. Dematté, L., Priami, C., Romanel, A.: The BlenX Language: a Tutorial. In: Bernardo, M., Degano, P., Zavattaro, G. (eds.) SFM 2008. LNCS, vol. 5016, pp. 313–365. Springer, Heidelberg (2008)

15. Deshpande, A., Gollu, A., Semenzato, L.: SHIFT Programming Language and Run-Time System for Dynamic Networks of Hybrid Automata. PATH Report, http://path.berkeley.edu/SHIFT/publications.html

16. Edelstein, S.J., Schaad, O., Henry, E., Bertrand, D., Changgeux, J.P.: A kinetic mechanism for nicotin acetylcholine receptors based on multiple allosteric transitions. Biol. Cybern. 75, 361–379 (1996)

17. Galpin, V., Hillston, J., Bortolussi, L.: HYPE applied to the modelling of hybrid biological systems. ENTCS, vol. 218, pp. 33–51 (2008); Also in Proceedings of MFPS 2008

18. Gillespie, D.T.: Exact stochastic simulation of coupled chemical reactions. Journal of Physical Chemistry 81, 2340–2361 (1977)

19. Henzinger, T.A.: The Theory of Hybrid Automata. In: The proceedings of the 11th Annual IEEE Symposium on Logic in Computer Science, LICS (1996)

20. Henzinger, T.A., Ho, P.-H., Wong-Toi, H.: HyTech: A Model Checker for Hybrid Systems. Software Tools for Technology Transfer 1, 110–122 (1997)

21. HybridSal home page, http://sal.csl.sri.com/hybridsal/

22. Hucka, M., Finney, A., Hoops, S., Keating, S., Le Novére, N.: Systems Biology Markup Language (SBML) Level 2: Structures and Facilities for Model Definitions, http://sbml.org/documents/

23. Lincoln, P., Tiwari, A.: Symbolic systems biology: Hybrid modeling and analysis of biological networks. In: Alur, R., Pappas, G.J. (eds.) HSCC 2004. LNCS, vol. 2993, pp. 660–672. Springer, Heidelberg (2004)
24. Le Novére, N., Bornstein, B., Broicher, A., Courtot, M., Donizelli, M., Dharuri, H., Li, L., Sauro, H., Schilstra, M., Shapiro, B., Snoep, J.L., Hucka, M.: BioModels Database: a Free, Centralized Database of Curated, Published, Quantitative Kinetic Models of Biochemical and Cellular Systems. Nucleic Acids Research 34, D689–D691 (2006)
25. Priami, C., Quaglia, P.: Beta-binders for biological interactions. In: Danos, V., Schachter, V. (eds.) CMSB 2004. LNCS (LNBI), vol. 3082, pp. 20–33. Springer, Heidelberg (2005)
26. Prism web site, http://www.prismmodelchecker.org/
27. Priami, C., Regev, A., Silverman, W., Shapiro, E.: Application of a stochastic name-passing calculus to representation and simulation of molecular processes. Information Processing Letters 80, 25–31 (2001)
28. Dematté, L., Priami, C., Romanel, A.: The Beta Workbench: a computational tool to study the dynamics of biological systems. Briefings in Bioinformatics 9(5), 437–449 (2008)

A   Appendix: Bio-PEPA System for the Acetylcholine Receptor Model

In this appendix we report the specification of the whole Acetylcholine receptor model in Bio-PEPA. Note that, in the definition of the species component, we use the following notation: $>>$ indicates a product (it corresponds to the operator $\uparrow$ in the Bio-PEPA syntax) and $<<$ indicates a reactant (it corresponds to the operator $\downarrow$). This is the syntax used in the Bio-PEPA tools [3].

```
\\Definition of compartments
comp1: 1e-16, 1;


\\Definition information about species components
[ B0: H= 1.66-5, N= 2, M= 1.66-5, V= comp1, unit = muM;
  B1: H= 1.66-5, N= 2, M= 1.66-5, V= comp1, unit = muM;
  B2: H= 1.66-5, N= 2, M= 1.66-5, V= comp1, unit = muM;
  A0: H= 1.66-5, N= 2, M= 1.66-5, V= comp1, unit = muM;
  A1: H= 1.66-5, N= 2, M= 1.66-5, V= comp1, unit = muM;
  A2: H= 1.66-5, N= 2, M= 1.66-5, V= comp1, unit = muM;
  I0: H= 1.66-5, N= 2, M= 1.66-5, V= comp1, unit = muM;
  I1: H= 1.66-5, N= 2, M= 1.66-5, V= comp1, unit = muM;
  I2: H= 1.66-5, N= 2, M= 1.66-5, V= comp1, unit = muM;
  D0: H= 1.66-5, N= 2, M= 1.66-5, V= comp1, unit = muM;
  D1: H= 1.66-5, N= 2, M= 1.66-5, V= comp1, unit = muM;
  D2: H= 1.66-5, N= 2, M= 1.66-5, V= comp1, unit = muM ]


\\Definition of parameters
[ kf_0= 3000;   kr_0= 8000;   kf_1= 1500;   kr_1= 16000;
  kf_2=30000;   kr_2= 700;   kf_3= 3000;   kr_3= 8.64;
  kf_4= 1500;   kr_4= 17.28;   kf_5= 0.54;   kr_5= 10800;
```

```
kf_6= 130;   kr_6= 2740;   kf_7= 3000;   kr_7= 4;
kf_8= 1500;   kr_8= 8;   kf_9= 19.7;   kr_9= 3.74;
kf_10= 19.85;   kr_10= 1.74;   kf_11= 20;   kr_11= 0.81;
kf_12= 3000;   kr_12= 4; kf_13= 1500;   kr_13= 8;
kf_14= 0.05;   kr_14= 0.0012; kf_15= 0.05; kr_15= 0.0012;
kf_16= 0.05;   kr_16=  0.0012 ]

\\Definition of functional rates
\\all kinetic laws are MA
[ f_alpha_f_0= fMA(kf_0);   f_alpha_r_0=  fMA(kr_0);
  f_alpha_f_1= fMA(kf_1);   f_alpha_r_1=  fMA(kr_1);
  f_alpha_f_2= fMA(kf_2);   f_alpha_r_2=  fMA(kr_2);
  f_alpha_f_3= fMA(kf_3);   f_alpha_r_3=  fMA(kr_3);
  f_alpha_f_4= fMA(kf_4);   f_alpha_r_4=  fMA(kr_4);
  f_alpha_f_5= fMA(kf_5);   f_alpha_r_5=  fMA(kr_5);
  f_alpha_f_6= fMA(kf_6);   f_alpha_r_6=  fMA(kr_6);
  f_alpha_f_7= fMA(kf_7);   f_alpha_r_7=  fMA(kr_7);
  f_alpha_f_8= fMA(kf_8);   f_alpha_r_8=  fMA(kr_8);
  f_alpha_f_9= fMA(kf_9);   f_alpha_r_9=  fMA(kr_9);
  f_alpha_f_10= fMA(kf_10);   f_alpha_r_10=  fMA(kr_10);
  f_alpha_f_11= fMA(kf_11);   f_alpha_r_11=  fMA(kr_11);
  f_alpha_f_12= fMA(kf_12);   f_alpha_r_12=  fMA(kr_12);
  f_alpha_f_13= fMA(kf_13);   f_alpha_r_13=  fMA(kr_13);
  f_alpha_f_14= fMA(kf_14);   f_alpha_r_14=  fMA(kr_14);
  f_alpha_f_15= fMA(kf_15);   f_alpha_r_15=  fMA(kr_15);
  f_alpha_f_16= fMA(kf_16);   f_alpha_r_16=  fMA(kr_16);
]


\\Species components
B0 = (alpha_f_0,1)<<B0 + (alpha_r_0,1)>>B0 + (alpha_f_5,1)<<B0 +
     (alpha_r_5,1)>>B0
B1 = (alpha_f_0,1)>>B1 + (alpha_r_0,1)<<B1 + (alpha_f_6,1)>>B1 +
     (alpha_r_6,1)<<B1+ (alpha_f_1,1)<<B1 + (alpha_r_1,1)>>B1
B2 = (alpha_f_2,1)<<B2 + (alpha_r_2,1)>>B2 + (alpha_f_1,1)>>B2 +
     (alpha_r_1,1)<<B2
A0 = (alpha_f_5,1)>>A0 + (alpha_r_5,1)<<A0 + (alpha_f_3,1)<<A0 +
     (alpha_r_3,1)>>A0 + (alpha_r_9,1)<<A0 + (alpha_f_9,1)>>A0
A1 = (alpha_f_3,1)>>A1 + (alpha_r_3,1)<<A1 + (alpha_f_4,1)<<A1 +
     (alpha_r_4,1)>>A1 + (alpha_f_6,1)<<A1 + (alpha_r_6,1)>>A1 +
     (alpha_r_10,1)<<A1 + (alpha_f_10,1)>>A1
A2 = (alpha_f_2,1)>>A2 + (alpha_r_2,1)<<A2 + (alpha_f_4,1)>>A2 +
     (alpha_r_4,1)<<A2 + (alpha_f_11,1)>>A2 + (alpha_r_11,1)>>A2
I0 = (alpha_f_7,1)<<I0 + (alpha_r_7,1)>>I0 + (alpha_f_9,1)>>I0 +
     (alpha_r_9,1)<<I0 + (alpha_f_14,1)<<I0 + (alpha_r_14,1)>>I0
I1 = (alpha_f_7,1)>>I1 + (alpha_r_7,1)<<I1 + (alpha_f_8,1)<<I1 +
     (alpha_r_8,1)>>I1 + (alpha_f_10,1)<<I1 + (alpha_r_10,1)>>I1 +
     (alpha_r_15,1)<<I1 + (alpha_f_15,1)>>I1
I2 = (alpha_f_8,1)>>I2 + (alpha_r_8,1)<<I2 + (alpha_f_11,1)>>I2 +
     (alpha_r_11,1)<<I2 + (alpha_r_16,1)<<I2 + (alpha_f_16,1)>>I2
D0 = (alpha_f_12,1)<<D0 + (alpha_r_12,1)>>D0 + (alpha_f_14,1)>>D0 +
```

```
       (alpha_r_14,1)>>D0
D1 = (alpha_f_12,1)>>D1 + (alpha_r_12,1)<<D1 + (alpha_f_13,1)<<D1 +
     (alpha_r_13,1)<<D1 + (alpha_f_15,1)>>D1 + (alpha_r_15,1)>>D1
D2 = (alpha_f_13,1)>>D2 + (alpha_r_13,1)<<D2 + (alpha_f_16,1)>>D2 +
     (alpha_r_16,1)<<D2
```

```
\\Model components
B0(1.66e-5) <kf_0,kr_0> B1(0) <kf_1,kr_1> B2(0) <kf_5,kr_5>
A0(0) <kf_3,kr_3,kf_6,kr_6> A1(0) <kf_4,kr_4,kf_2,kr_2> A2(0) <kf_9,kr_9>
I0(0) <kf_7,kr_7> I1(0) <kf_8,kr_8,kf_10,kr_10> I2(0)<kf_14,kr_14>
D0(0) <kf_12,kr_12,kf_15,kr_15> D1(0) <kf_13,kr_13,kf_16,kr_16> D2(0)
```

```
\\Event
[(event,  t = 20,  kf_0 <- 0; kf_1 <- 0; kf_3 <- 0; kf_4 <- 0;
  kf_7 <- 0; kf_8 <- 0; kf_12 <- 0; kf_{13} <- 0,   0)]
```

In Silico Modelling and Analysis of Ribosome Kinetics and aa-tRNA Competition

D. Bošnački[1,*], T.E. Pronk[2,**], and E.P. de Vink[3,***]

[1] Dept. of Biomedical Engineering, Eindhoven University of Technology
[2] Swammerdam Institute for Life Sciences, University of Amsterdam
[3] Dept. of Mathematics and Computer Science,
Eindhoven University of Technology
evink@win.tue.nl

Abstract. We present a formal analysis of ribosome kinetics using probabilistic model checking and the tool Prism. We compute different parameters of the model, like probabilities of translation errors and average insertion times per codon. The model predicts strong correlation to the quotient of the concentrations of the so-called cognate and near-cognate tRNAs, in accord with experimental findings and other studies. Using piecewise analysis of the model, we are able to give an analytical explanation of this observation.

1 Introduction

The translation mechanism that synthesizes proteins based on mRNA sequences is a fundamental process of the living cell. Conceptually, an mRNA can be seen as a string of codons, each coding for a specific amino acid. The codons of an mRNA are sequentially read by a ribosome, where each codon is translated using an amino acid specific transfer-RNA (aa-tRNA), building one-by-one a chain of amino acids, i.e. a protein. In this setting, aa-tRNA can be interpreted as molecules containing a so-called anticodon, and carrying a particular amino acid. Dependent on the pairing of the codon under translation with the anticodon of the aa-tRNA, plus the stochastic influences such as the changes in the conformation of the ribosome, an aa-tRNA, arriving by Brownian motion, docks into the ribosome and may succeed in adding its amino acid to the chain under construction. Alternatively, the aa-tRNA dissociates in an early or later stage of the translation.

Since the seventies a vast amount of research has been devoted, unraveling the mRNA translation mechanism and related issues. By now, the overall process of translation is reasonably well understood from a qualitative perspective. The translation process consists of around twenty small steps, a number of them being reversible. For the model organism *Escherichia coli*, the average frequencies of aa-tRNAs per cell have been collected, but regarding kinetics relatively little is

---

* Supported by FP6 LTR ESIGNET.
** Funded by the BSIK project Virtual Laboratory for e-Science VL-e.
*** Corresponding author.

C. Priami et al. (Eds.): Trans. on Comput. Syst. Biol. XI, LNBI 5750, pp. 69–89, 2009.
© Springer-Verlag Berlin Heidelberg 2009

known exactly. Over the past few years, Rodnina and collaborators have made good process in capturing the time rates for various steps in the translation process for a small number of specific codons and anticodons [21,23,24,12]. Using various advanced techniques, they were able to show that the binding of codon and anticodon is crucial at a number of places for the time and probability for success of elongation. Based on these results, Viljoen and co-workers started from the assumption that the rates found by Rodnina et al. can be used in general, for all codon-anticodon pairs as estimates for the reaction dynamics. In [9], a complete detailed model is presented for all 64 codons and all 48 aa-tRNA classes for *E. coli*, on which extensive Monte Carlo experiments are conducted. In particular, using the model, codon insertion times and frequencies of erroneous elongations are established. Given the apparently strong correlation of the ratio of so-called near-cognates vs. cognate and pseudo-cognates, and near-cognates vs. cognates, respectively, it is argued that competition of aa-tRNAs, rather than their availability decides both speed and fidelity of codon translation.

In the present paper, we propose to exploit abstraction and modelchecking of continuous-time Markov chains (CTMCs) with Prism [18,13] for the case of mRNA translation. The abstraction conveniently reduces the number of states and classes of aa-tRNA to consider. The tool provides built-in performance analysis algorithms and path chasing machinery, relieving its user from mathematical calculations. The outcomes are exact, unlike approximations obtained by simulation. More importantly, from a methodological point of view, the incorporated CSL-logic [2] allows to establish quantitative results for parts of the system, e.g. for first-passage time for a specific state. Such piecewise analysis proves useful when explaining the relationships suggested by the data collected from the model. Additionally, in our case, the Prism tool enjoys rather favourably response times compared to simulation.

Related work. Measurements in *E. coli* by Sørensen et al. [25] already suggested dependence of the availability of various codons, qualified as 'rare' or 'frequent', and translation rates. Wahab c.s. [27] noted that in *E. coli* strains expressing high levels of $tRNA_1^{Leu}$ isoacceptor, an increase of available *tRNA* led to a decrease of protein production. The present investigation started from the Monte-Carlo experiments of mRNA translation reported in [9]. A similar stochastic model, but based on ordinary differential equations, was developed in [14]. It treats insertion times, but no translation errors. The model of mRNA translation in [10] assumes insertion rates that are directly proportional to the mRNA concentrations, but assigns the same probability of translation error to all codons.

Currently, there exist various applications of formal methods to biological systems. A selection of recent papers from model checking and process algebra includes [22,5,6]. More specifically pertaining to the current paper, [4] applies the Prism model checker to analyze stochastic models of signaling pathways. Their methodology is presented as a more efficient alternative to ordinary differential equations models, including properties that are not of probabilistic nature. Also [13] employs Prism on various types of biological pathways, showing how the advanced features of the tool can be exploited to tackle large models.

In [3], we use the model presented in this paper to perform a formal analysis of amino acid replacement during mRNA translation. Building on the abstract stochastic model of arrival of tRNAs and their processing at the ribosome presented in the sequel, we compute probabilities of the insertion of amino acids into the nascent polypeptide chain. This allows us to construct the substitution matrix containing the probabilities of an amino acid replacing another. Finally in [3], we discuss the analogy of this matrix with standard mutation matrices, and analyze the mutual replacement of biologically similar amino acids. The main contribution of the present paper is the study of the underlying model of mRNA translation itself, exploiting probabilistic model checking and the approach of piecewise analysis.

Organization of the paper. Section 2 provides the biological background, discussing the mRNA translation mechanism. Its Prism model is introduced in Section 3. In Section 4, it is explained how error probabilities are obtained from the model and why they correlate with the near-cognate/cognate fraction. This involves adequate estimates of specific stochastic subbehaviour. Insertion times are the subject of Section 5. There too, it is illustrated how the quantitative information of parts of the systems is instrumental in deriving the relationship with the ratio of pseudo-cognate and near-cognates vs. cognates.[1]

2   A Kinetic Model of mRNA Translation

In nature, there is a fixed correspondence of a codon and an amino acid. This is the well-known genetic code, that couples all 61 relevant codons to 20 fundamental amino acids. The three codons not corresponding to an amino acid are so-called stop codons, that guide the termination of the translation process. Thus, an mRNA, as sequence of codons, codes for a unique sequence of amino acids, i.e. protein. However, the match of a codon and the anticodon of a tRNA is different from pair to pair. The binding influences the speed of the actual translation. Here, we give a brief overview of the translation mechanism. Our explanation is based on [23,17]. The basic idea is that mRNA is transcribed from the cellular DNA. A ribosome, an enzyme catalyzing translation, attaches to an individual mRNA and starts translating the sequence of codons into amino acids. The ribosome processes one codon at the time by recruiting aa-tRNA from the cell. Dependent on the match of the codon at the mRNA and the anticodon of the aa-tRNA, the amino acid carried by the aa-tRNA is transfer to the polypeptide chain under construction, i.e. the nascent protein. Two main phases can be distinguished: peptidyl transfer and translocation.

The peptidyl transfer phase runs through the following steps. aa-tRNA arrives at the A-site of the ribosome-mRNA complex by diffusion in a ternary complexation with elongation factor *Tu* (*EF-Tu*) and *GTP* at a rate determined by the interaction of *EF-Tu* and the ribosome. The initial binding is relatively

---

[1] An appendix presents supplementary data.

weak. Codon recognition comprises (i) establishing contact between the anti-codon of the aa-tRNA and the current codon in the ribosome-mRNA complex, and (ii) subsequent conformational changes of the ribosome. Given a codon, an anticodon can either be a cognate, a near-cognate or a non-cognate. As an aa-tRNA carries precisely one anticodon, we also speak of cognate, near-cognate and non-cognate aa-tRNA. The rates of confirmation of the ribosome are different for cognate and near-cognates. This does not apply to non-cognate anticodons; the aa-tRNA that carries it, dissociates from the ribosome almost immediately. GTPase-activation of the elongation factor *EF-Tu* is largely favoured, because of the conformational changes, in case of a strong complementary matching of the codon and a cognate anticodon. Otherwise, GTPase-activation is lessened. After GTP-hydrolysis, producing inorganic phosphate P_i and GDP, the affinity of the ribosome for the aa-tRNA reduces. The subsequent accommodation step also depends on the fit of the aa-tRNA. Accomodation happens rapidly for cognate aa-tRNA, whereas for near-cognate aa-tRNA this proceeds slower and the aa-tRNA is likely to be rejected.

The subsequent translocation phase will shift the peptide chain in nascent and the mRNA including the codon just processed, exposing a new codon for translation and releasing the A-site for the arrival of another aa-tRNA. The first step of the translocation phase involves the association of GTP. By GTP-hydrolysis of elongation factor *EF-G*, GDP and P_i are produced. This results in unlocking and movement of the aa-tRNA to the P-site of the ribosome. The latter step is preceded or followed by P_i-release, with GDP bound or unbound to P_i, respectively. Reconformation of the ribosome and release of *EF-G* moves the tRNA, that has transferred its amino acid to the polypeptide chain, into the E-site of the ribosome. Further rotation eventually leads to dissociation of the used tRNA.

Although overall qualitatively well understood, there is, at present, limited quantitative information regarding the translation mechanism and its individual steps. For *E. coli*, a number of specific rates have been collected by Rodnina and co-workers [23,12]. Some steps are known to be relatively rapid. The fundamental assumption of [9], that we also adopt here, is that experimental data found for the UUU and CUC codons, regarding their matching to *Phe-tRNA*, extrapolate to other codon-anticodon pairs as well. However, further assumptions are necessary to fill the overall picture. In particular, Viljoen proposes to estimate the delay due to so-called non-cognate aa-tRNA, that are blocking the ribosomal A-site, at 0.5ms. Also, accurate rates for the translocation phase are largely missing. Again following [9], we have chosen to assign, if necessary, high rates to steps for which data is lacking. This way these steps will not be rate limiting.

An overview of the reactions involving cognates and near-cognates and the corresponding rates are collected in Table 1. The upper and lower parts of Table 1 correspond respectively to reactions involving cognate and near-cognate tRNAs and they differ only in the reaction rates. The first reaction step represents the arrival of the ternary complex C (N) (aa-tRNA, EF-Tu, and GTP) to the ribosome $R1$ and the initial binding between those. Subsequent reactions

Table 1. Molecular reactions underlying the adapted model. Rates taken from the model in [9] based on experiments reported in [23]. See the main text for an explanation of the individual reactions. Rates for the reactions *CR4* and *NR4* are obtained by merging subsequent reactions, as discussion in Section 3.

correspond to various conformational changes of the ribosome-mRNA complex. The first selection step happens by means of the inverse reaction that transforms *CR3* (*NR3*) back to *CR2* (*NR2*). Because of the higher rate the near- and non-cognate aa-tRNAs have much greater chance to be rejected than the cognate ones. The one way reaction from *CR3* (*NR3*) to *CR4* (*NR4*) includes the GTP hydrolysis step which means that the *aa-tRNA* has passed the first selection test. Reactions from the second row correspond to the proofreading step in which either *aa-tRNA* is definitely accepted and the corresponding amino acid incorporated into the polypeptide chain —represented by conformation state *CR8*— or it is rejected which results in disassociation of the *aa-tRNA* from the ribosome.

We will fit the above model of protein synthesis in the language of the Prism model checker. The experiments confirm the main results of [9], viz. (i) insertion errors are proportional to the quotient of the frequencies of near-cognates and of cognates, (ii) *aa-tRNA* competition better predicts insertion times than *aa-tRNA* availability. In fact, we show in the latter case the stronger result that the ratio of near-cognate and cognate frequencies is an adequate estimate for insertion time. In addition, for the above results, we are able to actually derive the correlation with the quotient of near-cognates and cognates. It is the availability of an explicit model, together with the possibility to obtain, by model checking, quantitative information for parts of the system, that lead to a sharper analysis of the experimental data, that cannot obtained by simulation of a monolithic model.

3   The Prism Model

The model employed in the analysis below is an abstraction of the biological model as sketched in the previous section. The abstraction is twofold: (i) Instead of dealing with 48 individual classes of *aa-tRNA*, that are identified by the their anticodons, we restrict to four types of *aa-tRNA* distinguished by their matching with the codon under translation. (ii) We combine various detailed steps into one transition by accumulation of rates. The first reduction greatly simplifies the model, more clearly eliciting the essentials of the underlying process. The second abstraction is more a matter of convenience, though it helps in compactly presenting the model.

For a specific codon, we distinguish below four types of aa-tRNA: cognate, pseudo-cognate, near-cognate, non-cognate. Cognate aa-tRNAs have an anti-codon that strongly couples with the codon. The amino acid carried by the aa-tRNA is always the right one, according to the genetic code. The binding of the anticodon of a pseudo-cognate aa-tRNA or a near-cognate aa-tRNA is weaker, but sufficiently strong to occasionally result in the addition of the amino acid to the nascent protein. In case the amino acid of the aa-tRNA is, accidentally, the right one for the codon, we dubb the aa-tRNA of the pseudo-cognate type. If the amino acid does not coincide with the amino acid the codon codes for, we speak in such a case of a near-cognate aa-tRNA.[2] The match of the codon and the anticodon can be very poor too. We refer to such aa-tRNA as being non-cognate for the codon. This type of aa-tRNA does not initiate a translation step at the ribosome.

The Prism model can be interpreted as the superposition of four stochastic automata, each encoding the interaction of one of the types of aa-tRNA. The automata for the cognates, pseudo-cognates and near-cognates are very similar; the cognate type automaton only differs in its value of the rates from those for pseudo-cognates and near-cognates, while the automata for pseudo-cognates and for near-cognates only differ in their arrival process. The automaton for non-cognates is rather simple. See Figure 1.

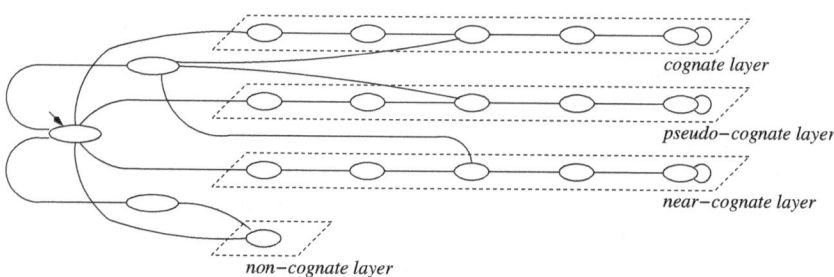

Fig. 1. Overview of Prism model as superposition of four aa-tRNA typed automata. Each layer models the processing of a specific type of aa-tRNA, viz. cognate, pseudo-cognate, near-cognate and non-cognate aa-tRNA.

Below, we are considering average transition times and probabilities for reachability based on exponential distributions. Therefore, following common practice

---

[2] The notion of a pseudo-cognate comes natural in our modeling. However, the distinction between a pseudo-cognate and a near-cognate is non-standard. Usually, a near-cognate refers to both types of tRNA.

in performance analysis, there is no obstacle to merge two subsequent sequential transitions with rates λ and μ, say, into a combined transition of rate $\lambda\mu/(\lambda+\mu)$. This way, a smaller model can be obtained that, although differently distributed, is equivalent to the original regarding expected values. However, it is noted, that in general, such a simplification is not compositional and should be taken with care.

We briefly discuss the Prism code implementing the abstract model. For the modeling of continuous-time Markov chains, Prism commands have the form

$$[\text{label}]\ \text{guard} \rightarrow \text{rate}:\ \text{update}\ ;$$

In short, from the commands whose guards are fulfilled in the current state, one command is selected proportional to its relative rate. Subsequently, the update is performed on the state variables. So, a probabilistic choice is made among commands. Executing the selected command results in a progress of time according to the exponential distribution for the particular rate. Labels are used to synchronize Prism commands, a feature not used in this paper. We refer to [18,13] for a proper introduction to the Prism model checker.

Initially, control resides in the common start state s=1 of the Prism model with four boolean variables cogn, pseu, near and nonc set to false.

```
s    : [0..8] init 1 ;
cogn : bool init false ;
pseu : bool init false ;
near : bool init false ;
nonc : bool init false ;
```

Next, an arrival process selects one of the booleans that is to be set to true. This is the initial binding of the ternary *aa-tRNA* complex at the ribosome. The continuation depends on the type of *aa-tRNA*: cognate, pseudo-cognate, near-cognate or non-cognate. In fact, a race is run that depends on the concentrations c_cogn, c_pseu, c_near and c_nonc of the four types of *aa-tRNA* and a kinetic constant k1f. Concentrations are taken from [7]. For example, following Markovian semantics, the probability in the race for cogn to be set to true (the others remaining false) is the relative concentration c_cogn/(c_cogn + c_pseu + c_near + c_nonc). The rates can therefore also be computed as relative frequencies per cell, as the volume of the cell cancels out. A small C program manipulating Table 4 in the appendix takes care of this. The values of the concentrations are provided to Prism via the command line, since they differ from codon to codon.

```
// initial binding
[ ] (s=1) -> k1f * c_cogn : (s'=2) & (cogn'=true) ;
[ ] (s=1) -> k1f * c_pseu : (s'=2) & (pseu'=true) ;
[ ] (s=1) -> k1f * c_near : (s'=2) & (near'=true) ;
[ ] (s=1) -> k1f * c_nonc : (s'=2) & (nonc'=true) ;
```

As the *aa-tRNA*, that has just arrived, may dissociate too, the reversed reaction
is in the model as well. However, control does not return to the initial state
directly, but, for model checking purposes, first visits the special state s=0 representing dissociation. At the same time, the boolean that was true is reset.
Here, cognates, pseudo-cognates and near-cognates are handled with the shared
rate k2b. Non-cognates always dissociate as captured by the separate rate k2bx.

```
// dissociation
[ ] (s=2) &  ( cogn | pseu | near ) -> k2b :
          (s'=0) & (cogn'=false) &
             (pseu'=false) & (near'=false) ;
[ ] (s=2) &  nonc -> k2bx : (s'=0) & (nonc'=false) ;
```

An *aa-tRNA* that is not a non-cognate can continue from state s=2 in the codon
recognition phase, leading to state s=3. This is a reversible step in the translation
mechanism, so there are transitions from state s=3 back to state s=2 as well.
However, the rates for cognates vs. pseudo- and near-cognates, viz. k3bc, k3bp
and k3bn, differ significantly (see Table 2), which is essential to the fidelity of
the *mRNA*-translation mechanism. Note that the values of the booleans do not
change.

```
// codon recognition
[ ] (s=2) & ( cogn | pseu | near ) -> k2f : (s'=3) ;
[ ] (s=3) & cogn -> k3bc : (s'=2) ;
[ ] (s=3) & pseu -> k3bp : (s'=2) ;
[ ] (s=3) & near -> k3bn : (s'=2) ;
```

The next forward transition, from state s=3 to state s=4 in the Prism model, is
a combination of several detailed steps of the translation mechanism involving
the processing of *GTP*. The transition is one-directional, again with a significant
difference in the rate k3fc for a cognate *aa-tRNA* compared to the rates k3fp
and k3fn for pseudo-cognate and near-cognate *aa-tRNA*, that are equal.

```
// GTPase activation, GTP hydrolysis
// and  EF-Tu conformation change
[ ] (s=3) & cogn -> k3fc : (s'=4) ;
[ ] (s=3) & pseu -> k3fp : (s'=4) ;
[ ] (s=3) & near -> k3fn : (s'=4) ;
```

In state s=4, the *aa-tRNA* can either be rejected, after which control moves to
intermediate state s=5, or accommodates, i.e. the ribosome reconforms such that
the *aa-tRNA* can hand over the amino acid it carries, so-called peptidyl transfer.
In the latter case, control changes to state s=6. As before, rates for cognates and
those for pseudo-cognates and near-cognates are of different magnitudes. From
the intermediate rejection state s=5, with all booleans set to false again, control
returns to the start state s=1.

```
// rejection
[ ] (s=4) & cogn -> k4rc : (s'=5) & (cogn'=false) ;
[ ] (s=4) & pseu -> k4rp : (s'=5) & (pseu'=false) ;
[ ] (s=4) & near -> k4rn : (s'=5) & (near'=false) ;


// accommodation, peptidyl transfer
[ ] (s=4) & cogn -> k4fc : (s'=6) ;
[ ] (s=4) & pseu -> k4fp : (s'=6) ;
[ ] (s=4) & near -> k4fn : (s'=6) ;
```

After some movement back-and-forth between state s=6 and state s=7, the binding of the EF-G complex becomes permanent. In the detailed translation mechanism a number of (mainly sequential) steps follows, that are summarized in the Prism model by a single transition to a final state s=8, that represents elongation of the protein in nascent with the amino acid carried by the *aa-tRNA*. The synthesis is successful if the *aa-tRNA* was either a cognate or pseudo-cognate for the codon under translation, reflected by either cogn or pseu being true. In case the *aa-tRNA* was a near-cognate (non-cognates never pass beyond state s=2), an amino acid that does not correspond to the codon in the genetic code has been inserted. Thus, in this case, an insertion error has occurred.

```
// EF-G binding
[ ] (s=6) -> k6f : (s'=7) ;
[ ] (s=7) -> k7b : (s'=6) ;


// GTP hydrolysis, unlocking, tRNA movement
// Pi release, rearrangements of ribosome and EF-G
// dissociation of GDP
[ ] (s=7) -> k7f : (s'=8) ;
```

A number of transitions, linking the dissociation state s=0 and the rejection state s=5 back to the start state s=1, where a race of *aa-tRNAs* of the four types commences anew, and looping at the final state s=8, complete the Prism model. The transitions are deterministically taking, as no other transitions leave these states. Having no biological counterpart the transitions are assigned a high-rate making the time they take negligible.

```
// no entrance, re-entrance at state 1
[ ] (s=0) -> FAST : (s'=1) ;
// rejection, re-entrance at state 1
[ ] (s=5) -> FAST : (s'=1) ;
// elongation
[ ] (s=8) -> FAST : (s'=8) ;
```

Table 2 collects the rates as compiled from the biological literature and used in the Prism model above.

Table 2. Rates of the Prism model, adapted from [9,26]. Rate k2bx is based on the estimate of the average delay of non-cognate arrivals of 0.5ms. Rates k4fc, k4fp, k4fn and k7f are accumulative rates of sequentially composed transitions.

k1f	140	k3fc	260	k4rc	60	k6f	150
k2f	190	k3fp, k3fn	0.40	k4rp, k4rn	FAST	k7f	145.8
k2b	85	k3bc	0.23	k4fc	166.7	k7b	140
k2bx	2000	k3bp, k3bn	80	k4fp, k4fn	46.1		

In the next two sections, we will study the Prism model described above for the analysis of the probability for insertion errors, i.e. extension of the peptidyl chain with a different amino acid than the codon codes for, and of the average insertion times, i.e. the average time it takes to process a codon up to elongation.

4   Insertion Errors

In this section we discuss how the model checking features of Prism can be exploited to predict the misreading frequencies for individual codons. The translation of mRNA into a polypeptide chain is performed by the ribosome machinery with high precision. Experimental measurements show that on average, only one in 1,000 to 10,000 amino acids is added wrongly (cf. [12]).[3]

For a codon under translation, a pseudo-cognate anticodon carries precisely the amino acid that the codon codes for. Therefore, although different in codon-anticodon bound, successful matching of a pseudo-cognate does not lead to an insertion error, as –accidentally– the right amino acid has been used for elongation. In our model, the main difference of cognates vs. pseudo-cognates and near-cognates is in the kinetics. At various stages of the peptidyl transfer the rates for true cognates differ from those for pseudo-cognates and near-cognates up to three orders of magnitude.

Figure 2 depicts the relevant abstract automaton, derived from the Prism model discussed above. See also Table 1. In case a transition is labeled with two rates, e.g. 0.23/80, the leftmost number, viz. 0.23, concerns the processing of a cognate aa-tRNA, while the rightmost number, viz. 80, that of a pseudo-cognate or near-cognate. In three states a probabilistic choice has to be made: in state 2 leading to state 0 or 3, in state 3 leading back to state 3 or forward to state 4, and in state 4 leading to rejection in state 5 or eventually to success via state 6. The probabilistic choice in state 2 is the same for cognates, pseudo-cognates and near-cognates alike, the ones in state 3 and in state 4 depend on the type of aa-tRNA, cognates and pseudo-cognates vs. near-cognates.

A cognate aa-tRNA starting in state 1 will move to state 2 with probability 1. From here, it will dissociate with probability $85/(85 + 190) \approx 0.309$, moving to

---

[3] Our findings, see Table 5, based on the kinetic rates available and the assumptions made, are well within these boundaries.

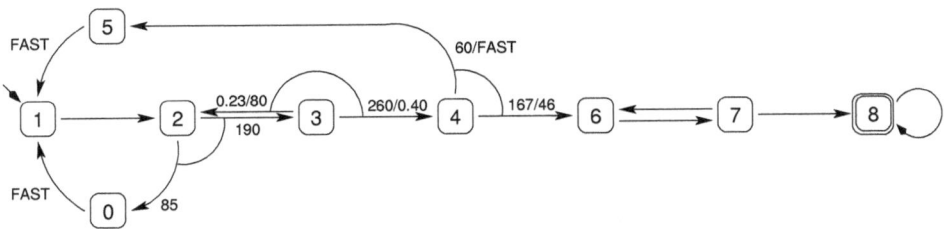

Fig. 2. Abstract automaton summarizing the Prism code. See also Table 1.

state 0, or will be recognized with the complementary probability $190/(85 + 190) \approx 0.691$, moving to state 3. The same holds for pseudo-cognate and near-cognate *aa-tRNA*. However, after recognition in state 3, a cognate *aa-tRNA* will go through the hydrolysis phase leading to state 4 for a fraction 0.999 of the cases (computed as $260/(0.23+260)$), a fraction being close to 1. In contrast, for a pseudo-cognate or near-cognate *aa-tRNA* this is $0.40/(0.40+80) \approx 0.005$ only. A similar difference can be noted in state 4 itself. Cognates will accommodate and continue to state 6 with probability 0.736, while pseudo-cognates and near-cognates will do so with the small probability 0.044, the constant FAST being set to 1000 in our experiments as in [9]. Since the transition from state 4 to state 6 is irreversible, the rates of the remaining transitions are not of importance here. For cognates, pseudo-cognates and near-cognates, the probability of reaching state 8 in one attempt can be easily computed, solving a small system of equations by hand or by using Prism. In the latter case, we have Prism evaluate the CSL-formula

$$\text{P=? [(s!=0 \& s!=5) U (s=8) \{(s=2) \& \mathit{cogn}\}]}$$

against our model. The formula asks to establish the probability for all paths where s is not set to 0 nor 5, until s have been set to 8, starting from the (unique) state satisfying s=2 & cogn. The expression $\{(s = 2)\&cogn\}$ is a so-called filter construction as supported by Prism. We obtain $p_s^c = 0.508$, $p_s^p = 0.484 \cdot 10^{-4}$ and $p_s^n = 0.484 \cdot 10^{-4}$, with p_s^c the probability for a cognate to end up in state 8 —and elongate the peptidyl chain— without going through state 0 nor state 5; p_s^p and p_s^n the analogues for success of pseudo- and near-cognates, respectively. Note that these values are the same for every codon.

Different among codons in *E. coli* are the concentrations of cognates, pseudo-cognates and near-cognates.[4] Ultimately, the frequencies f_c, f_p and f_n of the types of *aa-tRNA* in the cell, i.e. the actual number of molecules of the kind, determine the concentration of the *aa-tRNA*. Hence, under the usual assumption of homogeneous distribution, the frequencies determine the total rates for the arrival process of an anticodon. The probability for an anticodon arriving to be a cognate, pseudo-cognate or near-cognate can then be calculated from this.

---

[4] See Table 4 in the appendix.

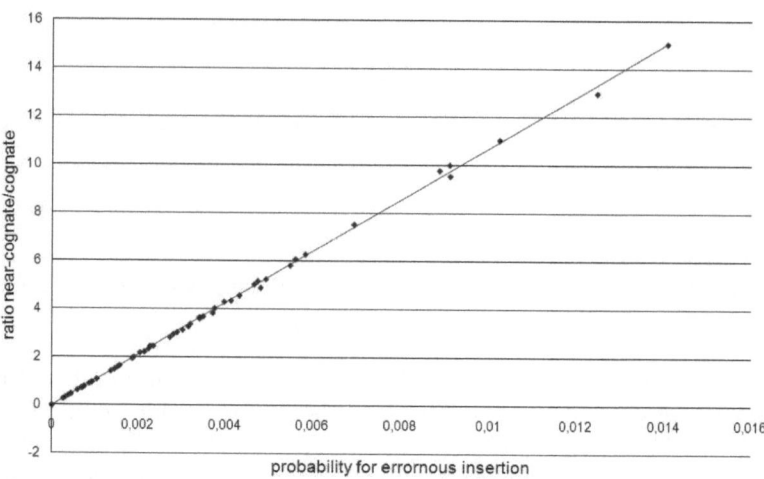

Fig. 3. Correlation of the ratio f_n/f_c of the frequency of near-cognates over the frequency of cognates vs. the probability of an insertion error. See also Table 5 in the appendix.

As concluded in [9] based on simulation results, the probability for an erroneous insertion, is strongly correlated with the quotient of the number of near-cognate anticodons and the number of cognate anticodons. See Figure 3.

As an advantage of the present setting, this correlation actually can be formally derived. This is as follows. We have that an insertion error occurs if a near-cognate succeeds to attach its amino acid. Note that we already have established $p_s^p, p_s^n \ll p_s^c$. Therefore,

$$P(\text{error}) = P(\text{near \& elongation} \mid \text{elongation})$$

$$= \frac{p_s^n \cdot (f_n/tot)}{p_s^c \cdot (f_c/tot) + p_s^p \cdot (f_p/tot) + p_s^n \cdot (f_n/tot)} \approx \frac{p_s^n \cdot f_n}{p_s^c \cdot f_c} \sim \frac{f_n}{f_c}$$

with $tot = f_c + f_p + f_n$, and where we have used that

$$P(\text{elongation}) = (f_c/tot) \cdot p_s^c + (f_p/tot) \cdot p_s^p + (f_n/tot) \cdot p_s^n .$$

Note, the ability to precalculate the probabilities p_s^c, p_s^p and p_s^n is instrumental in obtaining the above result. As such, it illustrates the approach of piecewise analysis, first establishing quantities for part of the system to obtain a quantity for the system as a whole.

5   Competition and Insertion Times

In this section, we continue the analysis of the Prism model for translation and discuss the correlation of the average insertion time for the amino acid specified by a codon, on the one hand, and and the *aa-tRNA* competition, i.e. the relative abundance of pseudo-cognate and near-cognate *aa-tRNAs*, on the other hand. The insertion time of a codon is the average time it takes to elongate the protein in nascent with an amino acid.

The average insertion time can be computed in Prism using the concept of *rewards*, also known as *costs* in Markov theory. Each state is assigned a value as its reward. Further, the reward of each state is weighted per unit of time. Hence, it is computed by multiplication with the average time spent in the state. The cumulative reward of a path in the chain is defined as a sum over all states in the path of such weighted rewards per state. Thus, by assigning to each state the value 1 as reward, we obtain the total average time for a given path. For example, in Prism the cumulative reward formula R=? [F (s=8)] which asks to compute the expected time to reach state s=8. Recall, in state s=8 the amino acid is added to the polypeptide chain. The formula returns the average reward of all the paths that lead from the initial state 1 to state 8. As explained above, in order to obtain the average time for insertion, we assign each state the value 1 as a reward, which in Prism can be done using the following code

```
rewards true: 1 endrewards
```

The construct expresses the fact that 1 is assigned to any state that satisfy the condition **true** (which is trivially satisfied by all states).

So, a script calling Prism for model checking the above formula then yields the expected insertion time per codon. Table 6 in the appendix lists the results. Although the correlation of cognate frequency and insertion times is limited, the qualitative claim of [25] of 'rare' codons being translated slow and 'frequent' codons being translated fast is roughly confirmed by the model. E.g., the codons AGC and CCA have amongst the lowest frequencies, 420 and 617, the lowest and two but lowest frequency, respectively, and translates indeed the slowest, 1.4924 and 1.5622 seconds, respectively. However, the codon CCA with an availability of 581, of one but lowest frequency, is translated at a moderate rate of 0.5525 seconds on the average. Thus, in line with our considerations, cognate availability per se does not sufficiently predict translation time. Comparably, the fastest insertion times, 52.7 and 64.5 milliseconds, are realized by the codons CUG and CGU, of the codons corresponding to amino acids the one and two but most abundant. The codon CUG of the highest frequency 5136, excluding stop codons, though has an average insertion time of 102.8 milliseconds.

A little bit more ingenuity is needed to establish average exit times, for example for a cognate to pass from state s=2 to state s=8. The point is that conditional probabilities are involved. However, since dealing with exponential distributions, elimination of transitions in favour of adding their rates to that of the remaining ones, does the trick. Various results, some of them used below, are collected in Table 3. (The probabilities of failure and success for the non-cognates are trivial, $p_f^x = 1$ and $p_s^x = 0$, with a time per failed attempt $T_f^x = 0.5 \cdot 10^{-3}$ seconds.)

There is a visible correlation between the quotient of the number of near-cognate *aa-tRNA* over the number of cognate *aa-tRNA* and the average insertion time. See Figure 4. In fact, the average insertion time for a codon is

Table 3. Exit probabilities and exit times (in seconds) for three types of aa-$tRNA$, superscripts c, p and n for cognate, pseudo-cognate and near-cognate aa-$tRNA$, respectively. Failure for exit to states $s=0$ or $s=5$, subscript f; success for exit to state $s=8$, subscript s.

p_s^c	0.5079	p_f^c	0.4921	T_s^c	0.03182	T_f^c	$9.342 \cdot 10^{-3}$
p_s^p	$4.847 \cdot 10^{-4}$	p_f^p	0.9995	T_s^p	3.251	T_f^p	0.3914
p_s^n	$4.847 \cdot 10^{-4}$	p_f^n	0.9995	T_s^n	3.251	T_f^n	0.3914

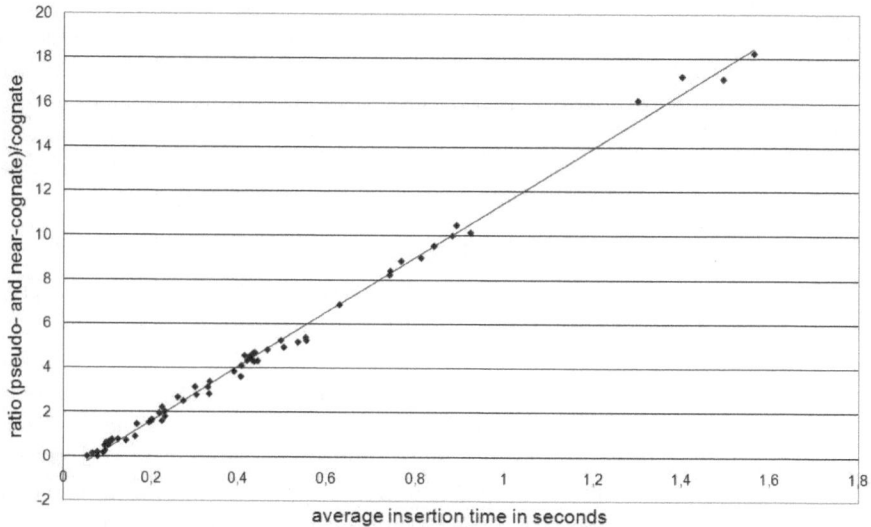

Fig. 4. Correlation of the ratio $(f_p + f_n)/f_c$ of total frequency of pseudo-cognates and near-cognates over the frequency of cognates vs. average insertion times. See also Table 6 in the appendix.

approximately proportional to the near-cognate/cognate ratio. This can be seen as follows. The insertion of the amino acid is completed if state $s=8$ is reached, either for a cognate, pseudo-cognate or near-cognate. As we have seen, the probability for either of the latter two is negligible, $p_s^p, p_s^n = 4.847 \cdot 10^{-4}$. Therefore, the number of cognate arrivals is decisive. With p_f^c and p_s^c being the probability for a cognate to fail, i.e. exit at state $s=0$ or $s=5$, or to succeed, i.e. reach of state $s=8$, the insertion time T_{ins} can be regarded as a geometric series. (Note the exponent i below.) Important are the numbers of arrivals of the other aa-$tRNA$ types per single cognate arrival, expressed in terms of frequencies.

An arrival occurring for the $(i+1)$st arrival of a cognate has spent $(i \times T_f^c) + T_s^c$ processing cognate aa-$tRNA$. The number of pseudo-cognate, near-cognate and non-cognate arrivals per individual cognate arrival are, on the average, the relative fractions $\frac{f_p}{f_c}$, $\frac{f_n}{f_c}$, and $\frac{f_x}{f_c}$, respectively (with f_p, f_n, and f_c as before in

(p)	(n)	(x) (cf) (p)	(n)	(x) (cf) (p)	(n)	(x) (cs)

Fig. 5. Accumulated delay after three cognate arrivals: (p) delay $(f_p/f_c)\cdot T_f^p$ for failing pseudo-cognates, (n) delay $(f_n/f_c)\cdot T_f^n$ for failing near-cognates, (x) delay $(f_x/f_c)\cdot T_f^x$ for non-cognates, (cf) exit time T_f^c for a failing cognate, (cs) exit time T_s^c for a successful cognate.

Section 4, and f_x the frequency of non-cognate *aa-tRNA*). See Figure 5. Summing over i, the number of failing cognate *aa-tRNA* for a successful cognate insertion, yields

$$T_{ins} = \sum_{i=0}^{\infty} (p_f^c)^i p_s^c \cdot (\text{delay for } i \text{ failing and 1 successful cognate arrivals})$$

$$= \sum_{i=0}^{\infty} (p_f^c)^i p_s^c \cdot \left((i+1) \cdot \left(\frac{f_p}{f_c} T_f^p + \frac{f_n}{f_c} T_f^n + \frac{f_x}{f_c} T_f^x \right) + i \cdot T_f^c + T_s^c \right)$$

$$\approx \frac{f_p + f_n}{f_c} p_s^c T_f^n \sum_{i=0}^{\infty} (i+1) \cdot (p_f^c)^i$$

$$\sim \frac{f_p + f_n}{f_c}.$$

Here, we have used that T_f^c and T_s^c are negligible, T_f^p equals T_f^n, and $\frac{f_x}{f_c} T_f^x$ is relatively small, from which it follows that $\frac{f_p + f_n}{f_c} T_f^n$ is the dominant summand. Note that the estimate is not accurate for small values of $f_p + f_n$. Nevertheless, closer inspection shows that for these values the approximation remains order-preserving. Again, the results obtained for parts of the systems are pivotal in the derivation.

6   Concluding Remarks

In this paper, we presented a stochastic model of the translation process based on presently available data of ribosome kinetics [12,9]. We used the model checking facilities of the Prism tool for continuous-time Markov chains. Compared to [9] that uses simulation, our approach is computationally more reliable (independent on the number of simulations) and has faster response times (taking seconds rather then minutes or hours). More importantly, model checking allowed us to perform piecewise analysis of the system, yielding better insight in the model compared to just observing the end-to-end results with a monolithic model. Based on this, we improved on earlier observations, regarding error probabilities and insertion times, by actually deriving the correlation suggested by the data.

In [7] a correlation was reported between the number of copies (concentrations) of cognate *tRNAs* and the frequency of usage of particular codons in the most abundant proteins in *E. coli*. It is suggested that this optimization is favorable for the cell growth: when they are urgently needed the most used proteins are translated with maximum speed and accuracy. On the other hand, we observed that there is a high correlation (0.86) between the cognate *tRNA* concentrations and the ration near-cognates vs. cognates which, according to our model, determines the error probabilities. Consequently, it would be interesting to check if there exists even better correlation between the near-cognates/cognates ratios and the codon usage frequencies than between the latter and the concentrations.

In conclusion, we have experienced *aa-tRNA* competition as a very interesting biological case study of intrinsic stochastic nature, falling in the category of the well known lambda-phage example [1]. Our model opens a new avenue for future work on biological systems that possess intrinsically probabilistic properties. It would be interesting to apply our method to processes which, similarly to translation, involve small numbers of molecules, like *DNA* replication [16,19], *DNA* repair [11,20], charging of the *tRNAs* with amino acids [8,15], etc., thus rendering approaches based on ordinary differential equations less attractive.

Acknowledgments. We are grateful to Timo Breit, Christiaan Henkel, Erik Luit, Jasen Markovski, and Hendrik Viljoen for fruitful discussions and constructive feedback.

References

1. Arkin, A., Ross, J., McAdams, H.H.: Stochastic kinetic analysis of developmental pathway bifurcation in phage lambda-infected. Escherichia coli cells. Genetics 149, 1633–1648 (1998)
2. Baier, C., Katoen, J.-P., Hermanns, H.: Approximate symbolic model checking of continuous-time Markov chains. In: Baeten, J.C.M., Mauw, S. (eds.) CONCUR 1999. LNCS, vol. 1664, pp. 146–161. Springer, Heidelberg (1999)
3. Bošnački, D., ten Eikelder, H.M.M., Steijaert, M.N., de Vink, E.P.: Stochastic analysis of amino acid substitution in protein synthesis. In: Heiner, M., Uhrmacher, A.M. (eds.) CMSB 2008. LNCS (LNBI), vol. 5307, pp. 367–386. Springer, Heidelberg (2008)
4. Calder, M., Vyshemirsky, V., Gilbert, D., Orton, R.: Analysis of signalling pathways using continuous time Markov chains. In: Priami, C., Plotkin, G. (eds.) Transactions on Computational Systems Biology VI. LNCS (LNBI), vol. 4220, pp. 44–67. Springer, Heidelberg (2006)
5. Chabrier, N., Fages, F.: Symbolic model checking of biochemical networks. In: Priami, C. (ed.) CMSB 2003. LNCS, vol. 2602, pp. 149–162. Springer, Heidelberg (2003)
6. Danos, V., Feret, J., Fontana, W., Harmer, R., Krivine, J.: Rule-based modelling of cellular signalling. In: Caires, L., Vasconcelos, V.T. (eds.) CONCUR 2007. LNCS, vol. 4703, pp. 17–41. Springer, Heidelberg (2007)
7. Dong, H., Nilsson, L., Kurland, C.G.: Co-variation of tRNA abundance and codon usage in *Escherichia coli* at different growth rates. Journal of Molecular Biology 260, 649–663 (1996)
8. Nureki, O., et al.: Enzyme structure with two catalytic sites for double-sieve selection of substrate. Science 280, 578–582 (1998)
9. Fluitt, A., Pienaar, E., Viljoen, H.: Ribosome kinetics and aa-tRNA competition determine rate and fidelity of peptide synthesis. Computational Biology and Chemistry 31, 335–346 (2007)

10. Gilchrist, M.A., Wagner, A.: A model of protein translation including codon bias, nonsense errors, and ribosome recycling. Journal of Theoretical Biology 239, 417–434 (2006)

11. Goodman, M.F.: Coping with replication 'train wrecks' in *Escherichia coli* using Pol V, Pol II and RecA proteins. Trends in Biochemical Sciences 25, 189–195 (2000)

12. Gromadski, K.B., Rodnina, M.V.: Kinetic determinants of high-fidelity tRNA discrimination on the ribosome. Molecular Cell 13(2), 191–200 (2004)

13. Heath, J., Kwiatkowska, M., Norman, G., Parker, D., Tymchyshyn, O.: Probabilistic model checking of complex biological pathways. In: Priami, C. (ed.) CMSB 2006. LNCS (LNBI), vol. 4210, pp. 32–47. Springer, Heidelberg (2006)

14. Heyd, A.W., Drew, D.A.: A mathematical model for elongation of a peptide chain. Bulletin of Mathematical Biology 65, 1095–1109 (2003)

15. Ibba, M., Söll, D.: Aminoacyl-tRNAs: setting the limits of the genetic code. Genes & Development 18, 731–738 (2004)

16. Johnson, K.A.: Conformational coupling in DNA polymerase fidelity. Annual Reviews in Biochemistry 62, 685–713 (1993)

17. Karp, G.: Cell and Molecular Biology, 5th edn. Wiley, Chichester (2008)

18. Kwiatkowska, M., Norman, G., Parker, D.: Probabilistic symbolic model cheking with Prism: a hybrid approach. Journal on Software Tools for Technology Transfer 6, 128–142 (2004), http://www.prismmodelchecker.org/

19. Martomo, S.A., Mathews, C.K.: Effects of biological DNA precursor pool asymmetry upon accuracy of DNA replication in vitro. Mutation Research 499, 197–211 (2002)

20. Ni, M., Wang, S.-Y., Li, J.-K., Ouyang, Q.: Simulating the temporal modulation of inducible DNA damage response in Escherichia coli. Biophysical Journal 93, 62–73 (2007)

21. Pape, T., Wintermeyer, W., Rodnina, M.: Complete kinetic mechanism of elongation factor Tu-dependent binding of aa-tRNA to the A-site of E. coli. EMBO Journal 17, 7490–7497 (1998)

22. Priami, C., Regev, A., Shapiro, E., Silverman, W.: Application of a stochastic name-passing calculus to representation and simulation of molecular processes. Information Processing Letters 80, 25–31 (2001)

23. Rodnina, M.V., Wintermeyer, W.: Ribosome fidelity: tRNA discrimination, proofreading and induced fit. Trends in Biochemical Sciences 26(2), 124–130 (2001)

24. Savelsbergh, A., et al.: An elongation factor G-induced ribosome rearrangement precedes tRNA–mRNA translocation. Molecular Cell 11, 1517–1523 (2003)

25. Sørensen, M.A., Kurland, C.G., Pedersen, S.: Codon usage determines translation rate in Escherichia coli. Journal of Molecular Biology 207, 365–377 (1989)

26. Viljoen, H.: Private communication (2008)

27. Wahab, S.Z., Rowley, K.O., Holmes, W.M.: Effects of $tRNA_1^{Leu}$ overproduction in Escherichia coli. Molecular Microbiology 7, 253–263 (1993)

A   Appendix: Suplementary Figures and Data

```
// translation model


stochastic
```

```
// constants
const double ONE=1;
const double FAST=1000;

// tRNA rates
const double c_cogn ;
const double c_pseu ;
const double c_near ;
const double c_nonc ;

const double k1f = 140;
const double k2b =  85;
const double k2bx=2000;
const double k2f = 190;
const double k3bc=   0.23;
const double k3bp=  80;
const double k3bn=  80;
const double k3fc= 260;
const double k3fp=   0.40;
const double k3fn=   0.40;
const double k4rc=  60;
const double k4rp=FAST;
const double k4rn=FAST;
const double k4fc= 166.7;
const double k4fp=  46.1;
const double k4fn=  46.1;
const double k6f = 150;
const double k7b = 140;
const double k7f = 145.8;

module ribosome

s : [0..8] init 1 ;
cogn : bool init false ;
pseu : bool init false ;
near : bool init false ;
nonc : bool init false ;

// initial binding
[ ] (s=1) -> k1f * c_cogn : (s'=2) & (cogn'=true) ;
[ ] (s=1) -> k1f * c_pseu : (s'=2) & (pseu'=true) ;
[ ] (s=1) -> k1f * c_near : (s'=2) & (near'=true) ;
[ ] (s=1) -> k1f * c_nonc : (s'=2) & (nonc'=true) ;
[ ] (s=2) &  ( cogn | pseu | near ) -> k2b : (s'=0) &
      (cogn'=false) & (pseu'=false) & (near'=false) ;
[ ] (s=2) &  nonc -> k2bx : (s'=0) & (nonc'=false) ;
```

```
// codon recognition
[ ] (s=2) & ( cogn | pseu | near ) -> k2f : (s'=3) ;
[ ] (s=3) & cogn -> k3bc : (s'=2) ;
[ ] (s=3) & pseu -> k3bp : (s'=2) ;
[ ] (s=3) & near -> k3bn : (s'=2) ;

// GTPase activation, GTP hydrolysis, reconformation
[ ] (s=3) & cogn -> k3fc : (s'=4) ;
[ ] (s=3) & pseu -> k3fp : (s'=4) ;
[ ] (s=3) & near -> k3fn : (s'=4) ;

// rejection
[ ] (s=4) & cogn -> k4rc : (s'=5) & (cogn'=false) ;
[ ] (s=4) & pseu -> k4rp : (s'=5) & (pseu'=false) ;
[ ] (s=4) & near -> k4rn : (s'=5) & (near'=false) ;

// accommodation, peptidyl transfer
[ ] (s=4) & cogn -> k4fc : (s'=6) ;
[ ] (s=4) & pseu -> k4fp : (s'=6) ;
[ ] (s=4) & near -> k4fn : (s'=6) ;

// EF-G binding
[ ] (s=6) -> k6f : (s'=7) ;
[ ] (s=7) -> k7b : (s'=6) ;

// GTP hydrolysis, unlocking,
// tRNA movement and Pi release,
// rearrangements of ribosome and EF-G,
// dissociation of GDP
[ ] (s=7) -> k7f : (s'=8) ;

// no entrance, re-entrance at state 1
[ ] (s=0) -> FAST*FAST : (s'=1) ;
// rejection, re-entrance at state 1
[ ] (s=5) -> FAST*FAST : (s'=1) ;
// elongation
[ ] (s=8) -> FAST*FAST : (s'=8) ;

endmodule

rewards
  true : 1;
endrewards
```

Table 4. Frequencies of cognate, pseudo-cognate, near-cognate and non-cognates for
E. coli as molecules per cell [7]. Stop codons UGA, UAG and UAA.

codon	cognate	pseudo-cognate	near-cognate	non-cognate	codon	cognate	pseudo-cognate	near-cognate	non-cognate
UUU	1037	0	2944	67493	GUU	5105	0	0	66369
UUC	1037	0	9904	60533	GUC	1265	3840	7372	58997
UUG	2944	0	2324	66206	GUG	3840	1265	1068	65301
UUA	1031	1913	2552	65978	GUA	3840	1265	9036	57333
UCU	2060	344	0	69070	GCU	3250	617	0	67607
UCC	764	1640	4654	64416	GCC	617	3250	8020	59587
UCG	1296	764	2856	66558	GCG	3250	617	1068	66539
UCA	1296	1108	1250	67820	GCA	3250	617	9626	57981
UGU	1587	0	1162	68725	GGU	4359	2137	0	64978
UGC	1587	0	4993	64894	GGC	4359	2137	4278	60700
UGG	943	0	4063	66468	GGG	2137	4359	0	64978
UGA*	6219	0	4857	60398	GGA	1069	5427	11807	53171
UAU	2030	0	0	69444	GAU	2396	0	4717	64361
UAC	2030	0	3388	66056	GAC	2396	0	10958	58120
UAG*	1200	0	5230	65044	GAG	4717	0	3464	63293
UAA*	7200	0	4576	59698	GAA	4717	0	10555	56202
CUU	943	5136	4752	60643	AUU	1737	1737	2632	65368
CUC	943	5136	1359	64036	AUC	1737	1737	6432	61568
CUG	5136	943	2420	62975	AUG	706	1926	4435	64407
CUA	666	5413	1345	64050	AUA	1737	1737	6339	61661
CCU	1301	900	4752	64521	ACU	2115	541	0	68818
CCC	1913	943	2120	66498	ACC	1199	1457	4338	64480
CCG	1481	720	5990	63283	ACG	1457	1199	4789	64029
CCA	581	1620	1430	67843	ACA	916	1740	2791	66027
CGU	4752	639	0	66083	AGU	1408	0	1287	68779
CGC	4752	639	2302	63781	AGC	1408	0	5416	64650
CGG	639	4752	6251	59832	AGG	420	867	6318	63869
CGA	4752	639	2011	64072	AGA	867	420	4248	65939
CAU	639	0	6397	64438	AAU	1193	0	1924	68357
CAC	639	0	3308	67527	AAC	1193	0	6268	64013
CAG	881	764	6648	63181	AAG	1924	0	6523	63027
CAA	764	881	1886	67943	AAA	1924	0	2976	66574

Table 5. Probabilities per codon for erroneous elongation

UUU 27.4e-4	CUU 46.7e-4	GUU 1.12e-10	AUU 14.4e-4
UUC 91.2e-4	CUC 13.6e-4	GUC 55.0e-4	AUC 35.0e-4
UUG 7.59e-4	CUG 4.49e-4	GUG 2.68e-4	AUG 58.3e-4
UUA 23.5e-4	CUA 18.9e-4	GUA 22.3e-4	AUA 34.4e-4
UCU 2.81e-10	CCU 34.1e-4	GCU 1.77e-10	ACU 2.73e-10
UCC 56.1e-4	CCC 10.4e-4	GCC 12.5e-4	ACC 34.2e-4
UCG 20.3e-4	CCG 37.6e-4	GCG 3.187e-4	ACG 31.7e-4
UCA 9.09e-4	CCA 22.8e-4	GCA 28.2e-4	ACA 29.1e-4
UGU 6.97e-4	CGU 1.21e-10	GGU 1.32e-10	AGU 8.70e-4
UGC 30.4e-4	CGC 4.59e-4	GGC 9.40e-4	AGC 37.2e-4
UGG 39.8e-4	CGG 88.7e-4	GGG 2.72e-10	AGG 140.7e-4
UGA 7.50e-4	CGA 3.98e-4	GGA 100.3e-4	AGA 48.1e-4
UAU 2.81e-10	CAU 91.1e-4	GAU 18.6e-4	AAU 15.2e-4
UAC 15.7e-4	CAC 47.5e-4	GAC 43.2e-4	AAC 49.3e-4
UAG 41.3e-4	CAG 69.4e-4	GAG 7.09e-4	AAG 32.1e-4
UAA 6.04e-4	CAA 22.7e-4	GAA 21.4e-4	AAA 14.6e-4

Table 6. Estimated average insertion time per codon in seconds

UUU 0.3327	CUU 0.8901	GUU 0.0527	AUU 0.2733
UUC 0.8404	CUC 0.6286	GUC 0.7670	AUC 0.4373
UUG 0.1245	CUG 0.1028	GUG 0.1041	AUG 0.8115
UUA 0.4436	CUA 0.9217	GUA 0.2604	AUA 0.4321
UCU 0.0893	CCU 0.4202	GCU 0.0756	ACU 0.0943
UCC 0.7409	CCC 0.1992	GCC 1.5622	ACC 0.4658
UCG 0.3035	CCG 0.4257	GCG 0.1010	ACG 0.4073
UCA 0.2313	CCA 0.5535	GCA 0.3002	ACA 0.5025
UGU 0.1432	CGU 0.0645	GGU 0.0924	AGU 0.1636
UGC 0.3296	CGC 0.1010	GGC 0.1673	AGC 0.3905
UGG 0.4360	CGG 1.3993	GGG 0.2308	AGG 1.4924
UGA 0.1098	CGA 0.0962	GGA 1.2989	AGA 0.5517
UAU 0.0758	CAU 0.8811	GAU 0.2180	AAU 0.2242
UAC 0.2008	CAC 0.5341	GAC 0.4144	AAC 0.4959
UAG 0.4319	CAG 0.7425	GAG 0.1106	AAG 0.3339
UAA 0.0963	CAA 0.4058	GAA 0.2243	AAA 0.1945

Qualitative and Quantitative Analysis of a Bio-PEPA Model of the Gp130/JAK/STAT Signalling Pathway

Maria Luisa Guerriero

Laboratory for Foundations of Computer Science,
The University of Edinburgh, UK

Abstract. Computational modelling of complex biochemical systems has grown in importance over recent years as a tool for supporting biological studies. Consequently, several formal languages have been recently proposed as modelling languages for biology. Among these, process algebras have been proved capable of providing researchers with new hypotheses on the behaviour of biochemical systems.

Bio-PEPA is a process algebra recently defined for the modelling and analysis of biochemical systems, which provides modellers with a wide range of analysis techniques: models can be analysed by stochastic simulation, model-checking, and mathematical methods based on ordinary differential equations.

In this work, we use Bio-PEPA for modelling the gp130/JAK/STAT signalling pathway, and we use both stochastic simulation and model-checking to analyse several qualitative and quantitative aspects of the system.

1 Introduction

Several modelling approaches have been used over recent years to analyse complex biological systems such as signaling pathways, ranging from traditional mathematical methods based on differential equations to computational methods based on stochastic simulation and model-checking. Each of these techniques can be more suitable than others in some context or to study some particular features of biological systems.

Process algebras are formal languages traditionally used to model distributed systems of concurrent computing devices. Starting from the biochemical π-calculus [1], several other process algebras have been recently adapted in order to model biochemical systems [2,3,4,5], following the "molecules as processes" paradigm introduced in the landmark paper [6]: molecules are modelled as concurrent processes, and biochemical reactions are represented by actions performed by synchronising processes.

Bio-PEPA [7,8] is a process algebra specifically defined to model and analyse biochemical networks. Compared to other process algebras, Bio-PEPA uses a more abstract view of biochemical systems, the so-called "species as processes" abstraction: processes represent molecular species instead of single molecules, and multi-way synchronisations of processes represent changes in the amounts of molecular species resulting from biochemical reactions. Such an abstract view enables modellers to deal with analysis techniques which are computationally infeasible when considering the "molecules as processes" abstraction.

C. Priami et al. (Eds.): Trans. on Comput. Syst. Biol. XI, LNBI 5750, pp. 90–115, 2009.

The main feature of Bio-PEPA is that it integrates several kinds of analysis techniques. Both discrete stochastic and continuous deterministic models can be automatically generated from Bio-PEPA models, thus allowing modellers to perform time-series analysis via stochastic simulation, Markovian analysis and ordinary differential equations (ODEs); in addition, system properties can be verified through model-checking and mathematical techniques such as bifurcation, stability and continuation analysis. Moreover, as for the other process algebras, Bio-PEPA is equipped with an operational semantics which supports various kinds of formal analysis (e.g. causality, equivalence, and reachability analysis).

In this work, we define a Bio-PEPA model of the gp130/JAK/STAT signalling pathway, a well-studied system which plays a major role in several biological processes both in human and other organisms. A lot of experimental data is available about the molecules in the pathway, and some mathematical and computational models have been already developed. For these reasons, the gp130/JAK/STAT pathway represents a good case study for exploiting some of the possible Bio-PEPA analysis methods in order to study different aspects (both qualitative and quantitative) of the system, and compare them with existing models.

The rest of the paper is structured as follows. First, the Bio-PEPA language is introduced in Sec. 2, while the pathway and the Bio-PEPA model are described in Sec. 3 and Sec. 4, respectively. The following three sections are devoted to the analysis of the model: in Sec. 5 several qualitative properties are analysed via model-checking, in Sec. 6 we present some stochastic simulation results, and in Sec. 7 model-checking is employed for quantitative analysis. Finally, Sec. 8 is an overview of the related work and Sec. 9 contains some concluding remarks.

2 Bio-PEPA

Bio-PEPA [7,8] is a process algebra which has been recently defined for the modelling and analysis of biochemical networks. It is a biologically-inspired language based on PEPA [9] and, differently from PEPA and other process algebras, it is able to explicitly represent details such as stoichiometric coefficients and the roles of species in reactions, and it supports the definition of general kinetic laws. Bio-PEPA models can be analysed by different techniques (stochastic simulation, analysis based on ODEs, numerical solution of the continuous-time Markov chain (CTMC), and probabilistic model-checking), since the mappings of Bio-PEPA models into specifications for those approaches have been defined [10].

The Bio-PEPA language is based on discrete levels of parameterised species: each component represents a species and its parameter may be interpreted as the number of molecules or discrete levels of concentration depending on the type of analysis to be applied. Parametric levels are considered for the definition of the transition system and for the derivation of a CTMC whose states represent the concentration levels of the species.

The syntax of Bio-PEPA is defined as:

$$S ::= (\alpha, \kappa) \text{ op } S \mid S + S \mid C \qquad P ::= P \underset{I}{\bowtie} P \mid S(x)$$

where op $= \downarrow \mid \uparrow \mid \oplus \mid \ominus \mid \odot$.

The component S is called a *species component* and abstracts a molecular species, whereas the component P, called a *model component*, describes the system and the interactions among components. The prefix term (α, κ) op S contains information about the role of the species in the reaction associated with the action type α: κ is the *stoichiometric coefficient* of the species and the *prefix combinator* "op" represents its role in the reaction. Specifically, $\downarrow$ indicates a *reactant*, $\uparrow$ a *product*, $\oplus$ an *activator*, $\ominus$ an *inhibitor* and $\odot$ a generic *modifier*. The operator "+" expresses the choice between possible actions and the constant C is defined by an equation $C \stackrel{def}{=} S$. The parameter $x \in \mathbb{R}^+$ in $S(x)$ represents the concentration of S. Finally, the process $P \bowtie_I Q$ denotes the cooperation between components: the set I determines those activities on which the operands are forced to synchronise. Reaction rates are defined as *functional rates* associated with actions.

Bio-PEPA supports a modelling style in terms of *concentration levels*: the species amounts are discretised into a number of levels, from level 0 (i.e. species not present) to a maximum level N (which depends on the maximum concentration of the species). Each level represents an interval of concentration and the granularity of the system is expressed in terms of the *step size H* (i.e. the length of the concentration interval).

Definition 1. *A Bio-PEPA system $\mathcal{P}$ is a 6-tuple $\langle \mathcal{V}, \mathcal{N}, \mathcal{K}, \mathcal{F}_R, Comp, P \rangle$, where: $\mathcal{V}$ is the set of compartments, $\mathcal{N}$ is the set of quantities describing the species (i.e. H and N), $\mathcal{K}$ is the set of parameter definitions, $\mathcal{F}_R$ is the set of functional rates, Components is the set of definitions of species components, P is the model component describing the system.*

For discrete state space analysis the behaviour of the system is defined in terms of an operational semantics. A *Stochastic Labelled Transition System (SLTS)* is defined for a Bio-PEPA system. From this we can obtain a *Continuous Time Markov Chain (CTMC)*. Both the SLTS and the CTMC derived from Bio-PEPA are defined in terms of levels of concentration, and the generated Markov chain is called *CTMC with levels*. For a full description of the language semantics see [10].

The Bio-PEPA language is supported by software tools such as the Bio-PEPA Workbench [11], which automatically processes Bio-PEPA models and generates other representations in forms suitable for simulation and model-checking. For instance, the generated simulation model can be executed using the Dizzy stochastic simulator [12]. The representation which is used for discrete state space generation and analysis by numerical solution of the underlying CTMC is expressed in the reactive modules language supported by the PRISM model-checker [13]. In addition, the Bio-PEPA Workbench generates reward structures and common CSL [14] formulae used in model-checking.

3   The Gp130/JAK/STAT Signalling Pathway

The gp130/JAK/STAT signalling pathway is a well-studied biological system, of great clinical interest because of its key role in human fertility, neuronal repair and haematological development [15,16,17]. Much experimental data is available on this pathway, and a few mathematical and computational models [18,19,20,21] have been developed.

The signalling cascade in the gp130/JAK/STAT pathway is triggered by members of the family of IL (interleukin)-6-type cytokines binding to plasma membrane receptor complexes containing the common signal transducing receptor chain gp130 (glycoprotein 130). Among the targets of gp130 signal transduction, we consider the transcription factors of the STAT (signal transducers and activators of transcription) family, in particular STAT3. A key feature of the pathway is the nuclear/cytoplasmic shuttling of STATs: upon activation, STATs can translocate into the nucleus and activate the transcription of downstream gene targets.

Different cytokines signal through the formation of different receptor complexes, all of them containing gp130 and another subunit. We focus here on two different cytokines: LIF (leukaemia inhibitory factor) and OSM (oncostatin M). LIF signals through an heterodimeric receptor complex gp130:LIFR. OSM exhibits the uncommon ability to signal through two different receptor complexes: the type I OSM receptor complex (gp130:LIFR), and the type II OSM receptor complex (gp130:OSMR).

Figure 1 is a graphical representation of the biochemical reactions occurring in the gp130/JAK/STAT pathway. In the inset the different types of receptor complexes are shown.

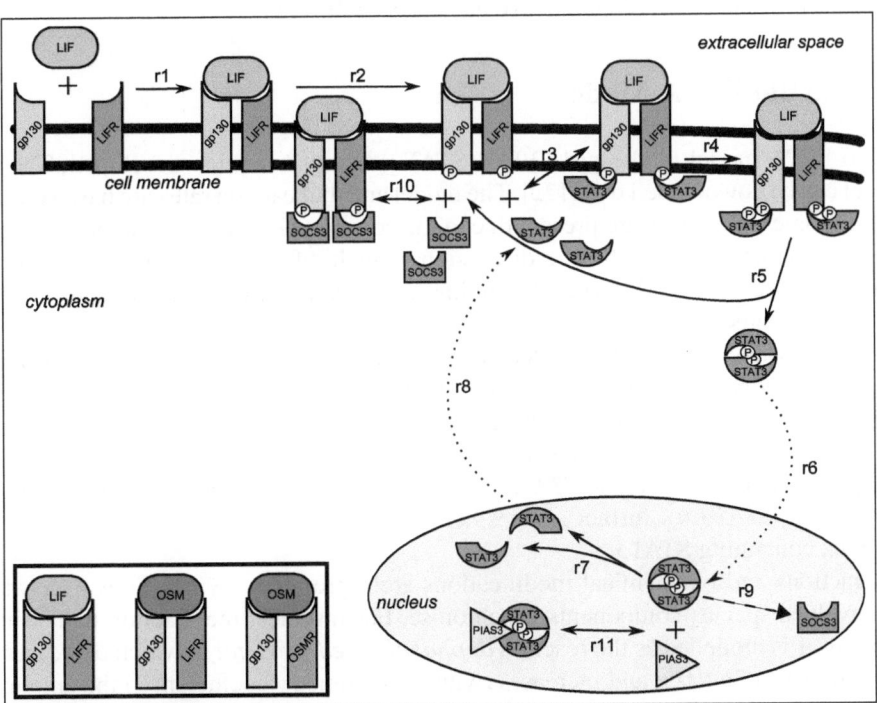

Fig. 1. Gp130/JAK/STAT pathway: graphical representation. Full arrows represent biochemical reactions, dotted arrows represent transports, dashed arrows represent syntheses.

The molecular species we consider in the model are: two ligands (LIF and OSM), three membrane-bound receptors (gp130, LIFR and OSMR), one effector (STAT3), and two inhibitors (SOCS3 and PIAS3). JAK kinase and TC-PTP phosphatase are implicitly modelled.

Four compartments are involved in the system: the exosol (the extracellular space, where the two ligands are located), the cell membrane (location of the receptors), the cytosol (initial location of STAT3), and the nucleus (in which STAT3 can translocate).

Receptors are activated by ligand bindings, and active receptors dimerise to form receptor complexes (gp130:LIFR or gp130:OSMR) (reaction r1 in Fig. 1). Once the receptor dimeric complex is formed, each receptor subunit (gp130, LIFR and OSMR) can undergo JAK-mediated phosphorylation (r2). STAT3 can bind on receptors' phosphorylated sites (r3), and the binding of STAT3 leads to its activation (phosphorylation) (r4).

Once phosphorylated, STAT3 dissociates from the receptor complex, and its phosphorylated site allows STAT3 to homodimerise (r5). When STAT3 is in dimeric form, it can translocate into the nucleus (r6) where it can carry out its specific functions (not modelled here): STAT3 binds to the DNA, thus activating the transcription of downstream gene targets. Nuclear STAT3 dimers are inactivated through TC-PTP -mediated dephosphorylation, which leads to the dimers' dissociation (r7) and to STAT3 export to the cytoplasm (r8), where STAT3 can undergo additional cycles of activation.

The two inhibition mechanisms considered are due to SOCS3 and PIAS3. SOCS3 is synthesised by STAT3 (r9) and it acts by competing with STAT3 in binding to receptors (r10). PIAS3 acts by binding to active nuclear STAT3 (r11).

4   The Bio-PEPA Model

A Bio-PEPA model of the gp130/JAK/STAT pathway has been developed. The full model can be downloaded from [22]. The model and the reaction rates are based on [21], though some differences are present due to the conceptual differences in the used modelling languages (see Sec. 8 for a discussion of such differences). All kinetic laws are assumed to be mass-action (i.e. depending on the amount of reactants and on given kinetic constants).

Each possible form of the molecular species is modelled as a distinct Bio-PEPA species component. For instance, STAT3 is modelled by four distinct species components representing, respectively, the cytoplasmic dephosphorylated monomeric form ($STAT3\_c$), the cytoplasmic phosphorylated dimeric form ($STAT3\text{-}PD\_c$), the nuclear phosphorylated dimeric form ($STAT3\text{-}PD\_n$), and the nuclear dephosphorylated monomeric form ($STAT3\_n$); further species components are defined for each state of each complex containing STAT3.

Reactions and biochemical modifications are represented by reactions over which the involved species components synchronise. For instance, the reaction representing r7 in Fig. 1 is modelled as the reaction $dephospho\_dedimer\_stat_{59}$, which decreases the amount of $STAT3\text{-}PD\_n$ and increases (with stoichiometry coefficient 2) the amount of $STAT3\_n$.

As an example, the definitions of the species *STAT3-PD_n* and *STAT3_n* are reported (here we use the simplified syntax of the Bio-PEPA Workbench, in which the trailing *S* in prefix terms (α, κ) op *S* can be omitted).

$$STAT3\text{-}PD\_n ::= (reloc\_stat\_cn_{58}, 1)\uparrow \; + \; (synth\_socs_{61}, 1)\oplus \; + (unbind\_pias_{80}, 1)\uparrow$$
$$+ \; (dephospho\_dedimer\_stat_{59}, 1)\downarrow \; + \; (bind\_pias\_stat_{80}, 1)\downarrow$$
$$STAT3\_n ::= (dephospho\_dedimer\_stat_{59}, 2)\uparrow \; + \; (reloc\_stat\_nc_{60}, 1)\downarrow$$

For each of the involved reactions, a functional rate specifying its kinetic rate law is defined. The ones used in the species definitions for *STAT3-PD_n* and *STAT3_n* are defined as follows.

$$reloc\_stat\_cn_{58} = \left[\frac{0.693}{k_{58}} \cdot STAT3\text{-}PD\_c\right];$$
//STAT3-PD_c relocation cytoplasm -> nucleus

$$dephospho\_dedimer\_stat_{59} = [k_{59} \cdot STAT3\text{-}PD\_n];$$
//STAT3-PD_n dephosphorylation & dedimerisation

$$reloc\_stat\_nc_{60} = \left[\frac{0.693}{k_{60}} \cdot STAT3\_n\right];$$
//STAT3-PD_n relocation nucleus -> cytoplasm

$$synth\_socs_{61} = [k_{61} \cdot STAT3\text{-}PD\_n];$$
//SOCS3 synthesis by STAT3-PD_n

$$bind\_pias\_stat_{80} = \left[\frac{k_{80}}{nucleus \cdot N_A} \cdot PIAS3 \cdot STAT3\text{-}PD\_n\right];$$
//PIAS3/STAT3-PD_n binding

$$unbind\_pias\_stat_{80} = [k_{-80} \cdot PIAS3{:}STAT3\text{-}PD\_n];$$
//PIAS3/STAT3-PD_n unbinding

As mentioned above, the Bio-PEPA Workbench [11] allows us to automatically generate representations of the Bio-PEPA model for different analysis tools. In the following sections we show some of the analyses performed using these generated models. In particular, we consider the PRISM [23,13] and Dizzy [12,24] models. We use the PRISM model-checker to verify that some desired properties of the system are satisfied, and the Dizzy simulation tool to perform time-series analysis via stochastic simulation.

5 Model-Checking Based Qualitative Analysis

As a first step in the analysis of the model we use the PRISM model-checker [23,13] to verify a number of qualitative properties of the system. Such properties are intended to be *consistency checks* on the model and they allow us to check for the presence of possible human errors in the modelling process. This kind of checks is particularly useful when modelling complex systems such as the pathway we consider here since, due to the size of the models, trivial typing errors are likely to occur and may be hard to identify.

5.1 PRISM Modelling and Specification Language

PRISM [23,13] is a probabilistic model-checker, which can be used to verify properties of CTMCs. Models are described using the state-based PRISM language, and it is possible to specify quantitative properties of the system using a property specification language which includes *CSL* (Continuous Stochastic Logic) [25,26]. The PRISM language is composed of *modules* and *variables*. A model is composed of a number of interacting modules and each module contains a number of local variables, whose values constitute the state of the module. The global state of the model is determined by the local state of all modules. The behaviour of the modules is given by a set of guarded commands, each describing a transition which is enabled when the guard is true. A command includes an update which gives new values to the variables.

PRISM properties are made up of *state properties* ϕ and *path properties* ψ. The syntax of PRISM properties is given by the following grammar.

$$\phi ::= \textbf{true} \mid \textbf{false} \mid expr \mid \phi \wedge \phi \mid \phi \vee \phi \mid \neg \phi \mid \phi \Rightarrow \phi \mid$$
$$\mathcal{P}_{\bowtie p}[\psi] \mid \mathcal{P}_{=?}[\psi] \mid \mathcal{S}_{\bowtie p}[\phi] \mid \mathcal{S}_{=?}[\phi]$$

$$\psi ::= \mathbf{X}\phi \mid \phi\, \mathbf{U}^I\, \phi \mid \phi\, \mathbf{U}\, \phi \mid \mathbf{F}^I\, \phi \mid \mathbf{F}\phi \mid \mathbf{G}^I\, \phi \mid \mathbf{G}\phi$$

Here *expr* is a boolean expression (containing literal values, identifiers and the standard arithmetic and relational operators), $\bowtie \in \{<, \leq, \geq, >\}$ is a relational parameter, $p \in [0, 1]$ is a probability, and I is an interval of $\mathbb{R}^+$.

The operators $\mathcal{P}_{\bowtie p}[\psi]$ and $\mathcal{P}_{=?}[\psi]$ are used to express transient properties (i.e. which depend on time) whereas the operators $\mathcal{S}_{\bowtie p}[\phi]$ and $\mathcal{S}_{=?}[\phi]$ are used to express steady state properties (i.e. which hold in the long run). The result of the verification of formulae $\mathcal{P}_{\bowtie p}[\psi]$ (resp. $\mathcal{S}_{\bowtie p}[\phi]$) is one of the boolean values **true** or **false** depending on whether ψ (resp. ϕ) is satisfied. The result of the verification of formulae $\mathcal{P}_{=?}[\psi]$ (resp. $\mathcal{S}_{=?}[\phi]$) is the expected probability with which ψ (resp. ϕ) is satisfied.

The operators $\mathbf{X}$, $\mathbf{U}$, $\mathbf{F}$, and $\mathbf{G}$ are used to express *neXt*, *Until*, *Finally*, and *Globally* properties, respectively. Time-bounded formulae are indexed by an interval I.

The PRISM language supports the specification and analysis of reward-based properties. Reward structures allow us to associate real values with certain states or transitions of the model. Such values, which can be thought of as "costs" of the specified states/transitions, are taken into account during the solution of the CTMC. In this way it is possible to reason about various quantitative measures such as "expected number of instances of processes", "expected number of occurrences of reactions", "expected time until a condition is satisfied", etc. The PRISM reward language supports the expression of both instantaneous and cumulative rewards.

5.2 Model-Checking the Bio-PEPA Model with PRISM

In the PRISM models generated by the Bio-PEPA Workbench, one module is defined for each species, and the module local variables are used to record the current quantity of each species. The transitions correspond to the activities of the Bio-PEPA model and the updates take the stoichiometry into account. Transition rates are specified in an auxiliary module which defines the functional rates corresponding to all the reactions.

Moreover, lower and upper bounds must be defined for each variable (i.e. for the amount of each species). The step size H in the Bio-PEPA model allows us to consider different PRISM models with different *granularity*, leading to systems with different numbers of levels.

As an example, we provide the PRISM definitions relative to the species $STAT3\text{-}PD\_n$ and $STAT3\_n$, which are obtained from the corresponding Bio-PEPA species definitions reported in Sec. 4.

First, the lower and upper levels for both species are computed from the defined step size H and the given bounds on species amounts.

$$MIN\_STAT3\text{-}PD\_n = MIN\_STAT3\_n = 0$$
$$MAX\_STAT3\text{-}PD\_n = MAX\_STAT3\_n = 1500$$

$$N_L\_STAT3\text{-}PD\_n = \left\lfloor \frac{MIN\_STAT3\text{-}PD\_n}{H} \right\rfloor \qquad N_U\_STAT3\text{-}PD\_n = \left\lfloor \frac{MAX\_STAT3\text{-}PD\_n}{H} \right\rfloor$$
$$N_L\_STAT3\_n \quad = \left\lfloor \frac{MIN\_STAT3\_n}{H} \right\rfloor \qquad N_U\_STAT3\_n \quad = \left\lfloor \frac{MAX\_STAT3\_n}{H} \right\rfloor$$

The specifications of the behaviour of $STAT3\text{-}PD\_n$ and $STAT3\_n$ are given by the two following modules. The third module contains the definition of the functional rates for all reactions.

module $STAT3\text{-}PD\_n$

$STAT3\text{-}PD\_n : [N_L\_STAT3\text{-}PD\_n .. N_U\_STAT3\text{-}PD\_n]$ **init** 0;

$[reloc\_stat\_cn_{58}] (STAT3\text{-}PD\_n + 1 \leq N_U\_STAT3\text{-}PD\_n) \rightarrow$
$\qquad\qquad 1 : (STAT3\text{-}PD\_n' = STAT3\text{-}PD\_n + 1);$

$[synth\_socs_{61}] (STAT3\text{-}PD\_n + 0 \leq N_U\_STAT3\text{-}PD\_n) \rightarrow$
$\qquad\qquad 1 : (STAT3\text{-}PD\_n' = STAT3\text{-}PD\_n + 0);$

$[dephospho\_dedimer\_stat_{59}] (STAT3\text{-}PD\_n \geq 1 + N_L\_STAT3\text{-}PD\_n) \rightarrow$
$\qquad\qquad 1 : (STAT3\text{-}PD\_n' = STAT3\text{-}PD\_n - 1);$

$[bind\_pias\_stat_{80}] (STAT3\text{-}PD\_n \geq 1 + N_L\_STAT3\text{-}PD\_n) \rightarrow$
$\qquad\qquad 1 : (STAT3\text{-}PD\_n' = STAT3\text{-}PD\_n - 1);$

$[unbind\_pias\_stat_{80}] (STAT3\text{-}PD\_n + 1 \leq N_U\_STAT3\text{-}PD\_n) \rightarrow$
$\qquad\qquad 1 : (STAT3\text{-}PD\_n' = STAT3\text{-}PD\_n + 1);$

endmodule

module $STAT3\_n$

$STAT3\_n : [N_L\_STAT3\_n .. N_U\_STAT3\_n]$ **init** 0;

$[dephospho\_dedimer\_stat_{59}] (STAT3\_n + 2 \leq N_U\_STAT3\_n) \rightarrow$
$\qquad\qquad 1 : (STAT3\_n' = STAT3\_n + 2);$

$[reloc\_stat\_nc_{60}] (STAT3\_n \geq 1 + N_L\_STAT3\_n) \rightarrow 1 : (STAT3\_n' = STAT3\_n - 1);$

endmodule

module *Rates*

$$[reloc\_stat\_cn_{58}] \left(\frac{\frac{0.693}{k_{58}} \cdot STAT3\text{-}PD\_c \cdot H}{H} > 0 \right) \rightarrow \left(\frac{\frac{0.693}{k_{58}} \cdot STAT3\text{-}PD\_c \cdot H}{H} \right) : \textbf{true};$$

$$[dephospho\_dedimer\_stat_{59}] \left(\frac{k_{59} \cdot STAT3\text{-}PD\_n \cdot H}{H} > 0 \right) \rightarrow \left(\frac{k_{59} \cdot STAT3\text{-}PD\_n \cdot H}{H} \right) : \textbf{true};$$

$$[reloc\_stat\_nc_{60}] \left(\frac{\frac{0.693}{k_{60}} \cdot STAT3\_n \cdot H}{H} > 0 \right) \rightarrow \left(\frac{\frac{0.693}{k_{60}} \cdot STAT3\_n \cdot H}{H} \right) : \textbf{true};$$

$$[synth\_socs_{61}] \left(\frac{k_{61} \cdot STAT3\text{-}PD\_n \cdot H}{H} > 0 \right) \rightarrow \left(\frac{k_{61} \cdot STAT3\text{-}PD\_n \cdot H}{H} \right) : \textbf{true};$$

$$[bind\_pias\_stat_{80}] \left(\frac{\frac{k_{80}}{nucleus \cdot N_A} \cdot PIAS3 \cdot H \cdot STAT3\text{-}PD\_n \cdot H}{H} > 0 \right) \rightarrow \left(\frac{\frac{k_{80}}{nucleus \cdot N_A} \cdot PIAS3 \cdot H \cdot STAT3\text{-}PD\_n \cdot H}{H} \right) : \textbf{true};$$

$$[unbind\_pias\_stat_{80}] \left(\frac{k_{-80} \cdot PIAS3 : STAT3\text{-}PD\_n \cdot H}{H} > 0 \right) \rightarrow \left(\frac{k_{-80} \cdot PIAS3 : STAT3\text{-}PD\_n \cdot H}{H} \right) : \textbf{true};$$

endmodule

The PRISM model generated from the Bio-PEPA model of the gp130/JAK/STAT pathway has 63 species and 118 reactions. Because of the well-known state space explosion problem of model-checking, even if we consider only a few levels for each species, the state space for this model is so huge that it makes the numerical solution of the CTMC nearly unmanageable. To overcome this problem, we consider a subdivision of the pathway into two distinct sub-models in such a way that the analysis of the individual sub-models becomes more feasible.

In order to find an appropriate modularisation, we adopt the approach proposed in [27,28], based on the identification of sub-systems with no retroactivity. For the considered model of the gp130/JAK/STAT pathway, two modules with low coupling can be easily identified.

In the first sub-model, which refers to the bindings of ligands to receptors and the activation of the receptor dimers, we consider all the distinct combinations of ligand/receptor complexes, and we describe in detail the formation of all possible types of active receptor dimers, considering the fact that different ligand-receptor pairs have different binding affinities.

In the second sub-model, which refers to the downstream signalling pathway, we instead consider as a starting point a single "generic" type of active receptor dimer (referred to as *rcpt-DP*), and we focus on the reactions involving the activation of STAT3 and its cytoplasmic/nuclear shuttling.

These two sub-models refer to sub-systems of the gp130/JAK/STAT pathway which act in a rather sequential way and, as a consequence, it is reasonable to assume that, for the downstream STAT3 signalling to occur, the receptor-complexes must have been already activated. The initial number of active receptor dimers in the second sub-model is defined as the sum of the steady-state quantities of all the active receptor dimers in the first sub-model. This assumption is justified by the fact that the activation of the receptors is fast compared with the following reactions, and therefore the amount of initially inactive receptors is negligible when considering the downstream pathway.

As discussed in [27,28], the absence of retroactivity ensures that the modularisation has no significant effect on the overall behaviour of the system. This, together with the fact that we use the output of the first sub-model as input of the second sub-model, ensures that the structural qualitative properties verified for the individual sub-models in the rest of this section also hold for the full model. Particular care should be taken when verifying quantitative temporal properties over sub-models. Here we only consider *semi-quantitative* analysis (Sec. 7) as we are interested in relative rather than absolute values. Therefore, in this particular case, the absence of retroactivity ensures the validity, in the full model, of the analysis results obtained in the sub-models. In general, however, the actual reaction rates in the composite model (and therefore the analysis results) might be different from the ones in the sub-models, and more advanced approaches for modularisation should be applied.

In the rest of this section we use $H = 200$ as the step size for the ligands-receptors sub-model, and $H = 300$ for the downstream sub-model. See Sec. 7 for a discussion of the choice of step size values.

Deadlock Detection. *Deadlock states* are the ones in which no transition is enabled. In some cases the presence of deadlock states is (correctly) due to the presence of irreversible reactions which lead to the transformation of all reactants into non-reactive proteins. In other cases deadlocks could be due to the scarcity of one of the reactants of a multimolecular reaction; in our model, for instance, all receptors are consumed (i.e. transformed into different forms, such as dimers) while still ligands are available. In other cases deadlocks could be caused by modelling errors.

PRISM automatically detects deadlock states when building the state space of models, and this feature can be considered the first step in the identification of potential modelling errors.

For instance, in the ligands-receptors sub-model, any state in which ligands are present while all *gp130* receptors have been consumed is a deadlock. This suggests that *gp130* is the bottleneck of the system.

Species Invariants. One simple and yet interesting property that can be verified is the presence of *invariants* in the amount of the involved proteins.

Species invariants are commonly present in biochemical systems because of the existence of basic constraints such as the law of conservation of mass, which states that the amount (i.e. mass) of reactants consumed by a reaction must be equal to the amount of products of the reaction.

For instance, given the conservation of mass and the absence of synthesis and degradation reactions, we expect that the sum of the amounts of *LIFR* receptor present in its various possible forms (free, as *gp130:LIF:LIFR* complex and as *gp130:OSM:LIFR* complex, with one or both of its subunits phosphorylated) is constant (and equal to the *LIFR* initial amount).

The satisfaction of the following properties confirms the existence of the expected invariants on the total amount of ligands and receptors (as an example, we report the ones for *LIF* and *LIFR*).

$\mathcal{P}_{\geq 1}[\mathbf{G}$ (*LIF* + *gp130:LIF:LIFR* + *gp130-P:LIF:LIFR* + *gp130:LIF:LIFR-P* + *gp130-P:LIF:LIFR-P* = N_U_LIF)] $\rightarrow$ **true**

$$\mathcal{P}_{\geq 1}[G\ (LIFR\ +\ gp130{:}LIF{:}LIFR\ +\ gp130{:}OSM{:}LIFR\ +\ gp130{-}P{:}LIF{:}LIFR\ +$$
$$gp130{:}LIF{:}LIFR{-}P\ +\ gp130{-}P{:}LIF{:}LIFR{-}P\ +\ gp130{-}P{:}OSM{:}LIFR\ +$$
$$gp130{:}OSM{:}LIFR{-}P\ +\ gp130{-}P{:}OSM{:}LIFR{-}P\ =\ N_U\_LIFR)]\ \rightarrow\ \textbf{true}$$

Here, and in the rest of the section, the notation $\mathcal{P}_{\bowtie p}[\psi] \rightarrow \textbf{true}$ (resp. **false**) means that ψ is satisfied (resp. is not satisfied), while the notation $\mathcal{P}_{=?}[\psi] \rightarrow p$ (with $p \in \mathbb{R}$) means that the result of ψ is the probability p.

Reachability Analysis. *Reachability* properties allow us to verify whether a given state is eventually reached. States of interest can be, for instance, the ones in which some species reaches a threshold or is totally consumed, or when the amounts of two species coincide.

We consider here the states in which a certain number of receptors are phosphorylated, and the ones in which a certain amount of active nuclear STAT3 (*STAT3-PD_n*) is present.

We consider first the ligands-receptors sub-model. The satisfaction of the first of the following properties guarantees that a state in which one fourth of the total amount of available receptors is phosphorylated is always reached at some time point. On the contrary, the second property, which is not satisfied, proves that we do not necessarily reach a state with one third of receptors phosphorylated.

$$\mathcal{P}_{\geq 1}[\textbf{F}\ (gp130{-}P{:}LIF{:}LIFR{-}P\ +\ gp130{-}P{:}OSM{:}LIFR{-}P\ +\ gp130{-}P{:}OSM{:}OSMR{-}P\ >$$
$$(N_U\_OSMR\ +\ N_U\_LIFR\ +\ N_U\_gp130)\ /\ 4)]\ \rightarrow\ \textbf{true}$$

$$\mathcal{P}_{\geq 1}[\textbf{F}\ (gp130{-}P{:}LIF{:}LIFR{-}P\ +\ gp130{-}P{:}OSM{:}LIFR{-}P\ +\ gp130{-}P{:}OSM{:}OSMR{-}P\ >$$
$$(N_U\_OSMR\ +\ N_U\_LIFR\ +\ N_U\_gp130)\ /\ 3)]\ \rightarrow\ \textbf{false}$$

The next property, instead, guarantees that in general we could reach a system where no *gp130:OSMR* receptor complex is activated.

$$\mathcal{P}_{\geq 1}[\textbf{F}\ (gp130{-}P{:}OSM{:}OSMR{-}P\ >\ 0)]\ \rightarrow\ \textbf{false}$$

Regarding the downstream sub-model, we check for the following properties, which guarantee that, at some time point, at least half the initial amount of STAT3 has been transported into the nucleus and activated, but not all of it.

$$\mathcal{P}_{\geq 1}[\textbf{F}\ (STAT3{-}PD\_n\ >\ N_U\_STAT3\_c\ /\ 2)]\ \rightarrow\ \textbf{true}$$

$$\mathcal{P}_{\geq 1}[\textbf{F}\ (STAT3{-}PD\_n\ >\ N_U\_STAT3\_c)]\ \rightarrow\ \textbf{false}$$

Reversibility. A system is called *reversible* if the initial state is reachable from any other state (i.e. the system is able to self-reinitialise). More generally, a state is called *reversible* if it can be reached again at some later time point.

The following property, if satisfied, guarantees the reversibility of the system: it states that it is always possible to return to the initial state (in the PRISM language "**init**" is a predefined formula which completely specifies the initial state).

$\mathcal{P}_{=?}[\mathbf{G} \ (\text{``init''} \Rightarrow \mathcal{P}_{\geq 1}[\mathbf{X} \ (!\text{``init''} \Rightarrow \mathcal{P}_{\geq 1}[\mathbf{F} \ (\text{``init''})])])]$

For the ligands-receptors sub-system the result of this property is 0, since we have considered bindings to be irreversible and, therefore, the system cannot return to the initial state in which all receptors and ligands are free.

The downstream sub-system, instead, is reversible (the result of the property is 1), thanks to the cytoplasmic/nuclear STAT3 shuttling, which enables the system to return to the initial state in which cytoplasmic STAT3 molecules are not phosphorylated and not bound to receptor dimers.

Liveness. The notion of *liveness* of a reaction in a given state refers to the possibility of it occurring in such a state. In particular, it is interesting to know which reactions are live in the initial state.

Since PRISM properties are state-based, it is not possible to explicitly check for the occurrence of a given reaction. However, knowing how each model component is affected by the occurrence of a given reaction, we can verify this kind of property by checking for the expected variations in the involved components.

We are interested, for instance, in verifying that in the initial state the binding reactions between ligands and receptors can occur, leading to the three possible types of ligand/receptor dimers (*gp130:LIF:LIFR*, *gp130:OSM:LIFR*, and *gp130:OSM:OSMR*).

The following three properties are satisfied, confirming that the three known types of complexes can be formed.

$\mathcal{P}_{\geq 1}[\mathbf{G} \ (\text{``init''} \Rightarrow \mathcal{P}_{>0}[\mathbf{X} \ (gp130 = N_U\text{-}gp130 - 1 \ \& \ LIF = N_U\text{\_}LIF - 1 \ \&$
$\quad LIFR = N_U\text{\_}LIFR - 1)])] \ \rightarrow \ \mathbf{true}$

$\mathcal{P}_{\geq 1}[\mathbf{G} \ (\text{``init''} \Rightarrow \mathcal{P}_{>0}[\mathbf{X} \ (gp130 = N_U\text{-}gp130 - 1 \ \& \ OSM = N_U\text{\_}OSM - 1 \ \&$
$\quad LIFR = N_U\text{\_}LIFR - 1)])] \ \rightarrow \ \mathbf{true}$

$\mathcal{P}_{\geq 1}[\mathbf{G} \ (\text{``init''} \Rightarrow \mathcal{P}_{>0}[\mathbf{X} \ (gp130 = N_U\text{-}gp130 - 1 \ \& \ OSM = N_U\text{\_}OSM - 1 \ \&$
$\quad OSMR = N_U\text{\_}OSMR - 1)])] \ \rightarrow \ \mathbf{true}$

The following property, instead, is not satisfied: it states, as desired, that *LIF* cannot bind to receptors to form *gp130:OSMR* dimers.

$\mathcal{P}_{\geq 1}[\mathbf{G} \ (\text{``init''} \Rightarrow \mathcal{P}_{>0}[\mathbf{X} \ (gp130 = N_U\text{-}gp130 - 1 \ \& \ LIF = N_U\text{\_}LIF - 1 \ \&$
$\quad OSMR = N_U\text{\_}OSMR - 1)])] \ \rightarrow \ \mathbf{false}$

Causality Analysis. *Causality relations* between given reactions can be expressed and verified by properties which relate the order of "appearance" of relevant molecules. This kind of property can be used, for instance, to verify the order in which intermediate products are formed within a cascade of events.

A form of causality relation can be expressed by using the sequence and consequence relations defined in [29]: specifically, while *sequence* formulae describe ordering relations between events (e.g. "in order to reach a given state, we must first reach another one"), *consequence* formulae describe causal relations (e.g. "if a given state occurs, it is necessarily followed by a second one").

For example, the ordering and causality relations between *STAT3* phosphorylation, homodimerisation and relocation into the nucleus can be verified by the following pairs of properties (assuming at system initialisation all *STAT3* is present in cytoplasmic monomeric form (*STAT3-P_c*).

When the result of the first property is 0, such a property states that it is not possible for a *STAT3-PD_c* molecule to be present if in all previous states we had no *rcpt-DP:STAT3-DP1* (a complex formed by a receptor dimer and a STAT3 molecule). Similarly, the following property (when it evaluates to 0) states that *STAT3-PD_c* must be produced before *STAT3-PD_n* appears.

$$\mathcal{P}_{=?}[(rcpt\text{-}DP{:}STAT3\text{-}DP1 = 0) \ \mathbf{U} \ STAT3\text{-}PD\_c > 0] \rightarrow 0$$

$$\mathcal{P}_{=?}[(STAT3\text{-}PD\_c = 0) \ \mathbf{U} \ STAT3\text{-}PD\_n > 0] \rightarrow 0$$

The following two properties complement the previous two, stating that if at least one complex *rcpt-DP:STAT3-DP1* is formed, then at least one *STAT3-PD_c* molecule will necessarily be formed.

$$\mathcal{P}_{=?}[\mathbf{G} \ (rcpt\text{-}DP{:}STAT3\text{-}DP1 > 0 \Rightarrow \mathcal{P}_{\geq 1}[\mathbf{F} \ (STAT3\text{-}PD\_c > 0)])] \rightarrow 1$$

$$\mathcal{P}_{=?}[\mathbf{G} \ (STAT3\text{-}PD\_c > 0 \Rightarrow \mathcal{P}_{\geq 1}[\mathbf{F} \ (STAT3\text{-}PD\_n > 0)])] \rightarrow 1$$

As another example, the following two properties verify that the transport of phosphorylated *STAT3* dimers can only occur from the cytoplasm to the nucleus, but not vice versa. The result of the first property is 0 (i.e. transport of *STAT3-PD* can occur from cytoplasm to nucleus), while the result of the second property is 1 (i.e. transport of *STAT3-PD* cannot occur from nucleus to cytoplasm) for all reachable values of i, j.

$$\mathcal{P}_{=?}[\mathbf{F} \ (STAT3\text{-}PD\_c = i \ \& \ STAT3\text{-}PD\_n = j \ \& \ \mathcal{P}_{\leq 0}[\mathbf{X} \ (STAT3\text{-}PD\_c = i - 1 \ \& \\ STAT3\text{-}PD\_n = j + 1)])] \rightarrow 0$$

$$\mathcal{P}_{=?}[\mathbf{F} \ (STAT3\text{-}PD\_c = i \ \& \ STAT3\text{-}PD\_n = j \ \& \ \mathcal{P}_{\leq 0}[\mathbf{X} \ (STAT3\text{-}PD\_c = i + 1 \ \& \\ STAT3\text{-}PD\_n = j - 1)])] \rightarrow 1$$

Conversely, the transport of dephosphorylated *STAT3* monomers can only occur from the nucleus to the cytoplasm.

$$\mathcal{P}_{=?}[\mathbf{F} \ (STAT3\_c = i \ \& \ STAT3\_n = j \ \& \ \mathcal{P}_{\leq 0}[\mathbf{X} \ (STAT3\_c = i - 1 \ \& \ STAT3\_n = j + 1)])] \rightarrow 1$$

$$\mathcal{P}_{=?}[\mathbf{F} \ (STAT3\_c = i \ \& \ STAT3\_n = j \ \& \ \mathcal{P}_{\leq 0}[\mathbf{X} \ (STAT3\_c = i + 1 \ \& \ STAT3\_n = j - 1)])] \rightarrow 0$$

6   Simulation Based Time-Series Analysis

In the previous section we have used model-checking in order to check for a number of simple formulae which guarantee us that some key properties of the gp130/JAK/STAT

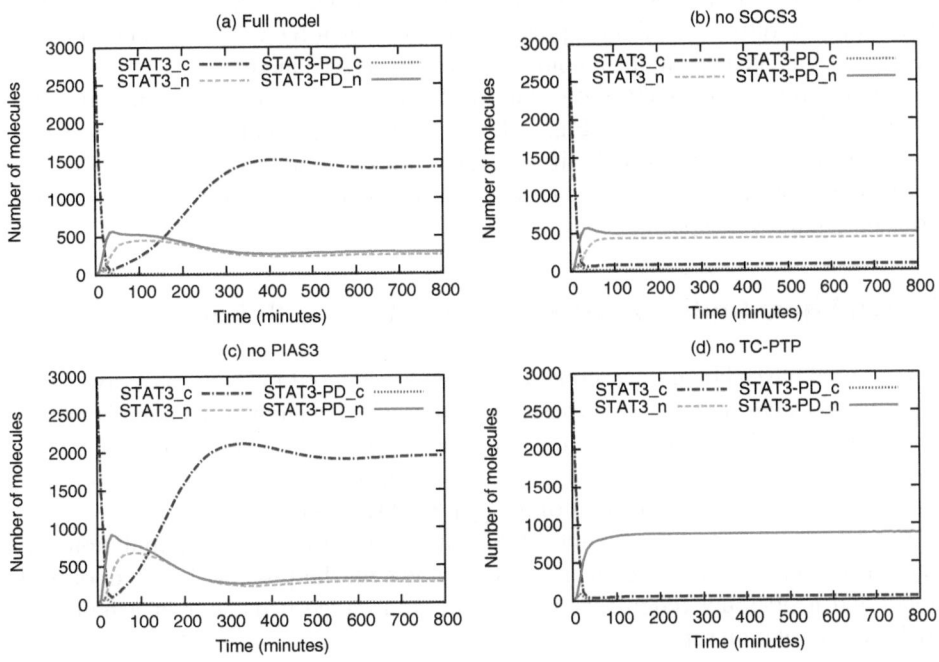

Fig. 2. Simulation results: full model vs. no inhibitors

model are satisfied. This analysis allows us to be more confident about the absence of modelling errors.

Now we progress our analysis of the model by means of stochastic simulation. We report here some results obtained by simulating the full model (comprising both the ligands-receptors and the downstream sub-systems) using the Gibson-Bruck [30] stochastic simulation engine implemented in Dizzy [12,24].

Figure 2 shows the time-series evolution produced by the model (Fig. 2(a)) versus the ones in which each of the three inhibitors has been removed (Fig. 2(b)–(d)). Each plot refers to average values computed over 1000 simulation runs, and the amounts of the four different forms of STAT3 are shown (cytoplasmic and nuclear dephosphorylated monomers, and cytoplasmic and nuclear phosphorylated dimers).

In all the performed simulations, at system initialisation STAT3 is only present in cytoplasmic monomeric form. As shown in Fig. 2(a), as time passes, STAT3 is phosphorylated, dimerised, and transported into the nucleus, until the systems reaches a state in which the inhibition of nuclear STAT3 by dephosphorylation and the nuclear/cytoplasmic shuttling lead nuclear and cytoplasmic STAT3 to be in equilibrium.

When the amount of nuclear STAT3 increases significantly, the inhibitory role of SOCS3 (which is under transcription control of STAT3) comes into play (Fig. 2(b)). SOCS3 is responsible for signal attenuation and, hence, after reaching a peak, nuclear STAT3 decreases.

PIAS3 slows down the production of active nuclear STAT3 by binding to it (Fig. 2(c)). Therefore, if PIAS3 is present, part of nuclear STAT3 is bound to it, while, if PIAS3 is knocked down, the amount of available STAT3 increases.

A third inhibitor, TC-PTP, allows nuclear STAT3 to translocate back into the cytoplasm, by dephosphorylating it (Fig. 2(d)). If TC-PTP is present, STAT3 nuclear/cytoplasmic shuttling occurs; instead, if TC-PTP is knocked out (i.e. if nuclear STAT3 is not dephosphorylated), STAT3 accumulates in the nucleus, whilst cytoplasmic STAT3 molecules quickly disappear.

7   Semi-quantitative Analysis of the CTMC with Levels

In Sec. 5 we have shown how model-checking can be used in order to discover modelling errors by checking for some basic properties which guarantee the model to behave as expected. In this section, instead, we use model-checking also for quantitative analysis, with the purpose of completing the simulation-based analysis in order to provide additional insight on the behaviour of the gp130/JAK/STAT pathway.

The main advantage of model-checking with respect to stochastic simulation is the fact that model-checking is exhaustive: it explores all the possible behaviours of the model and it does not require us to compute an average behaviour of a number of stochastic simulation runs.

As mentioned before, the main disadvantage of model-checking is the state space explosion problem, which implies that we cannot deal with too many levels for the model components without inducing an intractable model.

In has been shown (see [10]) that, as the number of levels increases, the behaviour of the CTMC with levels tends to the behaviour of ODEs (when the number of molecules is large enough to average out the randomness of the system); this result guarantees the theoretical correctness of the approach. However, if the number of levels is too small, the error introduced by the discretisation becomes significant and the numerical solution of the generated CTMC fails to reproduce the correct behaviour.

The number of levels for model components is related to the step size H and to the upper N_U and lower N_L bounds for each species. The step size H represents the granularity of the system, and it directly affects the accuracy of the results; the upper and lower bounds are also relevant to the accuracy, since imposing bounds on the numbers of molecules causes a state space truncation which might potentially have impact on the behaviour of the system.

Therefore, when performing CTMC analysis of Bio-PEPA models, the choice of the step size and of the upper and lower bounds is essential: they must be carefully selected so that the number of levels to be used for the model components is a suitable trade-off between accuracy and efficiency.

In the following sections we report some of the results obtained by using the PRISM model-checker to perform quantitative analysis. First we consider reward-based properties which allow us to observe the time-series for some of the species of the system (for comparison with the stochastic simulation), and we discuss the error introduced by discretising and bounding the model; afterwards, we define further properties in order to compute additional (semi-)quantitative measures.

Time-series Analysis Using State Rewards. A reward structure is automatically defined by the Bio-PEPA Workbench for each PRISM component, and it can be referred to either by the component name or by an integer value (implicitly assigned to reward structures based on the order in which they are defined). These reward structures associate an instantaneous reward equal to the current amount of the corresponding molecular species with each state. The evaluation of these reward-based properties corresponds to computing an average behaviour for the species at given time points.

As an example, the following reward is used to observe the time evolution of the receptor dimer *gp130:LIF:LIFR*.

$$\textbf{rewards } \text{``gp130:LIF:LIFR''}$$
$$\textbf{true} : gp130{:}LIF{:}LIFR \cdot H;$$
$$\textbf{endrewards}$$

Figure 3 reports the results obtained by verifying on the ligands-receptors sub-system the reward-based property

$$\mathcal{R}^i_{=?}[I = T]$$

for time points $T \leq 30$ minutes, where i is an integer variable used to index the reward structure of interest.

Figure 4, instead, reports the results obtained by verifying the same reward-based property for time points $T \leq 800$ minutes on the downstream sub-system. In this figure, we also report the standard deviation of the number of molecules, which is computed by exploiting reward structures associating the square of the number of molecules of each species with each state: the standard deviation is calculated as the square root of the variance $E(Y)^2 - E(Y^2)$, where Y is the random variable representing a species in

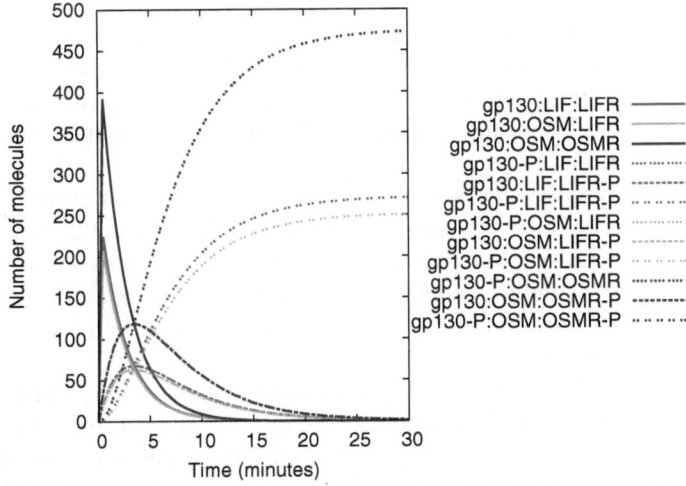

Fig. 3. Time-series by model-checking: ligands-receptors sub-model

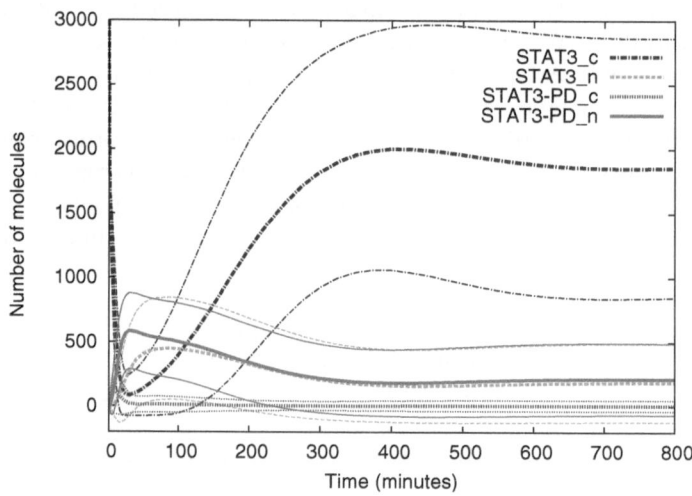

Fig. 4. Time-series by model-checking: downstream sub-model. Thick lines represent the expected numbers of molecules; thin lines represent their standard deviation.

the network, whereas $E(Y)$ and $E(Y^2)$ indicate the expected values for the amount of the species Y and for its square value.

Figures 3 and 4 have been obtained by analysing the sub-models with step sizes $H = 200$ and $H = 300$ respectively. In the next section we discuss the considerations which lead us to the choice of such values.

Three kinds of approximation errors could have been introduced by our analysis of the CTMC with levels due to, respectively, the discretisation of the amounts (H), their bounding (N_L and N_U), and the subdivision into modules.

In the next section we discuss the effect of varying the step size H on the behaviour of the system. Instead, we do not report results concerning the variation of the bounds N_L and N_U since, in this particular system, increasing the bounds does not have a significant effect: the reason for this is that no synthesis and degradation reactions are defined (with the single exception of SOCS3) and, as a consequence, the amount of most molecular species is clearly bounded by the amounts of the molecules present at system initialisation.

The choice of how to modularise the system has been carried out in order to minimise the interaction between the two modules. However, the modularisation has certainly an impact on the quantitative behaviour. In the whole system, for instance, STAT3 and SOCS3 molecules can bind to receptor dimers as soon as they start being phosphorylated; in the downstream sub-model, instead, we had to fix an initial amount of phosphorylated receptor dimers.

Despite these possible sources of approximation, comparing Fig. 4 and Fig. 2, we notice that the results obtained by analysing the downstream sub-model using PRISM instantaneous rewards do not differ significantly from the behaviour observed by averaging the results obtained by 1000 stochastic simulation runs of the whole model. Both the time-scale and the relative amounts of molecules are the same in both figures, and the only significant difference regarding the absolute amounts is the amount of cytoplasmic monomeric STAT3, which is higher in Fig. 4. We can also observe that the standard deviation reported in Fig. 4 is quite high, due to the stochastic noise which has been introduced by using a small number of levels.

Experimenting with Step Sizes. As previously stated, the choice of the step size has a great impact on both accuracy and performance of the analysis: the smaller the step size is, the larger the CTMC state space and, hence, the smaller the discretisation error introduced, but also the longer the time needed for solving the CTMC.

Before choosing the values to be used for the step size H in the analysis of the models, we have performed a number of experiments varying H in order to find values representing a good trade-off between accuracy and performance of the analysis. In Fig. 5 and Fig. 6 we report some results which show how changing the step size affects the behaviour of the system (in ligands-receptors and downstream sub-systems, respectively).

In Fig. 5, we compare the results obtained by using six different values for H (1000, 500, 300, 250, 200, 150) in the analysis of the ligands-receptors sub-model, and we can observe that H in this case does not have a big impact on the results.

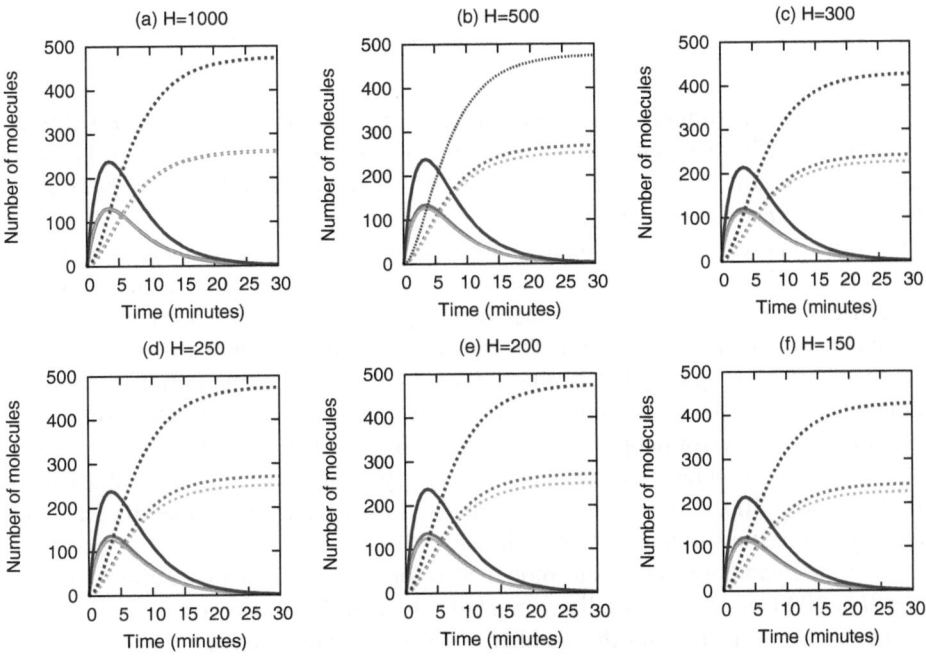

Fig. 5. Time-series by model-checking: ligands-receptors sub-model. The three types of receptor complexes are shown, gp130:LIF:LIFR (red), gp130:OSM:LIFR (green), and gp130:OSM:OSMR (blue), in the stage when one (full line) or both (dashed line) receptors are phosphorylated.

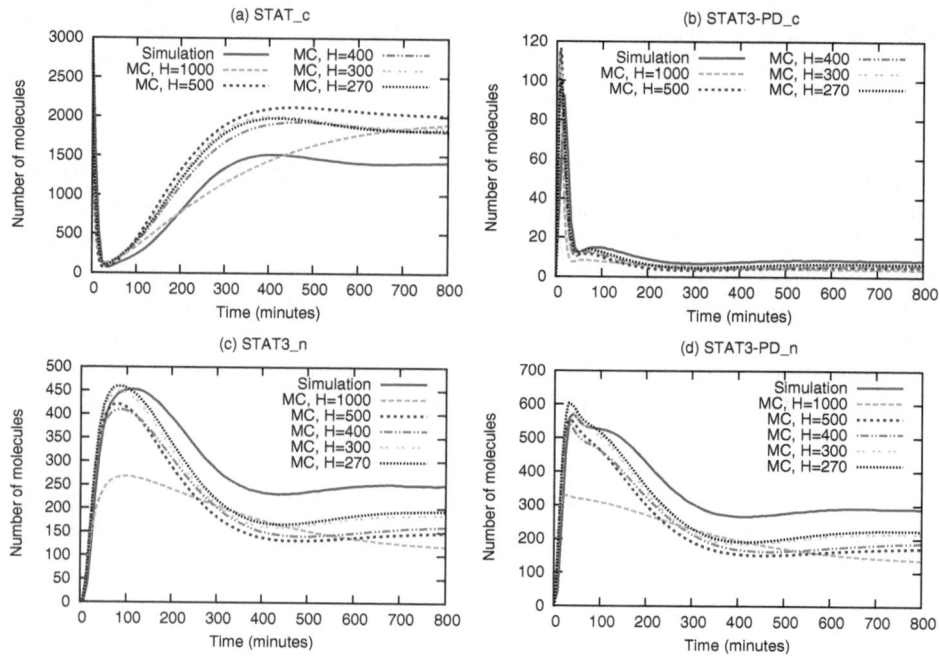

Fig. 6. Time-series by model-checking: STAT3 sub-model

The first notable difference is that in Fig. 5(a) the amounts of gp130:LIF:LIFR and gp130:OSM:LIFR are equal: as expected, with $H = 1000$ (i.e. one single level for each ligand and receptor) we are not able to observe the fact that $LIFR$ has an higher binding affinity with LIF than with OSM.

The other interesting thing is that, contrary to what we expected, there is no noticeable increase of accuracy when decreasing H. Instead, after observing the similarities between Fig. 5(b), (d) and (e), and between Fig. 5(c) and (f), respectively, we drew the conclusion that the first group is the "correct" one; the reason is the rounding error introduced when computing the number of levels starting from the initial amounts (remember that $N_L = \lfloor MIN/H \rfloor$ and $N_U = \lfloor MAX/H \rfloor$): when a small numbers of levels is used, this rounding error happens to be more significant than H itself.

In Fig. 6, we compare the results obtained by using five different values for H $(1000, 500, 400, 300, 270)$ in the analysis of the downstream sub-model; the value obtained by stochastic simulation is also shown.

As for the ligands-receptors sub-model, also for this sub-model we notice that when using $H = 1000$ we obtain a totally wrong behaviour, and we observe a general increase in accuracy when increasing the number of levels. For the smallest values of H, the relative values and the trends for the considered species are correctly reflected compared to the stochastic simulation results: for instance, both the peaks' amplitude and the time at which they occur are reproduced quite accurately.

Though not exact with respect to the stochastic simulation, these results are satisfactory enough for the kinds of semi-quantitative properties we are interested in analysing in the next section. Exact quantitative analysis via CTMC, instead, is infeasible for systems such as the model we consider here. Indeed, the time needed for obtaining the results shown in Fig. 6 ranges from a couple of seconds to hours, and for $H = 250$ the size of the CTMC already becomes prohibitively large to analyse.

Semi-quantitative Properties. Using again the "trade-off" step sizes $H = 200$ and $H = 300$, we consider here a few more *semi-quantitative* properties of the two sub-models.

For instance, we are interested in analysing the impact that the different affinities of ligand/receptor pairs have on the consumption of the different ligands and receptors and on the relative amount of type I and type II receptors formed.

Though this kind of analysis is clearly quantitative (since it involves calculating probabilities and, hence, numbers of molecules), we consider such properties *semi-quantitative* because we are not interested in computing absolute values, but rather in knowing relative values with respect to each other.

The following properties measure the probability with which the amount of each molecular species never changes from the initial amount.

$$\mathcal{P}_{=?}[\mathbf{G}\ (LIF = N_U\_LIF)] \rightarrow 7.53 \cdot 10^{-2}$$

$$\mathcal{P}_{=?}[\mathbf{G}\ (OSM = N_U\_OSM)] \rightarrow 1.45 \cdot 10^{-6}$$

$$\mathcal{P}_{=?}[\mathbf{G}\ (gp130 = N_U\_gp130)] \rightarrow 0$$

$$\mathcal{P}_{=?}[\mathbf{G}\ (LIFR = N_U\_LIFR)] \rightarrow 1.24 \cdot 10^{-4}$$

$$\mathcal{P}_{=?}[\mathbf{G}\ (OSMR = N_U\_OSMR)] \rightarrow 4.56 \cdot 10^{-4}$$

From the obtained results we notice, for instance, that *gp130* is always used (indeed, it is necessary to form all receptor dimers), and that it is more likely for *OSM* to be consumed than *LIF* (indeed, *LIF* is only used in the formation of one type of receptor dimers).

We measure also the probability with which the amount of each molecular species reaches its lower bound.

This group of properties shows that *gp130* is totally consumed in any possible evolution of the system, that *LIF* and *OSM* are never totally consumed, and that the probability of *LIFR* being totally consumed is equal to the probability of *OSMR* not being used at all. These results mean that *gp130* is the bottleneck of the system, while *LIF* and *OSM* are present in abundance.

$$\mathcal{P}_{=?}[\mathbf{F}\ (LIF = N_L\_LIF)] \rightarrow 0$$

$$\mathcal{P}_{=?}[\mathbf{F}\ (OSM = N_L\_OSM)] \rightarrow 0$$

$$\mathcal{P}_{=?}[\mathbf{F}\ (gp130 = N_L\_gp130)] \rightarrow 1$$

$\mathcal{P}_{=?}[\mathbf{F} \ (LIFR = N_L\_LIFR)] \rightarrow 4.56 \cdot 10^{-4}$

$\mathcal{P}_{=?}[\mathbf{F} \ (OS\,MR = N_L\_OS\,MR)] \rightarrow 1.24 \cdot 10^{-4}$

Finally, we consider the reward-based property

$\mathcal{R}^i_{=?}[C \le T]$

and we verify it on the downstream sub-model for time points $T \le 800$ minutes, where i is an integer variable referring to a transition reward structure.

In addition to state rewards, in fact, PRISM allows for the definition of reward structures which associate with each transition a cumulative reward equal to its expected number of occurrences up to the considered time.

In Fig. 7 and Fig. 8 the expected number of occurrences for some of the reactions of the downstream sub-model is shown.

In particular, in Fig. 7 we compare the number of occurrences of receptors/STAT3 and receptors/SOCS3 binding reactions, which shows intuitively the different binding affinities of STAT3 and SOCS3 to the receptor dimers.

In Fig. 8, instead, we consider the number of occurrences of transport reactions of STAT3 molecules from cytoplasm to nucleus and back. In Fig. 8(a) we compare the number of occurrences of transport in the two directions: we count each transport from cytoplasm to nucleus twice since STAT3 molecules are translocated in the nucleus in dimeric form and, hence, a pair of STAT3 molecules is moved at each reaction occurrence. We observe that, since at system initialisation no STAT3 molecule is present in the nucleus, the difference between the two curves in Fig. 8(a) (multiplied by the step

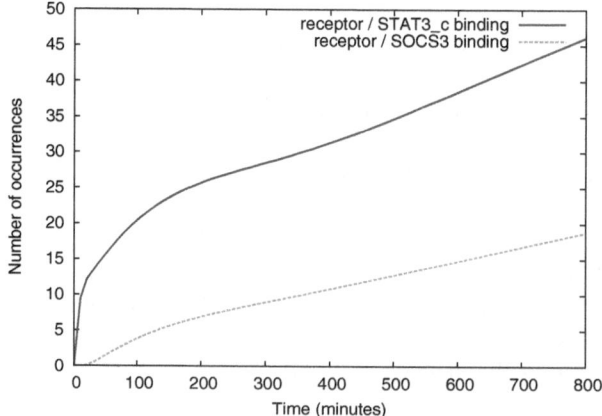

Fig. 7. Expected number of occurrences of receptor binding reactions in the downstream sub-model. The full red line plots the number of occurrences of reactions *bind_rcpt_DP_stat*$_{27}$ and *bind_rcpt_DP_stat*$_{28}$, while the dashed green line plots the number of occurrences of reactions *bind_rcpt_DP_socs*$_{62}$ and *bind_rcpt_DP_socs*$_{63}$.

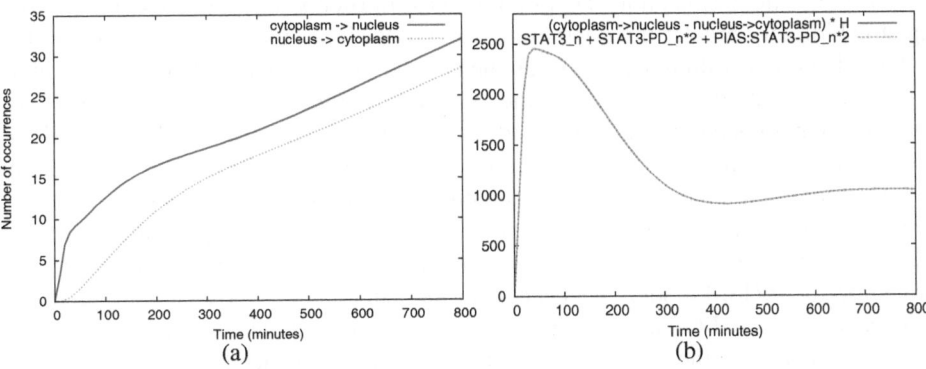

Fig. 8. Expected number of occurrences of transport reactions in the downstream sub-model. In (a) the full red line plots twice the number of occurrences of reaction $reloc\_stat\_cn_{58}$, while the dashed green line plots the number of occurrences of reaction $reloc\_stat\_nc_{60}$. In (b) the full red line is the difference between the lines in (a) multiplied by H, while the dashed green line is the total current amount of STAT3 molecules in the nucleus ($STAT3\_n + STAT3\text{-}PD\_n \cdot 2 + PIAS{:}STAT3\text{-}PD\_n \cdot 2$).

size H) must be the number of STAT3 molecules present in the nucleus. This consideration is confirmed by the perfect agreement of the two curves in Fig. 8(b).

8   Related Work

Given its significant impact on various cellular processes, the gp130/JAK/STAT pathway has been subject of numerous studies, both experimental and computational. Consequently, a few variants of the pathway model have been developed in order to analyse different aspects of it.

In [18] the focus is on the shuttling of STATs from nucleus to cytoplasm and back. A more complete model is developed in [19], which also reports the results of a global sensitivity analysis of parameter interaction. The role of inhibitory mechanisms is instead studied in [20]. These three works are based on mathematical modelling and the analysis is performed by ODEs solvers.

In [21], a process algebra based computational model of the gp130/JAK/STAT pathway is presented and analysed using the BetaWB tool [31], a stochastic simulator for the BlenX language [32]. The Bio-PEPA model we present here is strongly based on the BlenX model described in [21], and the simulation results of the two models match well. This agreement is particularly interesting in view of the conceptual differences existing in the two process algebras. One of these differences concerns the treatment of complexes, which in BlenX are considered as molecular species consisting of the individual molecules composing them, while in Bio-PEPA they are considered as different species not explicitly related to the sub-components. Secondly, immediate reactions can be defined in BlenX, while they are not admitted in Bio-PEPA because of Bio-PEPA's underlying CTMC semantics. Finally, stoichiometric information can be specified in

Bio-PEPA, while they cannot be explicitly coded in BlenX (requiring reactions involving stoichiometry greater than one to be decomposed into multiple steps). In addition to these theoretical differences between the languages, we mention that the focus in the two works is quite different. In [21] the effects of a number of experiments involving quantitative parameters are analysed and compared with experimental data. The aim of the present work, instead, is to exploit model-checking, in addition to stochastic simulation, to analyse both qualitative and quantitative properties of the model behaviour.

A few works have recently been published regarding the application of model-checking techniques to the analysis of biochemical systems. In [33] the authors demonstrate how the PRISM model-checker can be adopted to model and analyse biochemical pathways, using the FGF pathway as a case study. The approach proposed in this work differs from ours in the level of abstraction considered. Instead of taking a variable number of levels into account, the authors of [33] consider an abstraction in which one single copy of each involved molecular species is present and such that module variables represent changes in state of the molecules. This approach has the evident advantage of reducing the CTMC state space, though it might not be quantitatively correct in general: it can be seen as a level of abstraction equivalent to ours when one single level is used for each species. In the same work, the authors also consider a number of state space reduction techniques, some of which (based on lumpability and symmetry reduction) are exact, meaning that the behaviour of the reduced CTMC is preserved.

The notion of CTMC with levels of concentrations has been introduced in [34], in which the ERK signalling pathway was used as a case study, and in [35] the PRISM model-checker is used to analyse it. Following these works, the notion of discrete levels of concentrations has been adopted also in IDD-CSL [36], an Interval Decision Diagram based model-checker for stochastic Petri nets, which allows for the verification of CSL properties.

In [37] the authors propose a framework, based on Petri nets, in which qualitative and quantitative (stochastic and continuous) analysis of biochemical pathways are integrated. Qualitative properties such as boundedness, liveness and reversibility are considered, in addition to the possibility to check for P- and T-invariants, and behavioural properties are verified by probabilistic model-checking.

Finally, BIOCHAM [38,39] is a framework for modelling, simulating and analysing biochemical systems, in which different semantics (differential, stochastic, discrete, and boolean) are considered. BIOCHAM allows for the verification of temporal properties expressed in the Computation Tree Logic (CTL) by using the NuSMV model-checker [40].

9   Conclusions and Future Work

In this work we have used the gp130/JAK/STAT signalling pathway as a case study for modelling and analysis using the Bio-PEPA process algebra. Among the possible analysis methods made available by the Bio-PEPA Workbench, we have considered stochastic simulation and model-checking.

The results obtained by simulation agree well with existing mathematical and computational models. The application of the model-checking approach to the analysis of the pathway model, though limited by the state space explosion problem, provided us with some useful insight. First, it can be used for consistency checking, in order to guarantee the satisfaction of essential properties and, therefore, the absence of modelling errors. Second, it allows us to check for the satisfaction of semi-quantitative behavioural properties over the whole model, without the need for computing average values over a number of stochastic simulation runs.

In order to deal with the computational complexity of model-checking, we have subdivided the pathway model into two distinct sub-models. The time-series analysis obtained by analysing the sub-models individually via model-checking shows a reasonably good agreement with the behaviour obtained via stochastic simulation. The issue of modularisation of models of biochemical systems is a complex one. In this work we have adopted a simple approach which is adequate for this particular case study. A general approach for modularisation of models deserves additional study, in particular in view of the possible performance improvement which this technique could bring in model-checking.

Finally, in order to fully exploit the framework provided by Bio-PEPA further analysis could be performed on the MATLAB model generated by the Bio-PEPA Workbench using ODEs based methods to perform, for instance, bifurcation, stability, and continuation analysis.

Acknowledgments. The author wishes to thank Jane Hillston for her helpful comments. This research is supported by the EPSRC grant EP/E031439/1 "Stochastic Process Algebra for Biochemical Signalling Pathway Analysis".

References

1. Regev, A., Silverman, W., Shapiro, E.: Representation and simulation of biochemical processes using the π-calculus process algebra. In: Proceedings of Pacific Symposium on Biocomputing (PSB 2001), vol. 6, pp. 459–470 (2001)
2. Regev, A., Panina, E.M., Silverman, W., Cardelli, L., Shapiro, E.Y.: BioAmbients: an Abstraction for Biological Compartments. Theoretical Computer Science 325(1), 141–167 (2004)
3. Cardelli, L.: Brane Calculi - Interactions of Biological Membranes. In: Danos, V., Schachter, V. (eds.) CMSB 2004. LNCS (LNBI), vol. 3082, pp. 257–278. Springer, Heidelberg (2005)
4. Priami, C., Quaglia, P.: Operational patterns in Beta-binders. In: Priami, C. (ed.) Transactions on Computational Systems Biology I. LNCS (LNBI), vol. 3380, pp. 50–65. Springer, Heidelberg (2005)
5. Danos, V., Laneve, C.: Formal molecular biology. TCS 325(1) (2004)
6. Regev, A., Shapiro, E.: Cells as Computation. Nature 419(6905), 343 (2002)
7. Ciocchetta, F., Hillston, J.: Bio-PEPA: An extension of the process algebra PEPA for biochemical networks. In: Proc. of FBTC 2007. ENTCS, vol. 194, pp. 103–117 (2008)
8. Ciocchetta, F., Hillston, J.: Bio-PEPA: A Framework for the Modelling and Analysis of Biological Systems. Theoretical Computer Science 410(33-34), 3065–3084 (2009)
9. Hillston, J.: A Compositional Approach to Performance Modelling. Cambridge University Press, Cambridge (1996)
10. Ciocchetta, F., Hillston, J.: Calculi for Biological Systems. In: Formal Methods for Computational Systems Biology (SFM 2008). LNCS, vol. 5016, pp. 265–312. Springer, Heidelberg (2008)

11. Bio-PEPA Workbench Home Page:
    `http://www.dcs.ed.ac.uk/home/stg/software/biopepa/`
12. Ramsey, S., Orrell, D., Bolouri, H.: Dizzy: stochastic simulation of large-scale genetic regulatory networks. J. Bioinf. Comp. Biol. 3(2), 415–436 (2005)
13. PRISM Home Page: `http://www.prismmodelchecker.org`
14. Aziz, A., Sanwal, K., Singhal, V., Brayton, R.: Model-checking continuous-time Markov chains. ACM Trans. Comput. Logic 1(1), 162–170 (2000)
15. Underhill-Day, N., Heath, J.: Oncostatin M (OSM) Cytostasis of Breast Tumor Cells: Characterization of an OSM Receptor β-Specific Kernel. Cancer Research 66(22), 10891–10901 (2006)
16. Heinrich, P., Behrmann, I., Haan, S., Hermanns, H., Müller-Newen, G., Schaper, F.: Principles od interleukin (IL)-6-type cytokine signalling and its regulation. Biochem. J. 374, 1–20 (2003)
17. Kisseleva, T., Bhattacharya, S., Braunstein, J., Schindler, C.: Signaling through the JAK/STAT pathway, recent advances and future challenges. Gene 285, 1–24 (2002)
18. Swameye, I., Müller, T., Timmer, J., Sandra, O., Klingmüller, U.: Identification of nucleocytoplasmic cycling as a remote sensor in cellular signaling by databased modeling. PNAS 100, 1028–1033 (2003)
19. Mahdavi, A., Davey, R.E., Bhola, P., Yin, T., Zandstra, P.W.: Sensitivity Analysis of Intracellular Signaling Pathway Kinetics Predicts Targets for Stem Cell Fate Control. PLoS Computational Biology 3(7), 1257–1267 (2007)
20. Singh, A., Jayaraman, A., Hahn, J.: Modeling Regulatory Mechanisms in IL-6 Transduction in Hepatocytes. Biotechnology and Bioengineering 95(5), 850–862 (2006)
21. Guerriero, M.L., Dudka, A., Underhill-Day, N., Heath, J.K., Priami, C.: Narrative-based computational modelling of the Gp130/JAK/STAT signalling pathway. BMC Systems Biology 3(1), 40 (2009)
22. Bio-PEPA Home Page: `http://www.biopepa.org/`
23. Hinton, A., Kwiatkowska, M., Norman, G., Parker, D.: PRISM: A tool for automatic verification of probabilistic systems. In: Hermanns, H., Palsberg, J. (eds.) TACAS 2006. LNCS, vol. 3920, pp. 441–444. Springer, Heidelberg (2006)
24. Dizzy Home Page: `http://magnet.systemsbiology.net/software/Dizzy`
25. Aziz, A., Kanwal, K., Singhal, V., Brayton, V.: Verifying continuous time Markov chains. In: Alur, R., Henzinger, T.A. (eds.) CAV 1996. LNCS, vol. 1102, pp. 269–276. Springer, Heidelberg (1996)
26. Baier, C., Katoen, J.P., Hermanns, H.: Approximate Symbolic Model Checking of Continuous-Time Markov Chains. In: Baeten, J.C.M., Mauw, S. (eds.) CONCUR 1999. LNCS, vol. 1664, pp. 146–161. Springer, Heidelberg (1999)
27. Saez-Rodriguez, J., Kremling, A., Gilles, E.: Dissecting the puzzle of life: modularization of signal transduction networks. Computers and Chemical Engineering 29, 619–629 (2005)
28. Conzelmann, H., Saez-Rodriguez, J., Sauter, T., Bullinger, E., Allgöwer, F., Gilles, E.: Reduction of mathematical models of signal transduction networks: simulation-based approach applied to EGF receptor signalling. Systems Biology 1(1), 159–169 (2004)
29. Monteiro, P., Ropers, D., Mateescu, R., Freitas, A., de Jong, H.: Temporal logic patterns for querying dynamic models of cellular interaction networks. ECCB 24, 227–233 (2008)
30. Gibson, M., Bruck, J.: Efficient Exact Stochastic Simulation of Chemical Systems with Many Species and Many Channels. The Journal of Chemical Physics 104, 1876–1889 (2000)
31. Dematté, L., Priami, C., Romanel, A.: The Beta Workbench: a computational tool to study the dynamics of biological systems. Briefings in Bioinformatics 9(5), 437–449 (2008), `http://www.cosbi.eu/Rpty_Soft_BetaWB.php`

32. Dematté, L., Priami, C., Romanel, A.: The BlenX Language: A Tutorial. In: Bernardo, M., Degano, P., Zavattaro, G. (eds.) SFM 2008. LNCS, vol. 5016, pp. 313–365. Springer, Heidelberg (2008)
33. Heath, J., Kwiatkowska, M., Norman, G., Parker, D., Tymchyshyn, O.: Probabilistic Model Checking of Complex Biological Pathways. Theoretical Computer Science 319, 239–257 (2008)
34. Calder, M., Gilmore, S., Hillston, J.: Modelling the Influence of RKIP on the ERK Signalling Pathway Using the Stochastic Process Algebra PEPA. In: Priami, C., Ingólfsdóttir, A., Mishra, B., Riis Nielson, H. (eds.) Transactions on Computational Systems Biology VII. LNCS (LNBI), vol. 4230, pp. 1–23. Springer, Heidelberg (2006)
35. Calder, M., Vyshemirsky, V., Gilbert, D., Orton, R.: Analysis of signalling pathways using continuous time Markov chains. In: Priami, C., Plotkin, G. (eds.) Transactions on Computational Systems Biology VI. LNCS (LNBI), vol. 4220, pp. 44–67. Springer, Heidelberg (2006)
36. The Idd-CSL Home Page: http://www-dssz.informatik.tu-cottbus.de/software/software.html
37. Heiner, M., Gilbert, D., Donaldson, R.: Petri Nets for Systems and Synthetic Biology. In: Bernardo, M., Degano, P., Zavattaro, G. (eds.) SFM 2008. LNCS, vol. 5016, pp. 215–264. Springer, Heidelberg (2008)
38. The BIOCHAM Home Page: http://contraintes.inria.fr/BIOCHAM/
39. Fages, F., Soliman, S., Chabrier-Rivier, N.: Modelling and querying interaction networks in the biochemical abstract machine BIOCHAM. Journal of Biological Physics and Chemistry 4(2), 64–73 (2004)
40. NuSMV Home Page: http://nusmv.irst.itc.it/

Rule-Based Modelling and Model Perturbation

Vincent Danos[1], Jérôme Feret[2], Walter Fontana[3],
Russ Harmer[4], and Jean Krivine[3,5]

[1] University of Edinburgh
[2] INRIA–ENS–CNRS
[3] Harvard Medical School
[4] CNRS–Université Paris Diderot
[5] Institut des Hautes Etudes Scientifiques

Abstract. Rule-based modelling has already proved to be successful for taming the combinatorial complexity, typical of cellular signalling networks, caused by the combination of physical protein-protein interactions and modifications that generate astronomical numbers of distinct molecular species. However, traditional rule-based approaches, based on an unstructured space of agents and rules, remain susceptible to other combinatorial explosions caused by mutated and/or splice variant agents, that share most but not all of their rules with their wild-type counterparts; and by drugs, which must be clearly distinguished from physiological ligands.

In this paper, we define a syntactic extension of Kappa, an established rule-based modelling platform, that enables the expression of a structured space of agents and rules that allows us to express mutated agents, splice variants, families of related proteins and ligand/drug interventions uniformly. This also enables a mode of model construction where, starting from the current consensus model, we attempt to reproduce *in numero* the mutational—and more generally the ligand/drug perturbational—analyses that were used in the process of inferring those pathways in the first place.

1 Introduction

In recent years, there has been extensive development in the use of modelling to understand cellular signalling networks (see [1, 2, 3, 4] among many others). To date, this line of work has focussed almost exclusively on describing wild-type behaviours, i.e. it deals with the interactions between proteins that take place in a normal healthy cell. This is already highly non-trivial since these signalling networks employ a strategy of binding, modification and unbinding between proteins that generates astronomical numbers of non-isomorphic molecular species. This poses an essentially unsolvable scalability problem for any modelling approach, such as ODE-based chemical kinetics or Petri nets, based on exhaustively enumerating *reactions* between fully-specified molecular species.

In recent years, a new modelling approach has been used to tame this combinatorial explosion, namely agent- or rule-based modelling [5]. In this setting,

C. Priami et al. (Eds.): Trans. on Comput. Syst. Biol. XI, LNBI 5750, pp. 116–137, 2009.

molecular species are left implicit; instead, *agents* are used to represent not complexes but their constituent proteins. Each type of agent has a name and a set of *sites*. Instead of reactions, we write *rules* that mention names of agent types and some, but not necessarily all, of their respective sites. In this way, and unlike reactions, a rule need only make explicit those aspects of the agents upon which it acts that are actually relevant to the interaction being described by the rule. So reaction-based models leave agents implicit, considering them at best as an aggregation of molecular species, whereas rule-based models make agents explicit but the reactions implicit, instead considering their rules to be aggregations of reactions.

It should be noted that such wild-type models, be they reaction- or rule-based, can already handle situations where, typically as a result of gene amplification or ablation, a protein is either over- or under-expressed. Under such circumstances, the response of a cell to external conditions may be exaggerated or attenuated as a consequence of the induced perturbation of mass-action kinetics and of the nature and numbers of complexes that exist in the cell's resting state (cf. [6]). This does not bring about *new* protein-protein interactions, it only affects the relative importance of the wild-type interactions, e.g. if protein X has a binding partner Y that is over-expressed, X will be attracted to the greater than usual mass of Ys to the detriment of its binding with other partners.

However, many disease states are the result of genetic mutations that build incorrect proteins, with aberrant behaviour, rather than the straightforward modulation of protein expression levels (although in some cases the two defects co-exist and synergize). Such mutant proteins may only differ by one or two amino acids from their wild-type cousins and yet have radically different behaviour, e.g. erbB1 with the single substitution L858R, which exists in many kinds of solid tumour, has a constitutively active kinase domain, as does B-Raf with the single V600E mutation. The flip side of this is that much of the wild-type behaviour of a protein is actually shared with such mutants, for instance a binding domain far from any site of mutation will quite likely retain its usual functionality. This poses a further serious challenge to modelling since mutated proteins therefore duplicate large chunks of an already highly combinatorial wild-type network, while also potentially adding interactions.

To tackle these issues, we introduce a syntactic extension of Kappa that allows the definition of a structured space of agents. Agents can either be declared *ab initio* or derived from existing agents in a manner reminiscent of object-oriented programming (particularly the prototype-based approach). In the latter case, the new agent can gain, lose, rename, mutate or duplicate sites of the agent from which it is derived. This organizes the space of agents hierarchically and thus enables us to write *generic* rules that mention agents that have many descendants in the hierarchy. These generic rules act as shorthand for sets of normal Kappa rules; they capture behaviours *shared* by splice variants (e.g. p46, p52 and p66 Shc), genetically related proteins (e.g. ERK1 and ERK2) or mutated proteins. In particular, the conciseness of generic rules enables us to write and analyze large Kappa models far more easily. We illustrate this with a small generic rule set (15 rules) for the erbB receptor network that, once expanded into Kappa, has over 300 rules and which grows considerably larger still if we add in drug interventions and mutated erbB agents.

In summary, our agent hierarchy allows us to write large models in a comfortable way, to navigate the perturbation space of the model (ligands, mutations and drugs) and investigate the consequences of chosen perturbations, i.e. those for which we have experimental data, with the static and causal analyses of Kappa. This is particularly interesting for mutational perturbations as these enable us to reproduce, *in numero*, biochemical experiments that employ engineered mutations. In this way, our rules—a formalization of the *consensus* pathway assembled by many biochemical experiments—can be tested by checking, *in numero*, whether perturbing them with mutated agents—representing the engineered mutations—matches those experimental results. Of course, this procedure can never "prove" that a rule is correct but it can be used to reject rules that lead to behaviour incompatible with experimental results. It can also point to the existence of missing links in a model if it throws up false negatives with respect to the experimental data, e.g. it predicts some but not all experimentally observed phosphorylated sites. In other words, it enables us to put our assumptions under the microscope and verify that the consensus wild-type pathway behaves as expected when subjected to perturbations—and if it doesn't, we will need to change our consensus model.

Contribution and relation to existing work. Rule-based modelling is one branch of a rich literature based on the idea of representing proteins and their interactions as concurrent processes, thereby viewing a signalling network as a kind of massively distributed system. This was initially expressed in the formalism of π-calculus [7,8] but, since then, a number of variants of π-calculus [9,10] and of other languages for distributed systems [11,12,13,14,15,16,17] have also been proposed for representing various aspects of biological processes, notably the importance of causality and compartments.

Rule-based modelling, rather like the BWB/BlenX system [18], was developed out of these ideas but, instead of being based on some prior formalism for general distributed systems, is a domain-specific modelling language for biological processes. Our language Kappa is particularly closely related to BioNetGen (BNG) [19]. Although the original aims of BNG were rather different—it was conceived as a language for describing systems of ODEs in a higher-level fashion, rather than as a modelling language in its own right—the two approaches have much in common and, in particular, our agent hierarchy proposal would work just as well in BNG as in Kappa.

Despite these many advances, to the best of our knowledge none of the above-cited approaches, including Kappa and BNG, can deal with *all* the potential sources of combinatorial explosion in signalling models. Our extension of Kappa with agent hierarchies directly addresses this problem in the specific context of rule-based modelling. Given that mutating agents, via small changes in their sites and thus interaction capabilities, is central to our proposal, it would be interesting to investigate the possible connections of this work with the recent use of mutations on the structure of BlenX programs in order to evolve networks via genetic algorithms [20]. However, it should be stressed that our work was originally intended to facilitate the construction (and documentation) of large models in a way that makes explicit any underlying uniformities, rather than in directly enabling an evolutionary analysis of networks.

2   Kappa and Agent Variants

A Kappa [21] model consists of a collection of concrete agents and rules. Each agent, or more properly agent type, has a *name*, an associated set of *sites*, each with an optional internal state, and a *copy number*. An atomic rule falls into one of five classes—a binding between two agents, an unbinding, the modification of an agent, the creation of an agent or the deletion of an agent—but a rule can also be non-atomic, combining several actions.

Given a Kappa model, its *contact map*, which is computed statically from the rules, specifies which agents can bind and on which sites. (See e.g. Figs. 1, 4.) On the other hand its *influence map*, also computed statically, specifies the causal relations of activation and inhibition between rules, that is to say a rule activates (inhibits) another if its application may add (subtract) from the set of instances of the other one. We will make use of the static analysis of rule accessibility [22] which identifies whether a rule is dead, i.e. cannot be applied, or is potentially applicable; in the latter case, we will use the story sampler [23] to extract, from stochastic simulations [24] of the model, the chains of rule firings that can lead to an actual application of the rule. If we find such a story, this confirms that the static analysis didn't produce a false positive.

The concrete syntax we use to present agents, agent variants and rules should be self-explanatory (although we stress that it can be formalized). One key thing to remember, as said earlier, is that in the definition of a rule one has the option of not mentioning some sites of an agent. In situations where agents have up to a dozen different sites (e.g. the members of the EGF receptor family), this is key to obtaining concise models. This, combined with the ability to mention generic agents, allows us to express enough uniformities for also obtaining concise descriptions of perturbed models.

2.1   Agent Variants

A variant on an agent always introduces a new name and can arise in several different ways: it can lose or mutate an existing site, gain a new site or rename/duplicate an existing site. To represent these possibilities formally, we need only introduce two perturbation operations on agents, one to add a site, the other to replace a site with a set of sites. The latter operation subsumes site deletion (by replacing a site with the empty set), site renaming (replacing with a singleton set) and duplication.

For example,

```
%gen: A(s,t)
%gen: B = A[+u s\{} t\{t1,t2}]
```

declares the agent A with sites s and t and derives from it an agent B with sites
t1, t2 and u. This defines a tree of *agent variants*; most nodes of the tree are
labelled 'gen' for generic but leaves of the tree can be labelled 'conc' for concrete
which signals that that agent can be used in a Kappa model. Note that we have
a second tree structure that traces site linkages: any site can be traced back to
either a site addition or to a site declared ab initio; and conversely, following
the linkages the other way, a site in agent A maps to a set of sites in any given
descendant agent B (empty if the site has been deleted, singleton if it has just
been renamed). This is important for compiling generic rules into a bona fide
Kappa model.

Mutation of a site is represented by the compound operation of deleting the
original site and, if desired, adding a new site to "replace" it. If the desired
result of the mutation is simply the loss of certain wild-type interactions, the
loss of the site is enough and no such new site need be added; but sometimes
mutations result in new interactions becoming possible in which case we would
need to introduce a new site in order to write the new rules expressing the
novel interactions of the mutated agent, e.g. the tyrosine kinase inhibitor er-
lotinib binds to the L858R mutated erbB1 with much higher affinity than to the
wild-type receptor.

2.2 A First Example

Let us make this more concrete with an example extracted from a larger model of
the MAPK cascade. We start with two basic agent types, MAP2K and MAPK,
from which we would like to derive some more specific agent types. Our first
declarations introduce the starting agents:

```
%gen: MAP2K(D,S~u,ST~u)
%gen: MAPK(CD,T~u,Y~u)
```

Formally, these declarations play a role analogous to that of the axioms in any
formal language and, as in that kind of setting, we use them as the starting point
to introduce more subtle objects. In this case, we wish to consider the three
common kinds of MAPK protein—ERKs, JNKs and p38s—and their respective
MAP2K upstream activators—MEKs, JNKKs and p38 kinases.

To do this, we first introduce three variants of MAPK and three of MAP2K:

```
%gen: ERK = MAPK[+FXFP]
%gen: JNK = MAPK
%gen: p38 = MAPK
```

Note that, while ERK gains a new site FXFP, an ERK-specific binding site
for immediate early gene products such as Fos and Jun [25], JNK and p38
simply inherit the sites of MAPK without making any changes. As we will see
shortly, the introduction of these three variants allows us to express concisely
the specificity of binding between these three distinct families of MAPKs and
their cognate upstream activators. Note also that these three agents are still

generic as they represent families of proteins: ERK covers two proteins (ERK1 and ERK2), JNK covers three (JNK1, JNK2 and JNK3) and p38 covers four (p38alpha/beta/gamma/delta); and several of those proteins have multiple splice variants.

We formalize this by a further layer of variants:

```
%conc: ERK1 = ERK[T\{T202} Y\{Y204}]
%conc: ERK2 = ERK[T\{T185} Y\{Y187}]
```

We show only the case of ERK1 and ERK2 as those of JNK and p38 are completely analogous. Recall that we use the 'conc' tag (rather than 'gen') to make explicit the fact that ERK1 and ERK2 are concrete, not generic, agents and, as such, can be used in a Kappa model.

Note that we have renamed (via singleton duplications) the sites of ERK to include specific information about the exact residue numbers of their phosphorylatable sites; this is not essential, of course, but does illustrate the documentary power of agent variants over and above their role of structuring the space of agents.

We must also introduce generic and concrete variants of MAP2K. Each variant covers two proteins: MEK1 and MEK2 for MEK; MEK4 and MEK7 for JNKK; and MEK3 and MEK6 for p38K. (Again, for the sake of simplicity, we only show the concrete variants of MEK.)

```
%gen: MEK = MAP2K
%gen: JNKK = MAP2K
%gen: p38K = MAP2K


%conc: MEK1 = MEK[S\{S218} ST\{S222}]
%conc: MEK2 = MEK[S\{S222} ST\{S226}]
```

Already, the simple fact of hierarchically structuring the agents under consideration yields a useful object in its own right that documents, in a completely formal way, a significant amount of biological knowledge (about exactly how related proteins relate to each other) that can easily be found in several online databases but which, in that medium, remains informal and purely descriptive, whereas, in this formalized setting, has already been subjected to an initial step of processing and structuring. It also includes a convenient documentation of the specific sites of interest, e.g. the precise identities of phosphorylation sites, that are otherwise rather cumbersome to keep track of.

Moreover, the creation of this agent hierarchy also facilitates the process of writing rules by enabling us to write them at the appropriately generic level. It eases the cognitive burden of writing rules by exposing clearly the similarities and differences between various agent types. More concretely, it allows us to avoid writing essentially the same rule many times for closely related agents and, as such, also eliminates the risk of forgetting cases (a very common mistake when developing large rule sets). We turn to this in the next subsection where we will complete the MAPK example.

2.3   Generic Rules

We have seen how we can structure agents hierarchically with concrete agents at the leaves and generic agents above them. In this context, a normal (or *concrete*) Kappa rule is a rule that only mentions concrete agents. A *generic* rule is syntactically just like a normal rule but mentions one or more generic agents. The purpose of such a rule is to be expanded into a set of concrete rules by replacing each generic agent G in the rule with all appropriate concrete agents C below it in the hierarchy. However, this expansion is modulated by the changes made to G's sites in C; notably, if site s of G is deleted in C, then no rule testing the existence of s can instantiate G to C. And we must also use the site linkages between C and G to deal with any renaming and duplication of G's sites in C. So, were we to write the single generic rule

```
MAP2K(D), MAPK(CD) <-> MAP2K(D!0), MAPK(CD!0)
```

this would "incorrectly", i.e. not as we wish, expand to a collection of concrete rules where all concrete descendants (in the agent hierarchy) of MAP2Ks can bind with all concrete descendants of MAPKs, e.g. JNK2 could bind ERK1. This is the reason why, in the previous section, we introduced a second layer of generic agents—ERK, JNK, p38; MEK, JNKK, p38K. Given that, we can write the following three generic rules that properly respect the desired specificity of binding between MAP2Ks and MAPKs.

```
MEK(D), ERK(CD) <-> MEK(D!0), ERK(CD!0)
JNKK(D), JNK(CD) <-> JNKK(D!0), JNK(CD!0)
p38K(D), p38(CD) <-> p38K(D!0), p38(CD!0)
```

These three generic rules expand to eighteen concrete rules if we take ERK1/2, JNK1/2/3 and p38α/β/γ/δ as concrete agents. If we included the many splice variants of the JNKs and p38s, the *same* three generic rules would expand to over thirty concrete rules. This illustrates the flexibility of our approach whereby a given generic rule can expand differentially depending on the background of concrete agent variants we select. In particular, a single rule set can be seen as existing at many levels of detail—and this is easily tunable by the modeller as a function of his/her current needs.

There is, however, an associated cost, over and above the obvious need to recompile one's generic rules, when changing the level of detail of a model: under certain circumstances, this will lead to a degradation in the performance of stochastic simulation. The reason for this is that the cost of an event in the simulator depends, in part, on the maximum outdegree of the "wake-up map", a graph derived from the rule set which keeps track of which rules are reactivated when a rule fires [24]. In the worst-case scenario, our generic rule expansion causes a "blow up" of the wake-up map with concomitant degradation in the simulator's performance.

More generally, our mechanism of using an agent hierarchy and generic rules to generate a concrete rule set allows the Kappa modeller a finer control of the granularity of his/her rules. Consider for example an agent A that can bind two agents, B1 or B2, and that binding with either is sufficient (and necessary) for A to bind a further agent C. To express this in Kappa, we would have to write two rules for A binding C; one for the case of B1, the other for B2. This isn't too bad—but if we have not two but a large number of activating ligands of A, it rapidly becomes tedious and error-prone to write the rule sets. By using a generic agent B, representing the class of A-activating ligands, we write just one generic rule that covers all cases (albeit requiring recompilation after the addition of new concrete descendants of B). Or to put it another way, we think of the generic agent B as generating a coarse-graining of the model's molecular species that no longer distinguishes between the various concrete descendants of B (i.e. B1, B2, etc).

With more complex agent hierarchies, one can express further, more subtle coarse-graining effects such as the MAP2K-MAPK binding specificity example above. However, it should be admitted that the example of MAPK is particularly conducive to a treatment of this kind (which is why we use it as our initial example!) and that not all signalling pathways exhibit the same degree of sharing of structure found here, as expressed by the highly generic nature of the rules. This in itself is a useful aspect of our language extension in that it enables us to recognize, formally, the fact that a pathway is highly generic or, on the contrary, particularly obtuse and dependent on many specific details. Indeed, the purpose of this extension is not to obtain a maximal "compression" of a concrete rule set into as few generic rules as possible; rather it is to illuminate the structure of a model by expressing it at an appropriate level of abstraction.

3   The Perturbation Space

Now that we have shown, with the MAPK example, how our parsimonious language extension enables rapid development of large rule sets via the mechanism of generic rules, let us turn to the main problem of interest here which is to build realistic models incorporating multiple erbB ligands and receptors, mutated forms of those receptors and monoclonal antibodies (mAbs) and tyrosine kinase inhibitors (TKIs) targeting those receptors. Unlike the previous MAPK model where the use of agent variants was convenient but hardly indispensable, in this case it would be a nightmarish process to write the rules directly in Kappa. As we will see, the use of agent variants not only helps to structure the model in a human-understandable manner, it also radically tames the combinatorial explosions caused by having multiple ligands and receptors and by the introduction of mutations.

We first define our agent hierarchy. It has two roots, erbB for the receptors and erbL for the ligands, each with four children.

```
%gen: erbB(L,CR,N,atp,AS,C,Y~u)
%gen: erbBL(L)
```

The next layer of agents splits the space of ligands into four, each with a different repertoire of receptors to which it binds.

```
%gen: erbBL1 = erbBL
%gen: erbBL14 = erbBL
%gen: erbBL34 = erbBL
%gen: erbBL4 = erbBL
```

Note that a hierarchical presentation of a model has a degree of intensionality and, in particular, is of course not unique—indeed, the compiled model is actually a presentation of itself. This begs the remark that a presentation is both a way to achieve compactness of description and to document knowledge about relationships between agents that *disappears* in the compilation process.

We also need variants for the four erbB receptors. Note that we introduce them as generic agents and only later specialize them as wild-types and mutant ones (only erbB1-WT is shown here).

```
%gen: erbB1 = erbB[Y\{Y1016, Y1092, Y1110, Y1172, Y1197}]
%gen: erbB2 = erbB[L\{}]
%gen: erbB3 = erbB[N\{}]
%gen: erbB4 = erbB
%conc: erbB1_WT = erbB1
```

To keep our presentation uncluttered, we have only shown the full repertoire of phosphorylation sites for erbB1; in the full model, the other receptors also have a similar complement of Y sites. Note though that erbB2 loses the site L and erbB3 the site N. Finally, let us note the concrete ligands. In what follows, we will in fact only consider EGF and HRG.

```
%conc: EGF = erbBL1
%conc: TGFalpha = erbBL1
%conc: AR = erbBL1

%conc: BTC = erbBL14
%conc: HB-EGF = erbBL14
%conc: ER = erbBL14

%conc: HRG = erbBL34
%conc: NRG2 = erbBL34

%conc: NRG3 = erbBL4
%conc: NRG4 = erbBL4
```

3.1   The Consensus Model

We build our consensus model on the basis of a conservative reading of the literature; see e.g. [26, 27]. Specifically, we consider that ligands bind monomer receptors which can then externally (on the trans-side of the plasma membrane) dimerize; this in turn enables the formation of asymmetric dimers that lead to receptor binding on the cis-side and cross-phosphorylation.

```
erbBL1(L), erbB1(L,CR) -> erbBL1(L!0), erbB1(L!0,CR)
erbBL14(L), erbB1(L,CR) -> erbBL14(L!0), erbB1(L!0,CR)
erbBL14(L), erbB4(L,CR) -> erbBL14(L!0), erbB4(L!0,CR)
erbBL34(L), erbB3(L,CR) -> erbBL34(L!0), erbB3(L!0,CR)
erbBL34(L), erbB4(L,CR) -> erbBL34(L!0), erbB4(L!0,CR)
erbBL4(L), erbB4(L,CR) -> erbBL4(L!0), erbB4(L!0,CR)
```

These six generic rules expand into a significant number of concrete Kappa rules in a manner that depends on the level of detail requested in the identities of ligands. For example, although there are three ligands of type erbBL1 and three of type erbBL14, the two ligands of type erbBL34 actually exist in multiple splice variants, as do those of type erbB4. In any case, at the very least these six generic rules give rise to fifteen concrete rules; and, of course, we also need the unbinding rule:

```
erbBL(L!0), erbB(L!0) -> erbBL(L), erbB(L)
```

Note that this generic unbinding rule will generate concrete rules that will never apply, e.g. a descendant of erbBL1 unbinding erbB3. However, these dead rules are detected by our static analysis and so can be removed (if desired) from the generated rule set.

```
erbBL(L!1), erbB(L!1,CR), erbBL(L!2), erbB(L!2,CR) -> \
 erbBL(L!1), erbB(L!1,CR!0), erbBL(L!2), erbB(L!2,CR!0)
erbBL(L!1), erbB(L!1,CR), erbB2(CR) -> \
 erbBL(L!1), erbB(L!1,CR!0), erbB2(CR!0)
```

The first of these generic rules deals with most of the cases of (external) dimerization. The only difficulty comes from the fact that erbB2 has no ligand binding site L and, as such, cannot ever match the generic erbB agent in that rule which mentions L. For this reason, we must explicitly include the second rule that covers the case of erbB2 dimerizing with a different erbB receptor type; we do not consider erbB2 homodimerization. These three rules generate a very large number (well over 150) of concrete rules due not only to the fact that the erbB generic agent is capable of multiple matches but also because the concrete erbB agents can bind multiple ligand agents.

   We next have the rule for internal (or asymmetric) dimer formation. Here, a receptor binds its (external) dimer partner on a second site; this dimer is asymmetric because the bond is made between the N site of one receptor and the C site of the other. In a dimer not containing erbB3, this asymmetric dimer can flip states; this is the second rule.

```
erbB(CR!1,N,C), erbB(CR!1,C) -> erbB(CR!1,N!0,C), erbB(CR!1,C!0)
erbB(CR!1,N!2,C), erbB(CR!1,N,C!2) -> erbB(CR!1,N,C!3), erbB(CR!1,N!3,C)
```

The final rule is for trans phosphorylation of one erbB receptor by its dimer partner. In an asymmetric dimer, the receptor bound on site C is the *activator* whereas the receptor bound on N is the *activated*. It is thus the activator that gets phosphorylated.

`erbB(N!1,atp,AS), erbB(C!1,Y~u) -> erbB(N!1,atp,AS), erbB(C!1,Y~p)`

This one generic rule expands into many concrete rules for two independent reasons. Firstly, each erbB agent can be multiply instantiated: the first erbB can be anything but erbB3 (whose N site was deleted) and the second can be any of the four erbBs. Secondly, each receptor duplicates the site Y, so we get one concrete rule per duplicand. In the context of the purely wild-type model, neither the atp nor the AS site plays any role. This is because the rule tests only for the existence of these sites (which always succeeds) and that they are both unbound (which also always succeeds since we have no rules for binding to either of them). However, as we will see shortly, these two sites do play an important role once we take drug interventions and mutations into account.

We can neatly summarize the model so far with its contact map (Fig. 1).

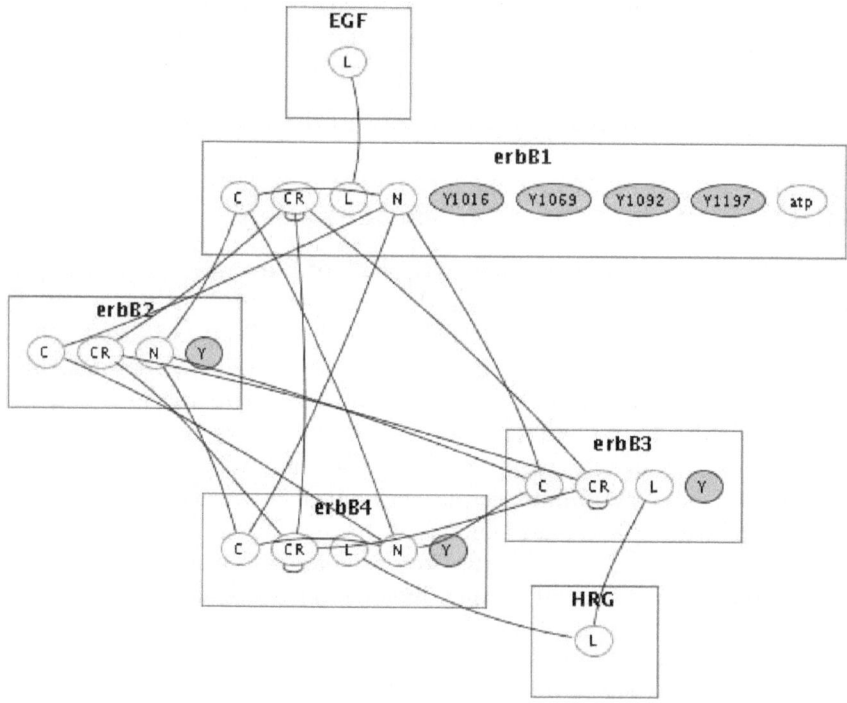

Fig. 1. Contact map of the consensus model: each concrete agent is represented once with all its sites; possible bindings are indicated by an edge joining two nodes, modifiable sites are indicated in grey. Only a restricted subset of known EGFR receptors ligands is shown, namely EGF and HRG.

3.2   Ligand Perturbations

The erbB receptor network clearly has a lot of flexibility in its response to ligands. In particular, the receptor dimers that form depend on the available ligands and the presented receptors. In addition, erbB3 has compromised capability to form asymmetric dimers: it can activate the catalytic activity of its dimer partner but cannot be activated by it. This phenomenon adds yet another layer of subtlety to erbB receptor activation. For example, in a cell line expressing erbB2, erbB3 and erbB4, one would expect HRG to promote phosphorylation of erbB2, via erbB2:erbB4 dimers, as well as phosphorylation of erbB3 and erbB4. On the other hand, were erbB4 not expressed, one would expect only erbB3 phosphorylation, via erbB2:erbB3 dimers. However, this kind of reasoning rapidly becomes highly complicated, particularly in the presence of multiple ligands, and we would like some way of deducing, from the rule set and a choice of expression levels of ligands and receptors, which receptors get phosphorylated (and, in some cases, on which sites).

We can do this using static analysis of the rule set. We first write dummy rules that detect typical molecular species of interest, e.g.

```
erbB2(Y~p) -> erbB2(Y~p)
```

We then ask the static analyser whether or not our dummy rules can fire. It responds in one of two ways: either a categorical 'no' or a tentative 'yes'. In the case of a 'no', we know (since the static analysis never produces false negatives) that our rule set cannot create the molecular species in question—starting from the declared initial solution. In the case of a 'yes', we have no certainly (since the analysis *can* give false positives) that the species can arise, but also no proof that it cannot. In an attempt to confirm the 'yes', we then use the story sampler to search for pathways leading to a dummy rule; if (at least) one exists, we have proof that the species can arise.

For example, the static analysis shows that, with our rule set, erbB2 phosphorylation cannot take place (categorical 'no') under the following conditions:

- HRG only; erbB2 and erbB3 only [erbB3 cannot phosphorylate erbB2]
- HRG only; erbB1, erbB2 and erbB3 only [erbB1 cannot bind HRG]

whereas it can potentially take place (tentative 'yes') under the following conditions:

- HRG only; erbB2, erbB3 and erbB4 only
- EGF and HRG; erbB1, erbB2 and erbB3 only.

To confirm this claim, we ask for stories leading to the appropriate observables. In both cases, we find indeed a story leading to phosphorylated erbB2 which confirms that the static analysis did not give us false positives (Figs. 2, 3).

This combination of static analysis and story sampling enables a powerful model development process where, starting from a consensus, perhaps overly restrictive, rule set, we investigate which observables of interest can arise under

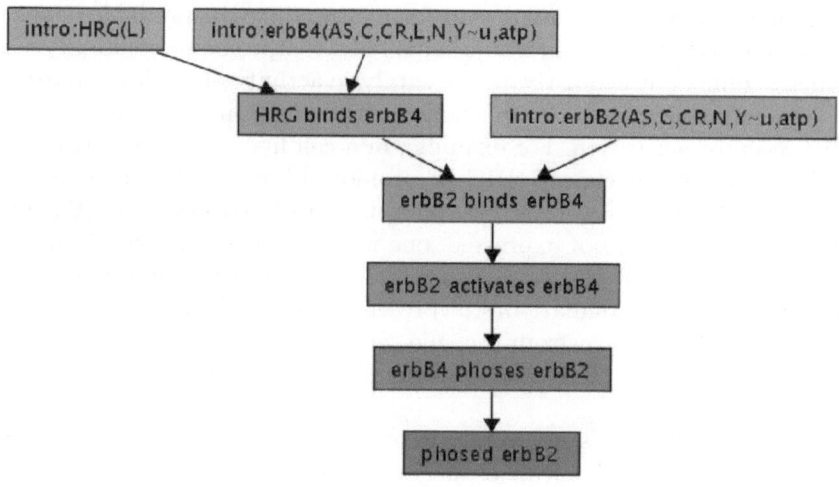

Fig. 2. Story leading to erbB2 phosphorylation by erbB4

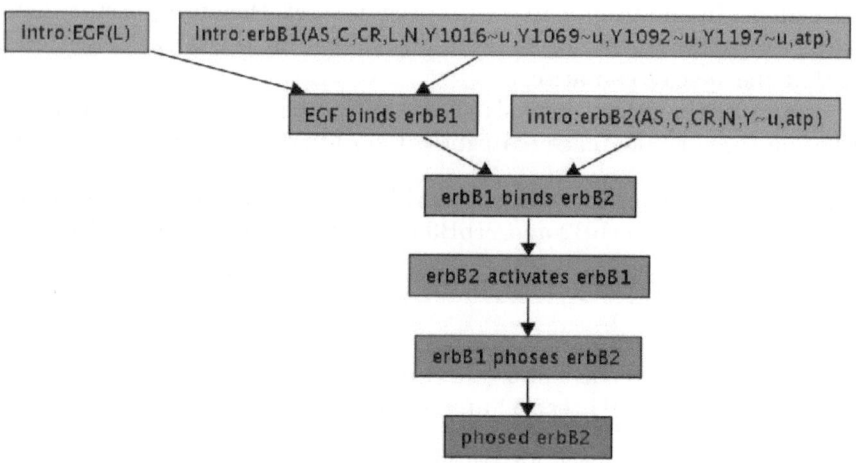

Fig. 3. Story leading to erbB2 phosphorylation by erbB1

which conditions. We then compare these predictions to experimental data in order to judge the accuracy and completeness of the model. If experimental data conflicts with the results of our analysis, this means one of two things: the consensus model either has fatal flaws or missing links. A 'fatal flaw' means that certain experimentally unobservable species can be generated by the rule set; in other words, that the mechanism described by the rules makes unwarranted assumptions. A 'missing link' corresponds to the more likely situation where an experimentally observed species remains inaccessible with our consensus model; this implies that the rule set lacks certain necessary rules.

For example, in the above discussion, we noted that, in our rule set with erbB1, erbB2 and erbB3 only, HRG stimulation leads only to erbB3 phosphorylation; whereas the combination of EGF and HRG leads to phosphorylation of all three receptors. This constitutes an experimentally refutable prediction. In the event of such a refutation, e.g. we observe erbB1 phosphorylation upon HRG stimulation, we could freely postulate various new rules, check that they do indeed open up the possibility of erbB1 phosphorylation and then compare and contrast their effects on other observables. If a new rule creates a 'fatal flaw', we can discount it; but in general this may still leave us with a choice between several proposed new mechanisms. To decide between these would require us to find a new, experimentally refutable prediction and do the experiment (or find it in the literature).

We stress that this remains a human-directed model development process—we do not consider automatically generated rules in any form—but one in which variant mechanisms can be built and evaluated in an organized fashion.

3.3 Drug Perturbations

In recent years, particularly with the realization that deregulated erbB signalling contributes to the development of multiple cancers, much research has focussed on finding ways of blocking the activity of this family of receptors via drug intervention. To date, two broad classes of drug have been developed: monoclonal antibodies (mAbs) and tyrosine kinase inhibitors (TKIs). Antibodies typically act as classical competitive inhibitors that exert their function by binding cell surface receptors in a way that physically obstructs their usual ligands from binding. On the other hand, TKIs behave as classical non-competitive inhibitors of kinase activity that do not prevent substrate binding but instead block the ATP binding site of the kinase domain, thus preventing substrate phosphorylation.

As an illustration of the ease with which we can incorporate these kinds of pharmaceutical intervention in our modelling framework, we include rules for C225 (cetuximab, a mAb) binding to erbB1's site L, ZD1839 (gefitinib, a TKI) binding to erbB1's atp site and 4D5 (trastuzumab, another mAb) binding to erbB2's dimerization site CR.

```
C225(L), erbB1(L) -> C225(L!0), erbB1(L!0)
ZD1839(L), erbB1(atp) -> C225(L!0), erbB1(atp!0)
4D5(L), erbB2(CR) -> C225(L!0), erbB2(CR!0)
```

As each of these molecules has a *reversible* inhibitory effect on the respective receptor, we also need the accompanying unbinding rules:

```
C225(L!0), erbB1(L!0) -> C225(L), erbB1(L)
ZD1839(L!0), erbB1(atp!0) -> ZD1839(L), erbB1(atp)
4D5(L!0), erbB2(CR!0) -> 4D5(L), erbB2(CR)
```

The contact map of the system (Fig. 4) including these inhibitors makes it clear that the antibodies (C225 and 4D5) act as competitive inhibitors: C225 competes with EGF for the ligand binding site of erbB1 and 4D5 competes for the dimerization binding site of erbB2.

The inhibitory effect of ZD1839 shows up only in the influence map: the rule binding ZD1839 to erbB1 inhibits all modification rules (concretely, the phosphorylations) dependent on erbB1. This is because ZD1839 binds to the site atp of erbB1 which must be free in order for erbB1 to modify its dimer partner. The presence of ZD1839 thus frustrates, without completely preventing, erbB1-dependent phosphorylation. We will return to this later.

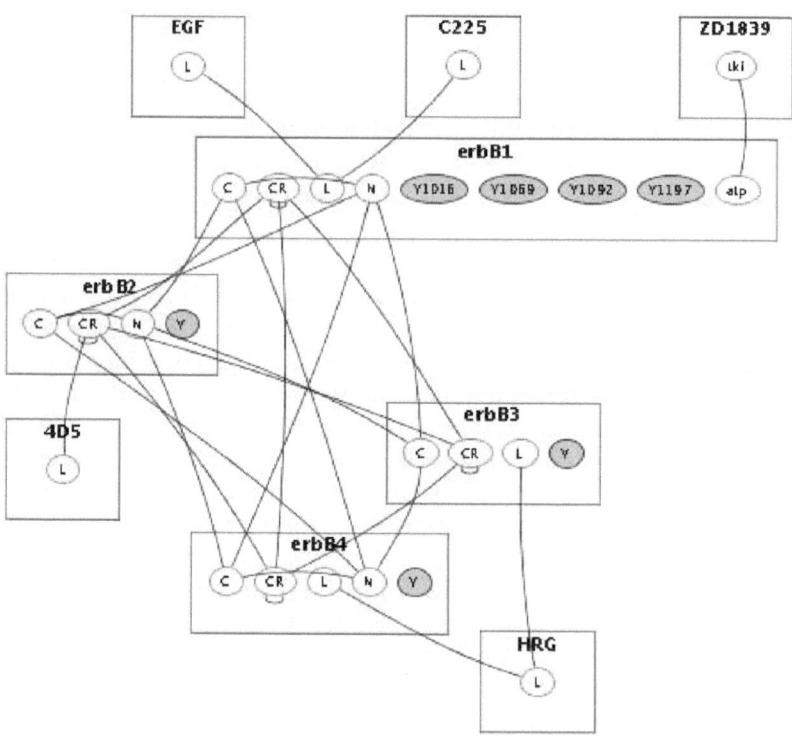

Fig. 4. Contact map of the consensus model with antibodies (C225 and 4D5) as competitive inhibitors, and ZD1839 as non-competitive inhibitor.

It should be noted that, in our examples of inhibitors, all agents are concrete. But, in general, drugs would also be organized, much like natural ligands, by an appropriate agent hierarchy.

3.4   The Uses of Mutational Perturbations

So far, we have seen how the use of agent variants allows us to organize agents hierarchically and thus write generic rules at a convenient level of granularity. In particular, this facilitates the development of models with families of related proteins, or proteins with multiple splice variants, that have overlapping functionality. However, agent variants also enable the treatment of mutated agents which likewise share a lot of the functionality of their wild-type cousins but which also potentially lose some of that functionality and/or gain new functionality.

This has two immediate applications. Firstly, it allows us to build models with a mixture of wild-type and mutated agents in order to investigate (statically or numerically) the consequences of mutations. This is particularly interesting in the context of models, as described previously, that also include drug interventions.

More subtly, it also allows us to cast a critical eye over the assumptions we make in building our wild-type model. After all, a lot of the experimental data from which consensus pathways have been deduced comes from mutation experiments. These typically eliminate one or more phosphorylation sites in a protein and investigate which, if any, pathways suffer from this perturbation. However, such data can be difficult to interpret and the deduced wild-type interactions may be incorrect.

For example, as explained in [28], one experiment showed that expressing a kinase-dead mutant of PI3K inhibited Ras activation upon EGF stimulation; this led the authors to propose a role for PI3K in activating Ras. But then, a second study demonstrated that a constitutively active (and membrane associated) mutant of PI3K did *not* promote Ras activation, which contradicted the conclusions of the first study. In the end, it turned out [28] that PI3K actually *inhibits* Ras *de*activation; so PI3K sensitizes Ras for activation but cannot by itself actually activate it. In more details: PI3K promotes Gab1 recruitment to the membrane which, on EGF stimulation, strongly recruits Shp2 to the membrane; Shp2 is a tyrosine phosphatase that dephoses the phospho-tyrosine binding sites for RasGAP (and for PI3K!) on erbB receptors and Gab1. So Shp2 inhibits Ras-GAP recruitment to the membrane which indirectly aids Ras activation (Sos which activates Ras has an easier job). This gives a measure of the daunting complexity of inferring a protein network, and as a consequence a measure of how helpful a methodology such as the one we illustrate here can prove.

Indeed, using agent variants, we can express the kinds of (artificially) mutated proteins used in biochemical studies and so replay numerically such experiments. We can therefore detect, *in numero*, if the hypothesized wild-type network is in fact incorrect, e.g. if we had a model for Ras activation including a rule for 'PI3K activates Ras', we would have been able to predict that a constitutively active PI3K mutant would activate Ras; the fact that, experimentally, this is not observed means that that rule must be wrong. This kind of perturbational analysis is not just useful for postdictive verification of inferences, it is also a discipline to build a model upon such data, and to build further data to refute predictions; this could be particularly interesting if two plausible molecular mechanisms (candidate consensus pathways) made divergent predictions.

3.5   Testing the Wild-Type Model

In two recent papers, Kuriyan and coworkers have developed a conceptual model of erbB receptor acivation that depends on the formation of an asymmetric dimer [27, 29]. We have used this when writing the above rules for the wild-type erbB network in the previous section. They developed their model using a combination of structural and mutational data and provide convincing evidence of its correctness by cotransfecting various artifical erbB constructs that lack one or more of the N, C and AS sites.

We can use agent variants, in combination with static analysis, to reproduce these kinds of results *in numero*. For example, kinase-dead erbB1 is obtained by defining a variant of erbB1 with the AS site deleted; this agent can no longer phosphorylate its dimer partner. Similarly, we can also introduce variants that delete either the N or the C site instead of, or in addition to, the AS site.

```
%conc: erbB1_KD = erbB1[AS/{}]
%conc: erbB1_noN = erbB1[N/{}]
%conc: erbB1_noC = erbB1[C/{}]
%conc: erbB1_KDnoN = erbB1[AS/{} ; N/{}]
%conc: erbB1_KDnoC = erbB1[AS/{} ; C/{}]
```

These agents inherit all rules from wild-type erbB1 that do not mention the sites that they lack. So erbB1_KD can freely form asymmetric dimers but phosphory-lates nothing, whereas erbB1_noN and erbB1_noC are partially compromised in their ability to form asymmetric dimers: the former can activate its partner and get phosphorylated, but cannot be activated and phosphorylate its partner; the latter can be activated by its partner and phosphorylate it, but cannot activate its partner and get phosphorylated.

We can now use static analysis, as in the previous section, to analyze the consequences of coexpressing pairs of these variant agents. We do this by checking the accessibility of the rules

```
erbB1(N!1), erbB1(C!1) -> erbB1(N!1), erbB1(C!1)
erbB1(Y1197~p) -> erbB1(Y1197~p)
```

(that respectively detect the possibility of an asymmetric dimer forming and an erbB1 receptor becoming phosphorylated) in an initial solution that includes EGF and a choice of any (one or) two of the erbB1 variants. In particular, we can recapitulate the results of [27] (see their Fig. 6; we use the same combination numbers) in completely automatic fashion:

1. wild-type erbB1 only: asymmetric homodimer accessible; phosphorylation accessible
2. erbB1_KD only: asymmetric homodimer accessible; phosphorylation inaccessible
3. erbB1_KD & erbB1_noN: one asymmetric heterodimer accessible; phosphorylation inaccessible
4. erbB1_KD & erbB1_noC: one asymmetric heterodimer accessible; phosphorylation accessible
5. erbB1_KDnoC only: asymmetric homodimer inaccessible; phosphorylation inaccessible
6. erbB1_KDnoC & erbB1_noN: one asymmetric heterodimer accessible; phosphorylation inaccessible
7. erbB1_KDnoC & erbB1_noC: asymmetric heterodimer inaccessible; phosphorylation inaccessible
8. erbB1_KDnoN only: asymmetric homodimer inaccessible; phosphorylation inaccessible
9. erbB1_KDnoN & erbB1_noN: asymmetric heterodimer inaccessible; phosphorylation inaccessible
10. erbB1_KDnoN & erbB1_noC: one asymmetric heterodimer accesible; phosphorylation accessible

In [27], this had to be done by hand, a task that soon begins to get rather subtle, particularly if you want to consider doubly-mutated agents and/or coexpression of more than two receptor constructs at a time. It is thus very useful to be able to express this situation in Kappa and rely on static analysis to detect the impossibility/possibility of phosphorylation. Moreover, if the static analysis announces that phosphorylation is not impossible, we can, as above, use the story sampler to search for ways in which this can actually take place. Again, in some cases, this is easy to do by hand but, beyond a certain degree of complexity, it is highly desirable to have an automatic method in order to avoid making mistakes.

These results demonstrate that our consensus model is indeed compatible with the experimental data of Kuriyan et al. and, as such, it passes the test. This comforts us, for now, in our choice of rules but of course provides no guarantee that future experimental data will not invalidate some of them.

3.6   The Limits of Perturbation Testing

We mentioned earlier that our language extension shields the modeller from the underlying rule set generated by generic rules. However, we should say that this is only true *qualitatively*—if we wish to manipulate the rate constants of our model in such a way that different concrete instantiations of one generic rule get different kinetics, this can only be done by examining and modifying directly the generated rule set.

More generally, the modelling methodology advocated above based on static analysis cannot be used to gauge the effect of perturbations, such as drugs, that restrict, but don't outlaw, the application of other rules. Or, to put it another way, a perturbation that operates entirely at the level of kinetics is undetectable by this method. We would however expect to observe the effects of such perturbations during stochastic simulation and/or story sampling. Indeed, it would be straightforward to observe the inhibition of erbB1's kinase activity by tracking that rule's activity in the absence and presence of drugs. More ambitiously, we could compare the relative strengths of each erbB's kinase activity and the way in which that is disturbed by drugs that target only one receptor; this would require running the story sampler many times at many time points to get a statistical picture of the model's activity profile over time. We leave this for future work.

4   Conclusions

Even if wild-type pathways are obviously central to a systemic view of molecular biology, modelling is not just about these. It is equally important to be able to navigate the space of derivatives of a model for two complementary reasons. Firstly, one needs to understand diseased conditions as natural perturbations of the wild-type; secondly, one also needs to represent synthetic perturbations (by genetic knock-outs, domain truncations, point mutations, etc) because they are key in the inference of the wild-type. This is a formidable challenge because the space of such model perturbations introduces a second kind of combinatorial explosion. The well-studied example of the EGF receptor family (see §3) is a powerful illustration of this fact. Now, we have to do something if we want our modelling vessel to stay afloat in the sea of perturbations. In other words, just as the passage from reactions to rules tames the first binding-caused explosion, we have to find a mechanism to tame what one might call the perturbation-caused explosion.

The fact that Kappa describes molecular interactions at the level of domain binding and modification seems a good start, since this is the granularity at which the engineering of perturbations in protein networks actually happens (e.g. Y to A mutations that disable a modification). But to tackle our representation problem, we need another ingredient, namely a syntactic extension of Kappa that enables a clean, uniform treatment of protein families, splice variants and mutated proteins. This is what we have proposed here. The idea is to structure agents hierarchically so that rules can be expressed at an appropriate level of abstraction, as generic rules, which are then automatically compiled into pure Kappa. This eases the pain (and pitfalls) of writing large rule sets (indeed the modeller has no need to ever look at the resulting concrete rule set, unless he/she wishes to modify its rate constants), and as we wanted, this give means to navigate their perturbation space.

Of course there is no magic: to work around the explosive generativity of wild-type pathways we capture postulated regularities by using rules (if the universe of reactions were lacking any regularity no method could describe them anyway, a rather grim perspective for systems biology); to work around the second source of complexity, again we capture regularities of another kind, namely that much of the wild-type behaviour of a protein is actually shared with its mutants and isoforms.

We have shown that this strategy works well with our EGF example, as we were able to neatly set a wild-type model together with a selection of derivatives. With this model in place, one can bring the usual analysis tools of Kappa to bear on the rule set. As we have shown further, even in the absence of quantitative information about rates and copy numbers, one can obtain qualitative predictions about the induced perturbed behaviours and thus support on a full-scale the traditional informal inferences that are commonplace in the experimental investigation of protein networks.

Acknowledgements. Jean Krivine is supported via grants from the Agence Nationale de la Recherche (ANR-07-PHYSIO-013-01) and the Génopole Evry held by A. Benecke of the IHES.

References

1. Kholodenko, B.N., Demin, O.V., Moehren, G., Hoek, J.B.: Quantification of Short Term Signaling by the Epidermal Growth Factor Receptor. J. Biol. Chem. 274(42), 30169–30181 (1999)
2. Kiyatkin, A., Aksamitiene, E., Markevich, N.I., Borisov, N.M., Hoek, J.B., Kholodenko, B.N.: Scaffolding protein GAB1 sustains epidermal growth factor-induced mitogenic and survival signaling by multiple positive feedback loops. J. Biol. Chem. 281, 19925–19938 (2006)
3. Orton, R.J., Sturm, O.E., Vyshemirsky, V., Calder, M., Gilbert, D.R., Kolch, W.: Computational modelling of the receptor tyrosine kinase activated MAPK pathway. Biochemical Journal 392(2), 249–261 (2005)
4. Schoeberl, B., Eichler-Jonsson, C., Gilles, E.-D., Müller, G.: Computational modeling of the dynamics of the map kinase cascade activated by surface and internalized EGF receptors. Nature Biotechnology 20, 370–375 (2002)
5. Hlavacek, W.S., Faeder, J.R., Blinov, M.L., Posner, R.G., Hucka, M., Fontana, W.: Rules for Modeling Signal-Transduction Systems. Science's STKE 2006(344) (2006)
6. Maslov, S., Ispolatov, I.: Propagation of large concentration changes in reversible protein-binding networks. Proceedings of the National Academy of Sciences 104(34), 13655–13660 (2007)
7. Regev, A., Silverman, W., Shapiro, E.: Representation and simulation of biochemical processes using the π-calculus process algebra. In: Altman, R.B., Dunker, A.K., Hunter, L., Klein, T.E. (eds.) Pacific Symposium on Biocomputing, vol. 6, pp. 459–470. World Scientific Press, Singapore (2001)
8. Regev, A., Shapiro, E.: Cells as computation. Nature 419 (September 2002)
9. Priami, C., Regev, A., Shapiro, E., Silverman, W.: Application of a stochastic name-passing calculus to representation and simulation of molecular processes. Information Processing Letters (2001)

10. Baldi, C., Degano, P., Priami, C.: Causal π-calculus for biochemical modeling. In: Proceedings of the AI*IA Workshop on BioInformatics 2002, pp. 69–72 (2002)
11. Priami, C., Quaglia, P.: Beta Binders for Biological Interactions. In: Danos, V., Schachter, V. (eds.) CMSB 2004. LNCS (LNBI), vol. 3082, pp. 20–33. Springer, Heidelberg (2005)
12. Cardelli, L.: Brane Calculi Interactions of Biological Membranes. In: Danos, V., Schachter, V. (eds.) CMSB 2004. LNCS (LNBI), vol. 3082, pp. 257–278. Springer, Heidelberg (2005)
13. Regev, A., Panina, E.M., Silverman, W., Cardelli, L., Shapiro, E.: BioAmbients: an abstraction for biological compartments. Theoretical Computer Science 325, 141–167 (2004)
14. John, M., Ewald, R., Uhrmacher, A.M.: A Spatial Extension to the π Calculus. Electronic Notes in Theoretical Computer Science, vol. 194(3), pp. 133–148 (2008)
15. Calder, M., Gilmore, S., Hillston, J.: Modelling the influence of RKIP on the ERK signalling pathway using the stochastic process algebra PEPA. In: Priami, C., Ingólfsdóttir, A., Mishra, B., Riis Nielson, H. (eds.) Transactions on Computational Systems Biology VII. LNCS (LNBI), vol. 4230, pp. 1–23. Springer, Heidelberg (2006)
16. Ciocchetta, F., Hillston, J.: Bio-PEPA: an extension of the process algebra PEPA for biochemical networks. Electronic Notes in Theoretical Computer Science, vol. 194(3), pp. 103–117 (2008)
17. Calzone, L., Fages, F., Soliman, S.: BIOCHAM: an environment for modeling biological systems and formalizing experimental knowledge. Bioinformatics 22(14), 1805–1807 (2006)
18. Dematte, L., Priami, C., Romanel, A.: The BlenX language: a tutorial. In: Bernardo, M., Degano, P., Zavattaro, G. (eds.) SFM 2008. LNCS, vol. 5016, pp. 313–365. Springer, Heidelberg (2008)
19. Blinov, M.L., Faeder, J.R., Hlavacek, W.S.: BioNetGen: software for rule-based modeling of signal transduction based on the interactions of molecular domains. Bioinformatics 20, 3289–3292 (2004)
20. Dematté, L., Priami, C., Romanel, A., Soyer, O.: Evolving BlenX programs to simulate the evolution of biological networks. Theoretical Computer Science 408(1), 83–96 (2008)
21. Danos, V., Laneve, C.: Formal molecular biology. Theoretical Computer Science 325(1), 69–110 (2004)
22. Danos, V., Feret, J., Fontana, W., Krivine, J.: Abstract Interpretation of Cellular Signalling Networks. In: Logozzo, F., Peled, D.A., Zuck, L.D. (eds.) VMCAI 2008. LNCS, vol. 4905, pp. 83–97. Springer, Heidelberg (2008)
23. Danos, V., Feret, J., Fontana, W., Harmer, R., Krivine, J.: Rule-Based Modelling of Cellular Signalling. In: Caires, L., Vasconcelos, V.T. (eds.) CONCUR 2007. LNCS, vol. 4703, pp. 17–41. Springer, Heidelberg (2007)
24. Danos, V., Feret, J., Fontana, W., Krivine, J.: Scalable Simulation of Cellular Signaling Networks. In: Shao, Z. (ed.) APLAS 2007. LNCS, vol. 4807, pp. 139–157. Springer, Heidelberg (2007)
25. Murphy, L.O., Smith, S., Chen, R.H., Fingar, D.C., Blenis, J.: Molecular interpretation of ERK signal duration by immediate early gene products. Nat. Cell Biol. 4(8), 556–564 (2002)
26. Burgess, A.W., Cho, H.S., Eigenbrot, C., Ferguson, K.M., Garrett, T.P.J., Leahy, D.J., Lemmon, M.A., Sliwkowski, M.X., Ward, C.W., Yokoyama, S.: An Open-and-Shut Case? Recent Insights into the Activation of EGF/ErbB Receptors. Molecular Cell 12(3), 541–552 (2003)

27. Zhang, X., Gureasko, J., Shen, K., Cole, P.A., Kuriyan, J.: An Allosteric Mechanism for Activation of the Kinase Domain of Epidermal Growth Factor Receptor. Cell 125(6), 1137–1149 (2006)
28. Sampaio, C., Dance, M., Montagner, A., Edouard, T., Malet, N., Perret, B., Yart, A., Salles, J., Raynal, P.: Signal strength dictates phosphoinositide 3-kinase contribution to Ras/extracellular signal-regulated kinase 1 and 2 activation via differential Gab1/Shp2 recruitment: consequences for resistance to epidermal growth factor receptor inhibition. Mol. Cell Biol. 28(2), 587–600 (2008)
29. Zhang, X., Pickin, K.A., Bose, R., Jura, N., Cole, P.A., Kuriyan, J.: Inhibition of the EGF receptor by binding of MIG6 to an activating kinase domain interface. Nature 450(7170), 741 (2007)

Extended Stochastic Petri Nets
for Model-Based Design of Wetlab Experiments

Monika Heiner[1], Sebastian Lehrack[1], David Gilbert[2], and Wolfgang Marwan[3]

[1] Department of Computer Science, Brandenburg University of Technology
Postbox 10 13 44, 03013 Cottbus, Germany
{monika.heiner,slehrack}@informatik.tu-cottbus.de
[2] School of Information Systems, Computing and Mathematics
Brunel University, Uxbridge, Middlesex UB8 3PH, UK
david.gilbert@brunel.ac.uk
[3] Otto von Guericke University & Magdeburg Centre for Systems Biology
c/o Max Planck Institute for Dynamics of Complex Technical Systems,
Sandtorstr. 1, 39106 Magdeburg, Germany
marwan@mpi-magdeburg.mpg.de

Abstract. This paper introduces extended stochastic Petri nets to model wetlab experiments. The extentions include read and inhibitor arcs, stochastic transitions with freestyle rate functions as well as several deterministically timed transition types: immediate firing, deterministic firing delay, and scheduled firing. The extensions result into non-Markovian behaviour, which precludes analytical analysis approaches. But there are adapted stochastic simulation analysis (SSA) methods, ready to deal with the extended behaviour. Having the simulation traces, we apply simulative model checking of PLTL, a linear-time temporal logic (LTL) in a probabilistic setting.

We present some typical model components, demonstrating the suitability of the introduced Petri net class for the envisaged application scenario. We conclude by looking briefly at a classical example of prokaryotic gene regulation, the lac operon case.

1 Motivation

This paper extends the Markovian stochastic Petri nets $\mathcal{SPN}_{Bio}$ as introduced in [GHL07] to model and analyse biochemical networks. Related application scenarios are discussed in [BGHO08], [GBHD09]. Case studies demonstrating a unifying framework to integrate the qualitative, stochastic and continuous paradigms can be found in [HGD08], [GHR+08], [HDG10]. Thus, $\mathcal{SPN}_{Bio}$ have been proven to be useful in systems and synthetic biology. However, there are limitations in expressivity.

Generally, biologists face the problem to design wetlab experiments to validate or contradict the current understanding of the biochemical network under investigation. In order to be better able to do so, they ask for the following advanced features:

C. Priami et al. (Eds.): Trans. on Comput. Syst. Biol. XI, LNBI 5750, pp. 138–163, 2009.
© Springer-Verlag Berlin Heidelberg 2009

- stochastic and deterministic firing behaviour within one model,
- relative and absolute timing of the transitions' firing,
- construction of arbitrary schedules of programmed interventions.

Therefore, we are going to extend $\mathcal{SPN}_{Bio}$ belonging to the Markovian world by several features supporting the comfortable modelling of wetlab experiments. The extentions lead to the definition of biochemically interpreted Generalised Stochastic Petri nets $\mathcal{GSPN}_{Bio}$ and Deterministic and Stochastic Petri nets $\mathcal{DSPN}_{Bio}$. They include read and inhibitor arcs, stochastic transitions with freestyle rate functions as well as several deterministically timed transition types: immediate firing, deterministic firing delay, and scheduled firing.

The extension go beyond the Markov property, which precludes analytical analysis approaches; but there are adapted stochastic simulation analysis (SSA) methods, ready to deal with the extended behaviour. Having the simulation traces we apply simulative model checking of linear-time temporal logic (LTL) in a probabilistic setting (PLTL). Simulative model checking approximates the probability of a given temporal logic formula by considering finite sets of finite paths through the state space. Thus, it works even for systems with infinite state spaces.

We discuss in detail some typical model components, demonstrating the suitability of the introduced Petri net class $\mathcal{DSPN}_{Bio}$ for the envisaged application scenario. These components will be analysed by checking sets of stochastic simulation traces against PLTL properties. In doing so, a special category of properties, the so-called invariant properties, will be used to prove at the same time the plausibility of the applied simulation algorithm.

We conclude by looking briefly at a classical example of prokaryotic gene regulation, the lac operon case.

2   Stochastic Modelling

We assume basic knowledge of the standard notions of qualitative place/transition Petri nets, see e.g. [Mur89], [Rei82], [HGD08]. To be self-contained we start with recalling the fundamentals of (biochemically interpreted) stochastic Petri nets, belonging to the Markovian world, before introducing the extended notions resulting finally into non-Markovian Petri nets.

2.1   The Markovian Case - Stochastic Petri Nets ($\mathcal{SPN}_{Bio}$)

As with a qualitative Petri net, a stochastic Petri net maintains a discrete number of tokens on its places. But contrary to the time-free case, a firing rate (waiting time) is associated with each transition t, which are random variables $X_t \in [0, \infty)$, defined by probability distributions. Therefore, all reaction times can theoretically still occur, but the likelihood depends on the probability distribution. Consequently, the system behaviour is described by the same discrete state space, and all the different execution runs of the underlying qualitative

Petri net can still take place. This allows the use of the same powerful analysis techniques for stochastic Petri nets as they are applied for qualitative Petri nets.

For a better understanding we describe the general procedure of a particular simulation run for a stochastic Petri net. Each transition gets its own local timer. When a particular transition becomes enabled, meaning that sufficient tokens arrive on its preplaces, then the local timer is set to an initial value, which is computed at this time point by means of the corresponding probability distribution. In general, this value will be different for each simulation run. The local timer is then decremented at a constant speed, and the transition will fire when the timer reaches zero. If there is more than one enabled transition, a race for the next firing will take place. After the firing of the winning transition, the timers of the others still enabled transitions keep their values or are reset, depending on the specific type of the net.

Technically, various probability distributions can be chosen to determine the random values for the local timers. Biochemical systems are the prototype for exponentially distributed reactions. Thus, for our purposes, the firing rates of all transitions follow an exponential distribution, which can be described by a single parameter λ, and each transition needs only its particular, generally marking-dependent parameter λ to specify its local time behaviour. The following definition summarises this informal introduction.

Definition 1 (Stochastic Petri net, Syntax). *A biochemically interpreted stochastic Petri net is a quintuple* $\mathcal{SPN}_{Bio} = (P, T, f, v, m_0)$, *where*

- *P and T are finite, nonempty, and disjoint sets. P is the set of* places, *and T is the set of* transitions.
- $f : ((P \times T) \cup (T \times P)) \to \mathbb{N}_0$ *defines the set of directed* arcs, *weighted by nonnegative integer values.*
- $v : T \to H$ *is a function, which assigns a* stochastic hazard function h_t *to each transition t, whereby*
  $H := \bigcup_{t \in T} \left\{ h_t \mid h_t : \mathbb{N}_0^{|\bullet t|} \to \mathbb{R}^+ \right\}$ *is the set of all stochastic hazard functions, and* $v(t) = h_t$ *for all transitions* $t \in T$.
- $m_0 : P \to \mathbb{N}_0$ *gives the* initial marking.

The stochastic hazard function h_t defines the marking-dependent transition rate $\lambda_t(m)$ for the transition t, i.e. $h_t = \lambda_t(m)$. The domain of h_t is restricted to the set of preplaces of t, denoted by $\bullet t$ with $\bullet t := \{p \in P \mid f(p, t) \neq 0\}$, to enforce a close relation between network structure and hazard functions. Therefore, $\lambda_t(m)$ actually depends on a sub-marking only.

Stochastic Petri net, Semantics. Transitions become enabled as usual, i.e. if all preplaces are sufficiently marked. However there is a time, which has to elapse, before an enabled transition $t \in T$ fires. The transition's firing delay (waiting time) is an exponentially distributed random variable X_t with the *probability density function:*

$$f_{X_t}(\tau) = \lambda_t(m) \cdot e^{(-\lambda_t(m) \cdot \tau)}, \qquad \tau \geq 0.$$

The firing itself does not consume time and follows the standard firing rule of qualitative Petri nets. The semantics of a stochastic Petri net (with exponentially distributed firing delays for all transitions) is described by a continuous time Markov chain (CTMC). The CTMC of a stochastic Petri net without parallel transitions is isomorphic to the reachability graph of the underlying qualitative Petri net, while the arcs between the states are now labelled by the transition rates. For more details see [MBC+95], [BK02], [HGD08].

Based on this general $\mathcal{SPN}_{Bio}$ definition, specialised biochemically interpreted stochastic Petri nets can be defined by specifying the required kind of stochastic hazard function more precisely. In this paper, we are going to use the molecule semantics with mass action transition rates. Therefore we deploy the *stochastic mass-action hazard function*, which tailors the general $\mathcal{SPN}_{Bio}$ definition to biochemical mass-action networks, where tokens correspond to molecules:

$$h_t := c_t \cdot \prod_{p \in {}^\bullet t} \binom{m(p)}{f(p,t)} .$$

The constant c_t is the transition-specific stochastic rate constant, and $m(p)$ is the current number of tokens on a preplace p of the transition t. The binomial coefficient describes the number of non-ordered combinations of the $f(p,t)$ molecules, required for the reaction, out of the $m(p)$ available ones. In the following we abbreviate this formula by $BioMassAction(c_t)$.

See [GHL07] for another example, reading the tokens as concentration levels.

2.2   The Non-markovian Case - Extended Stochastic Petri Nets

We start off with an overview and brief biochemical motivation before introducing two classes of extended stochastic Petri nets.

There are quite a number of various extensions based on the fundamental stochastic Petri net class $\mathcal{SPN}$, see e.g. [MBC+95], [Ger01]. The most important additional features concern deterministically timed transitions, or *deterministic transitions* for short, which come along in different types. The crucial point is that the firing delay (waiting time) before an enabled transition fires does not depend anymore on a random variable, but is specified by a fixed time duration. To avoid confusion, we will call the transitions with a probabilistic firing delay, as introduced in the former subsection, *stochastic transitions*, if necessary. In summary, our extended stochastic Petri nets support the following features:

- read and inhibitor arcs,
- programmed transitions (freestyle rate functions),
- deterministic firing delay,
- scheduled transitions.

Read and inhibitor arcs. are popular add-ons enhancing modelling comfort. Read arcs (often also called test arcs) allow to specify positive side-conditions, e.g., if the occurrence of a subunit depends on the conformation of a protein complex, or if a cell's reaction to a given stimulus depends on the specific physiological conditions of the cell. Contrary, inhibitor arcs allow to specify negative side-conditions in an abstract way, e.g., if the presence of a given protein or condition inhibits a specific reaction.

Speaking in technical terms, read and inhibitor arcs are directed arcs, going always from places to transitions. The standard firing rule needs to be adapted accordingly. The enabling condition is extended in the following way: if there is an arc a with a weight $w = f(p, t)$ connecting a place p with a transition t, then t can be enabled in a marking m if the following conditions are also satisfied:

- a is a read arc $\land m(p) \geq w$,
- a is an inhibitor arc $\land m(p) < w$.

The token situation on p is not changed by the firing of t, i.e. $m'(p) = m(p)$ for $m \xrightarrow{t} m'$.

Programmed transitions are stochastic transitions with freestyle rate functions. The firing rate can be specified by arbitrary mathematical functions, stored in lookup tables, if necessary.

To give an example, a popular phenomenon in biology is cooperativity. A biochemical reaction may be controlled by an highly non-linear, cooperative mechanism. Simple versions of cooperativity may be represented by complicated Petri net structures, but there are limits. The kinetic mechanisms of a cooperative behaviour are often not completely understood. However, the acquired understanding must be included in the model to get a coherent system model.

Deterministic firing delay is the outstanding characteristics of deterministic transitions. The delay is always relative to the time point where the transition gets enabled. There is one popular special case, the zero delay, for which the *immediate transitions* are introduced. Immediate transitions have always highest priority, which creates a subtle difference between an immediate transition and a deterministic transition with zero firing delay: if there is a conflict between the two, the immediate transition gets priority.

We will use the function *TimedFiring(delay)* to assign the delay constant.

Scheduled transitions belong to the deterministic transitions. The deterministic firing occurs according to a schedule specifying absolute points of the simulation time. A schedule can specify just a single time point, or equidistant time points within a given interval, triggering the firing once or periodically. However, transitions only fire at their scheduled time points if they are enabled. Scheduled transitions can dramatically restrict the behaviour, as we will see in Section 4.3, example EX5.

Scheduled transitions allow to disturb the core model at well-defined time points as it is done experimentally with the actual biological system under investigation in the wetlab; see Section 5 for an example.

We will use two functions to assign the required values: *FixedTimedFiring_Single(time_point)*, *FixedTimedFiring_Periodic(begin_time_point, repetition, end_time_point)*.

2.3   Generalised Stochastic Petri Nets ($\mathcal{GSPN}_{Bio}$)

Generalised stochastic Petri nets ($\mathcal{GSPN}_{Bio}$) are stochastic Petri nets $\mathcal{SPN}_{Bio}$ extended by *inhibitor arcs* and *immediate transitions*.

Inhibitor arcs are a powerful modelling feature and are known to bring computational completeness. Consequently, Petri nets of the net class $\mathcal{GSPN}$ have the same expressivity as an universal Turing machine [PW03]. However, in terms of construction of the reachability graph (continuous-time Markov chain), they do not establish additional challenges for finite state spaces, i.e. bounded Petri nets.

Immediate transitions are a very special kind of deterministic transitions with zero firing delay, i.e. they fire immediately after getting enabled, and always prior to (general) deterministic and stochastic transitions. Consequently, getting enabled and the firing itself coincide for immediate transitions. A cyclic system behaviour involving only the firing of immediate transitions corresponds to an infinite behaviour without time progress; we get a *time deadlock*.

If a stochastic simulation encounters a situation with more than one immediate transition enabled, one is chosen randomly [Ger01]. However, an analysis approach will consider all possible choices.

In terms of the reachability graph (continuous-time Markov chain), induced by a $\mathcal{GSPN}_{Bio}$ Petri net, we distinguish between *transient* and *non-transient* states. A system never spends time in a transient state before changing into another state. Thus, the time spent (*sojourn time*) in transient states is always zero, and not exponentially distributed anymore.

Consequently, the underlaying semantics is not a continuous-time Markov chain anymore. However, the transient states can be removed such that the reduced reachability graph corresponds again to a continuous-time Markov chain. See [MBC+95] for a precise description of the reduction technique and related formal definitions. In summary this means that $\mathcal{GSPN}_{Bio}$ can still be analysed analytically, if the state space, i.e. the continuous-time Markov chain can be constructed.

2.4   Deterministic and Stochastic Petri Nets ($\mathcal{DSPN}_{Bio}$)

Deterministic and Stochastic Petri Nets ($\mathcal{DSPN}_{Bio}$) are generalised stochastic Petri nets ($\mathcal{GSPN}_{Bio}$) extended by deterministic transitions.

Deterministic transitions possess a deterministic firing delay (waiting time), specified by a nonnegative real value. When a deterministic transition gets enabled, a count-down timer is started, initialized with the transition's firing delay. If the transition gets disabled before the timer reaches zero, the timer is switched off, and the transition will not fire. Otherwise, the transition will fire as soon as the timer reaches zero. The firing itself does not consume time.

If we consider stochastic Petri nets without deterministic transitions, the probability of two transitions firing at the same time is practically zero. Contrary, in stochastic Petri nets with deterministic transitions, it is possible that two transitions want to fire simultaneously. We already discussed the special case of two concurrently enabled immediate transitions. To analyse such a system, all possible choices have to be considered, while in the simulation a random choice takes place.

Definition 2 (Deterministic and stochastic Petri net). *A biochemically interpreted deterministic and stochastic Petri net is a septuple* $DSPN_{Bio} = (P, T, f, g, v, d, m_0)$, *where*

- *P und T are finite, nonempty, and disjoint sets. P is the set of places, and T is the set of transitions.*
- *The set T is the union of three disjunctive transition sets, i.e.*
  $T := T_{stoch} \cup T_{im} \cup T_{timed}$ *with:*
  1. *T_{stoch}, the set of stochastic transitions with exponentially distributed waiting time,*
  2. *T_{im}, the set of immediate transitions with waiting time zero, and*
  3. *T_{timed}, the set of transitions with deterministic waiting time.*
- *$f : ((P \times T) \cup (T \times P)) \to \mathbb{N}_0$ defines the set of directed arcs, weighted by nonnegative integers.*
- *$g : (P \times T) \to \mathbb{N}_0$ defines the set of directed inhibitor arcs, weighted by nonnegative integers.*
- *$v : T_{stoch} \to H$ is a function, which assigns a stochastic hazard function h_t to each transition $t \in T_{stoch}$, whereby*
  $H := \bigcup_{t \in T_{stoch}} \left\{ h_t \mid h_t : \mathbb{N}_0^{|{}^\bullet t|} \to \mathbb{R}^+ \right\}$ *is the set of all stochastic hazard functions, and $v(t) = h_t$ for all transitions $t \in T_{stoch}$.*
- *$d : T_{timed} \to \mathbb{R}^+$ assigns to each deterministic transition $t \in T_{timed}$ a nonnegative deterministic waiting time.*
- *$m_0 : P \to \mathbb{N}_0$ gives the initial marking.*

The stochastic transitions correspond to the transitions of the net class $\mathcal{SPN}_{Bio}$, so they have an exponentially distributed waiting time following the definitions given in Section 2.1.

The net class $\mathcal{DSPN}_{Bio}$ is a subset of the class $e\mathcal{DSPN}$, introduced in [Ger01]. For details of the subset relation see [Leh07]. Therefore, the theory, which has been developed to analyse $e\mathcal{DSPN}$ Petri nets, see [Ger01] and [Haa03], can be deployed to analyse $\mathcal{DSPN}_{Bio}$, too.

The remaining two features *read arcs* and *scheduled transitions* are not explicitly mentioned in the definition above, because the just allow a simplified specification using the orthogonal basic concepts in $\mathcal{DSPN}_{Bio}$.

Read arcs do not extend the modelling power as long as an interleaving semantics is considered. A read arc and two opposite arcs are indistinguishable in terms of the reachability graph (continuous-time Markov chain).

Scheduled transitions can be replaced by net components consisting of immediate and deterministic transitions only; see [Leh07] for construction patterns. Thus, they do not extend the modelling power.

3   Stochastic Analysis

The non-Markovian behaviour of $DSPN_{Bio}$ precludes the standard analytical approaches belonging to the Markovian world. However, there are adapted stochastic simulation methods, ready to deal with the extended behaviour, see e.g. [Ger01], [Haa03], [Leh07], and many more. A detailed discussion of the necessary adaptions compared to the fundamental Gillespie algorithm [Gil77] is beyond the given space limitations of this paper. Having the simulation traces, we apply simulative model checking of linear-time temporal logic (LTL) in a probabilistic setting (PLTL).

Simulative model checking follows the idea of Monte Carlo sampling and handles large or even infinite state spaces through approximating results by analysing only a subset of the state space – a finite set of finite outputs (traces) from a stochastic simulation algorithm (SSA), e.g. Gillespie's exact SSA or any other suitable variations of it.

A natural choice of logic to describe properties of sets of traces is linear-time logic. A linear-time logic operates over sets of linear paths through the state space, equivalent to operating on simulation outputs. A given property holds if it holds in all possible paths. Consequently, there are no path quantifiers.

We apply PLTL, a probabilistic linear-time temporal logic [DG08], [MC208]. This logic extends standard Linear-time Temporal Logic (LTL) [Pnu81] to a stochastic setting with a probability operator and a filter construct, defining the initial state of the property. LTL is the fragment of full Computational Tree Logic (CTL*) [CGP01] without path quantifiers, implicitly quantifying universally over all paths. To be self-contained we briefly recall the PLTL basics.

Syntax. PLTL is a logic to create path formulae ϕ and to ask for their probabilities. The grammar given in Table 1 defines a PLTL formula ψ.

Semantics. The semantics is defined over finite sets of finite linear traces of temporal behaviour, in our case by stochastic simulation runs. Each trace is evaluated to a Boolean truth value, and the probability of a property holding true is computed by the fraction of true values in the set over the whole set. It goes without saying, the choice of simulator and simulation parameters used to compute the sequence of states can affect the semantics of the PLTL property and the correctness of the result.

$P_{\unlhd x}$ is any inequality comparison of the probability of the property holding true, for example $P_{\geq 0.5}$. The expression $P_{=?}$ returns the value of the probability of the property holding true. Equality testing of the probability, $P_{=x}$, is not supported for obvious reasons.

PLTL allows the use of filters over top-level LTL expressions, denoted by $\{AP\}$, similar to those used in Probabilistic Computational Tree Logic (PCTL) [HJ94] and Continuous Stochastic Logic (CSL) [ASSB96]. This permits specifications to refer to the state or states that the property is checked from, rather than default to the initial state. This means that for a query of the form $\phi\,\{AP\}$, ϕ is checked from the first state that AP is satisfied. This can be a different one for each stochastic run. The temporal operators follow the standard LTL semantics:

Table 1. PLTL syntax. Please note that the square and curly brackets are part of PLTL.

$$\psi \quad ::= \quad \mathbf{P}_{\trianglelefteq x}[\,\phi\,]$$
$$| \quad \mathbf{P}_{\trianglelefteq x}[\,\phi\,\{AP\}\,] \;.$$

$$\phi \quad ::= \quad \mathbf{X}\phi \mid \mathbf{G}\,\phi \mid \mathbf{F}\phi \mid \phi\,\mathbf{U}\,\phi \mid \phi\,\mathbf{R}\,\phi$$
$$| \quad \neg\,\phi \mid \phi \vee \phi \mid \phi \wedge \phi \mid \phi \Rightarrow \phi$$
$$| \quad AP \;.$$

$$AP \quad ::= \quad \neg\,AP \mid AP \vee AP \mid AP \wedge AP \mid AP \Rightarrow AP$$
$$| \quad value\ comp\ value$$
$$| \quad true \mid false \;.$$

$$comp \quad ::= \quad = \mid \neq \mid \geq \mid > \mid < \mid \leq \;.$$

$$value \quad ::= \quad value\ op\ value$$
$$| \quad variable \mid max(variable) \mid d(variable)$$
$$| \quad Int \mid Real \;.$$

$$op \quad ::= \quad + \mid - \mid * \mid /\,,$$

with $\trianglelefteq \in \{<, \leq, \geq, >\}$, $x \in [0,1]$. $\mathbf{P}_{\trianglelefteq x}$ can be replaced by $\mathbf{P}_{x=?}$.

- **Next (X)** - The property must hold true in the next time point.
- **Globally (G)** - The property must hold true always [1].
- **Finally (F)** - The property must hold true sometime in the future.
- **Until (U)** - The first property must hold true until the second property holds true.
- **Release (R)** - The second property can only ever not hold true if the first property becomes true.

The meta term *variable* stands for any variable in the model, *Int* is any integer number and *Real* is any real number. In our case of stochastic Petri net analysis, a *variable* is going to be a place name, and the formulae refer to the number of tokens on a place in a given state. Additionally, there is a predefined variable *time*, referring to the simulation time points. Thus we can, for example, express properties which occur after some simulation time has elapsed.

The function *max* operates over all the token values of a place to return the maximum in the given simulation runs, thus the peak of a species' concentration, modelled by a place, can be checked, e.g. $Protein = max(Protein)$. The function d operates on each place in each state individually to return the derivative, thus increasing token numbers can be checked, e.g. $d(Protein) > 0$.

This approach to simulative model checking incorporates two approximations. The truth value of a single trace is approximated by operating over a finite sequence of states only; and the probability of the property is approximated through sampling a finite number of traces only. Thus, a subset of the model's behaviour is considered only. However, there are two special categories

---

[1] To be precise, in the given setting of model checking by finite traces, globally means 'always –as far as known'.

of properties, where definitive, i.e. non-approximating answers are possible by simulative model checking.

— *Monotone properties* comply with the following condition: if the property is satisfied in any path through the state space, then it is satisfied in any extension of the path [HLMP04]. Formulae without the Globally operator are monotone properties. The Globally operator and semantically equivalent descriptions by the other operators are incompatible with the monotony property. Considering longer paths can only increase the probability.
— *Invariant Properties* have to hold true in every state in every path. Thus they comply with the following condition: if the property is satisfied in any path through the state space, then it is satisfied in any other path. Their probability is independent of the number of considered paths. They are often used as consistency checks, and so do we in this paper.

PLTL may be considered as a linear-time counterpart to CSL. It can easily be used to formalise the visual evaluation of diagrams as generated by deterministic/stochastic simulation runs or by recording experimental time series. In the following chapter we are going to use PLTL to analyse sets of stochastic simulation traces of extended stochastic Petri nets, which have been constructed to illustrate the expressiveness of $DSPN_{Bio}$.

4   Typical Components

We present some typical model components, controlling a network's inflow and outflow, and thus demonstrating the suitability of the introduced Petri net class $DSPN_{Bio}$ for the envisaged application scenarios of model-based design of wet-lab experiments. We use the following abbreviations introduced in Section 2:

— *BioMassAction(c_t)*,
— *TimedFiring(delay)*,
— *FixedTimedFiring_Single(time_point)*,
— *FixedTimedFiring_Periodic(begin_time_point, repetition, end_time_point)*;

and we apply the following drawing conventions:

— read arcs: identified by a black dot,
— inhibitor arcs: identified by a hollow dot,
— stochastic transition: hollow square,
— deterministically timed transition: black square,
— immediate transition: black rectangle.

We are going to examine the behaviour of each component by simulative PLTL model checking over 100 (1,000) simulation runs. The individual runs are independent, so generally different. We confine ourselves deliberately on introductory formulae to illustrate the key ideas, increasing at the same time our confidence in the accuracy of our simulation algorithm for the non-Markovian setting.

4.1   Time-Controlled Inflow/Outflow

EX1. In our first example we consider a closed system, consisting of one reversible reaction $A \leftrightarrow B$, modelled by the two transitions *t1* (*BioMass-Action(0.11)*) and *t2* (*BioMassAction(0.1)*). The two deterministically timed transitions *input* (*FixedTimedFiring_Periodic(11,1,20)*) and *output* (*Fixed-TimedFiring_Periodic(31,1,40)*) are responsible for the absolutely timed inflow and outflow of tokens, see Figure 1.

The transition *input* does not have preplaces, thus it fires for sure at the time points $11, 12, \ldots, 20$, producing each time 1,000 additional tokens on place A. Contrary, the transition *output* removes 1,000 tokens from place B at the time points $31, 32, \ldots, 40$, provided there are enough tokens to enable the firing. Figure 2 shows the first 100 time units of a single simulation run.

We give some introductory samples of temporal-logic formulae (queries), formalising the visual inspection of the simulation output as it might be done by the expert evaluating former or designing the next wetlab experiments. We apply these queries to a set of 100 stochastic (single) simulation traces. The ratio

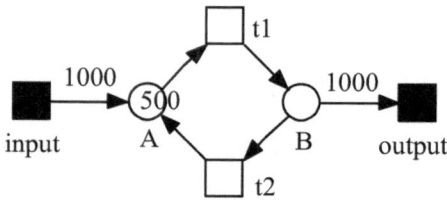

Fig. 1. First example of time-controlled inflow/outflow (EX1)

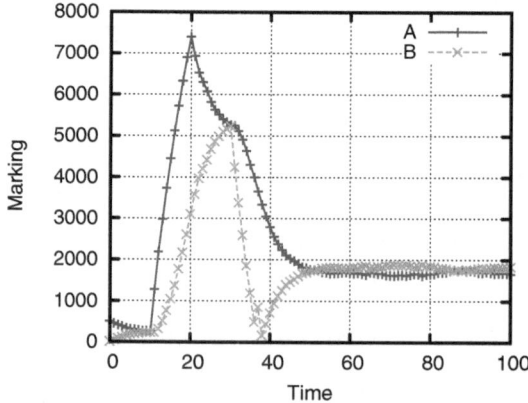

Fig. 2. Simulation result of the network given in Figure 1 (single run) (EX1)

of traces where the formula holds to the total number gives us a rough estimate of a formula's probability.

We check over exact Gillespie traces, i.e. all single events are logged. There are generally no "even" time points (like 30.000000 for 30). However, the firing of scheduled transitions at absolute time points (e.g. 20 in this example) causes exact time points in the simulation traces. We have to keep this in mind when refering to absolute time points in the following queries.

Please remember, all place names are read as integer variables in the following formulae; and the predefined variable *time* relates to the simulation time. The probabilities as computed by simulative model checking are given in brackets.

- Maxima *(probabilities: 1.0, 0.95)*.
  $\mathbf{P}_{=?}\,[\,\mathbf{G}(A < 7550)\,]$
  $\mathbf{P}_{=?}\,[\,\mathbf{G}(B < 5350)\,]$

- Peaks *(probabilities: 0.9, 1.0)*.
  $\mathbf{P}_{=?}\,[\,\mathbf{F}(time = 20 \wedge A > 0.9{\cdot}max(A) \wedge (3000 < B \wedge B < 3500))\,]$
  $\mathbf{P}_{=?}\,[\,\mathbf{F}((29 < time \wedge time < 30) \wedge (5000 < A \wedge A < 5400) \wedge B > 0.9max(B))\,]$

- Steady state, relative statements *(probabilities: 0.03, 0.59, 0.8, 0.91)*.
  $\mathbf{P}_{=?}\,[\,time \geq 50 \Rightarrow \mathbf{G}(A < B)\,]$
  $\mathbf{P}_{=?}\,[\,time \geq 55 \Rightarrow \mathbf{G}(A < B)\,]$
  $\mathbf{P}_{=?}\,[\,time \geq 60 \Rightarrow \mathbf{G}(A < B)\,]$
  $\mathbf{P}_{=?}\,[\,time \geq 70 \Rightarrow \mathbf{G}(A < B)\,]$

- Steady state, absolute statements *(probabilities: 0.39, 1.0)*.
  $\mathbf{P}_{=?}\,[\,time \geq 50 \Rightarrow \mathbf{G}((1500 < A \wedge A < 1800) \wedge (1600 < B \wedge B < 2000))\,]$
  $\mathbf{P}_{=?}\,[\,time \geq 60 \Rightarrow \mathbf{G}((1500 < A \wedge A < 1800) \wedge (1600 < B \wedge B < 2000))\,]$

EX2. We vary the pattern of our first example to remove repeatedly all currently available tokens on place B at equidistant time points, see Figure 3.

The immediate transition *output* consumes all tokens on place B, while there is a token on place *output_on*. The token on place *output_on* is controlled by the deterministically timed transition *switch_output_on* (*FixedTimedFiring_Periodic(20,20,_SimEnd)*) and the immediate transition *switch_output_off*. The transition *switch_output_on* initiates every 20 time units the cleaning process. The immediate transition *switch_output_off* switches off the outflow as soon as the place B is clean; otherwise each token arriving on B would be instantly removed and no token accumulation would be possible anymore. A single simulation run is given in Figure 4. We analyse a set of 100 of such stochastic traces by the following temporal-logic queries *(all yield probability 1.0)*.

- If the output is switched on, B is cleaned immediately.
  $\mathbf{P}_{=?}\,[\,\mathbf{G}(output\_on = 1 \Rightarrow B = 0)\,]$

- Cleaning of B at time point 20.
  $\mathbf{P}_{=?}\,[\,\mathbf{F}(time = 20 \wedge B = 0)\,]$

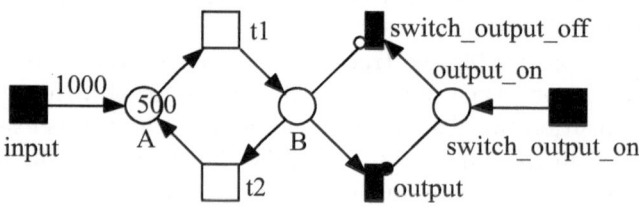

Fig. 3. Second example of time-controlled inflow/outflow (EX2)

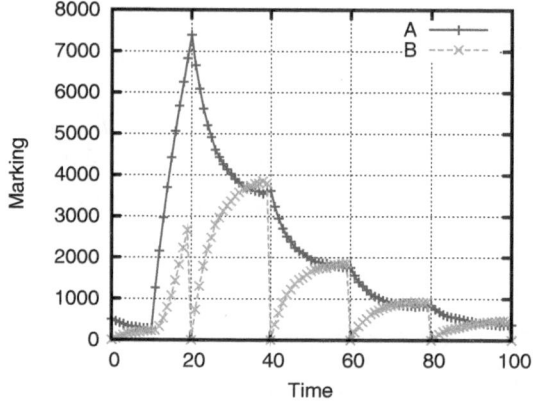

Fig. 4. Simulation result of the network given in Figure 3 (single run) (EX2)

- Cleaning of B at time point 20, ensuring that B does not get cleaned earlier.
  $\mathbf{P}_{=?} [\mathbf{F}(B > 0 \wedge (B > 0 \mathbf{U} (time = 20 \wedge B = 0)))]$
- Cleaning of B at time point 40, ensuring that B remains marked inbetween as soon as it got a token.
  $\mathbf{P}_{=?} [\mathbf{F}(time = 20 \wedge B = 0) \wedge \mathbf{F}(B > 0 \wedge (B > 0 \mathbf{U} (time = 40 \wedge B = 0)))]$

4.2   Token-Controlled Inflow

We discuss two examples and start again with a reversible reaction $A \leftrightarrow B$, modelled by the two stochastic transitions $t1$ ($BioMassAction(0.1)$) and $t2$ ($BioMassAction(0.005)$), which we consider as a closed system, challenged by experimental interventions.

EX3. In our first example of token-controlled inflow, the tokens on place A are raised by 50 as soon as the token amount drops below the threshold 30, see Figure 5. This behaviour is implemented by the immediate transition *input*, the firing of which is prevented by an inhibitor arc testing A. The weight 30

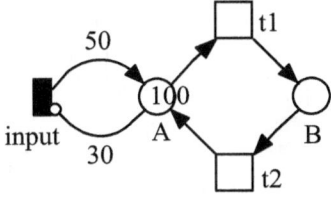

Fig. 5. First example of token-controlled inflow (EX3)

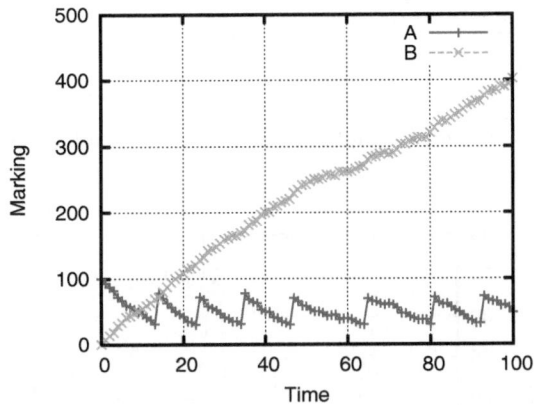

Fig. 6. Simulation result of the network given in Figure 5 (single run) (EX3)

of the inhibitor arc prevents the firing of *input* until the token amount drops below 30. 50 tokens are added to place A as soon as the inhibition condition becomes invalid, preventing again further inflow until the next drop occurs. Figure 6 shows a single simulation run. We analyse a set of 1,000 runs against the following formulae.

- The tokens on A never fall below the threshold 30 *(probability: 1.0)*.
  $\mathbf{P}_{=?}\,[\,\neg\,\mathbf{F}(A < 30)\,]$

- The transition *input* tries to keep the tokens on A between 30 and 80. But there are always some tokens on place B, which may return to A *(probabilities: 0.946, 0.996, 0.999)*.
  $\mathbf{P}_{=?}\,[\,\mathbf{F}(A = 30 \wedge \mathbf{G}(30 \leq A \wedge A \leq 80))\,]$
  $\mathbf{P}_{=?}\,[\,\mathbf{F}(A = 30 \wedge \mathbf{G}(30 \leq A \wedge A \leq 82))\,]$
  $\mathbf{P}_{=?}\,[\,\mathbf{F}(A = 30 \wedge \mathbf{G}(30 \leq A \wedge A \leq 84))\,]$

- There is a constant inflow due to the transition *input*, and the rate of $t1$ is (significantly) higher than of $t2$. Therefore, B increases permanently and

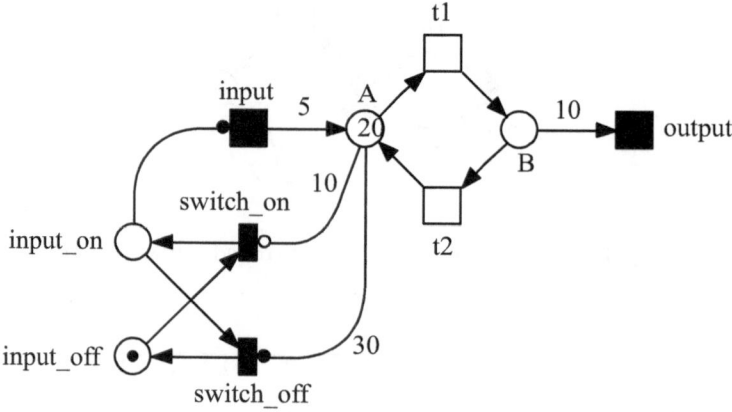

Fig. 7. Second example of token-controlled inflow (EX4)

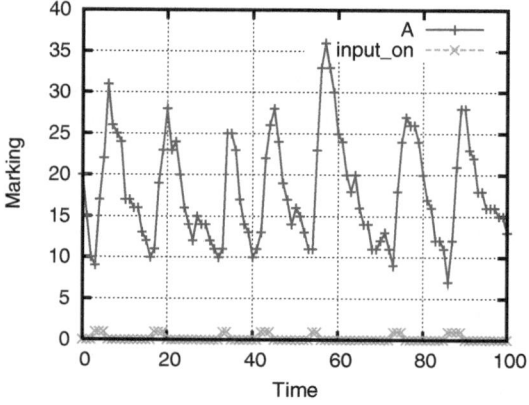

Fig. 8. Simulation results of the network given in Figure 7 (single run). The input is switched on/off (place *input_on*) in dependence on the token situation on A (EX4).

without limits. This is true in the averaged case only, e.g. 100 runs. $d(B)$ specifies the derivative.

$$\mathbf{P}_{=?} \; [\; \mathbf{G}(d(B) \geq 0) \;]$$

EX4. Our second example of token-controlled inflow is given in Figure 7. The transitions *t1* (*BioMassAction(0.2)*) and *t2* (*BioMassAction(0.1)*) form again the reversible reaction $A \leftrightarrow B$. We add the deterministically timed transition *output* (*FixedTimedFiring_Periodic(5,5,_SimEnd)*) to get a significant consumption of tokens. Each time *output* gets activated, it removes 10 tokens from B.

If the token amount on place A drops below 10, the deterministically timed transition *input* (*TimedFiring(0.5)*) starts working and adds by each firing 5

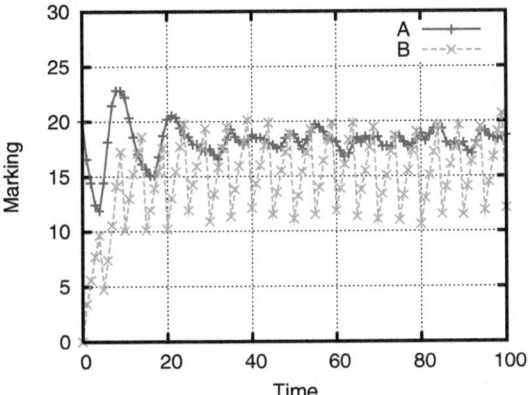

Fig. 9. Simulation results of the network given in Figure 7 (100 runs). A and B oscillate due to the repeated switching between inflow on/off (EX4).

tokens with 0.5 units waiting time inbetween, until there are at least 30 tokens on A. This behaviour is controlled by the immediate transitions *switch_on* and *switch_off*, and the two places *input_on* and *input_off*, forming a 1-P-invariant[1]. *Switch_on* can only fire if there are less than 10 tokens on A, and *switch_off* can only fire, if there are at least 30 tokens on A.

We give two related diagrams. The single run in Figure 8 shows how the input is switched on/off (place *input_on*) in dependence on the token situation on A. Figure 9 gives the average of 100 runs. It highlights the oscillation of A and B, caused by the repeated switching between inflow on/off. We analyse the token-controlled inflow component by the following formulae (1,000 runs) (*the first three yield probability 1.0*).

- The two places *input_on* and *input_off* form a 1-P-invariant.
  $\mathbf{P}_{=?}$ [$\mathbf{G}((input\_on = 1 \land input\_off = 0) \lor (input\_on = 0 \land input\_off = 1))$]

- The transition *input* is switched on/off if the token amount on A crosses the threshold 10 or 30, respectively.
  $\mathbf{P}_{=?}$ [$\mathbf{G}(A < 10 \Rightarrow input\_on = 1)$]
  $\mathbf{P}_{=?}$ [$\mathbf{G}(A \geq 30 \Rightarrow input\_off = 1)$]

- There is a delay of 0.5 time units between the on/off switch and the reaction of the actual inflow transition. E.g., after having switched off the input, 5 additional tokens will arrive by the already triggered firing of the transition *input*. Thus, even a weaker range than specified by the threshold values does not get probability 1 (*probability: 0.995*).
  $\mathbf{P}_{=?}$ [$\mathbf{G}(5 \leq A \land A \leq 40)$]

---

[1] Exactly one of both places carries a token at any point of time.

4.3   Switch between Deterministic and Stochastic Transitions

The following two networks demonstrate how to switch between deterministic and stochastic transitions. We start off with a time-controlled switch, before discussing a token-controlled switch.

In both cases we consider a non-reversible reaction $A \to B$, which is nevertheless modelled by two transitions: the stochastic transition t_stoch (*BioMassAction(0.1)*) and the deterministically timed transition t_det (*TimedFiring(0.25)*). The chosen net structure ensures that always one of these two transitions only is able to transfer tokens from place A to place B; with other words: the token flow occurs either stochastically or deterministically. The mutually exclusive firing is implemented by the two places *stochastic_on* and *det_on*, forming a 1-P-invariant and establishing side-conditions for t_stoch or t_det, respectively.

EX5. The actual time-controlled switch is performed by two deterministically timed transitions: *switch_to_det* (*FixedTimedFiring_Single(10)*) and *switch_to_stoch* (*FixedTimedFiring_Single(30)*), which fire (each once!) at the absolute time points 10 or 30, respectively, causing a switch in the other operation mode, see Figure 10. In summary, the modelled reaction $A \to B$ behaves deterministically between the time points 10 and 30, and stochastically else.

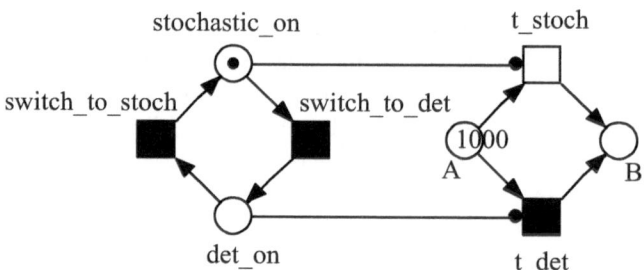

Fig. 10. Example of time-controlled switch between deterministic and stochastic behaviour. The semantic functions assigned to the transitions *switch_to_det* and *switch_to_stoch* allow them to fire only once (EX5).

EX6. We keep the basic principle for the token-controlled switch, but replace the transitions switching between the operation modi by immediate transitions, which depend on the token situation in place A. The immediate transitions *switch_to_det* and *switch_to_stoch* fire each once as soon as the token amount on place A drops below 700 or 500, see Figure 12. In summary, the modelled reaction $A \to B$ behaves deterministically for token amount between 500 and 700, and stochastically else.

The diagrams in Figure 11 and 13 show the behaviour of the two patterns for a single run each. For both we confirm the mutually exclusive operation mode of the stochastic and deterministic behaviour by the following query.

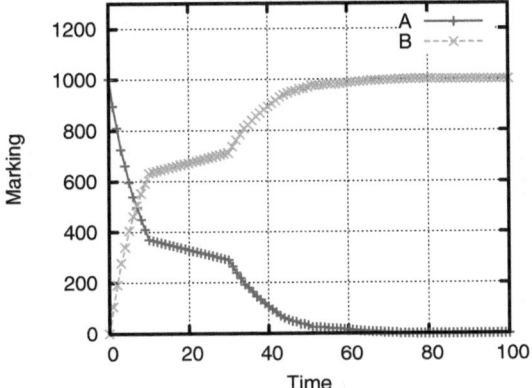

Fig. 11. Simulation result of the network given in Figure 10 (single run). There is a deterministic token flow from A to B between time points 10 and 30, and stochastic flow else (EX5).

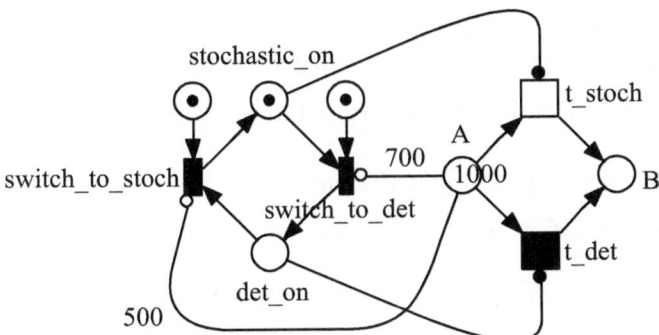

Fig. 12. Example of token-controlled switch between deterministic and stochastic behaviour. The additional preplaces of the immediate transitions bring the equivalence to the net component in Figure 10; i.e. the immediate transitions fire only once (EX6).

- The two places *stochastic_on* and *det_on* form a 1-P-invariant.

  $\mathbf{P}_{=?}$ [$\mathbf{G}((\textit{stochastic\_on} = 1 \wedge \textit{det\_on} = 0) \vee$
  $(\textit{stochastic\_on} = 0 \wedge \textit{det\_on} = 1))$]

We conclude the analyses with checking the range of deterministic versus stochastic behaviour for the two discussed patterns.

- Deterministic token flow from A to B between time points 10 and 30.

  $\mathbf{P}_{=?}$ [$(10 \leq \textit{time} \wedge \textit{time} < 30) \Rightarrow \textit{det\_on} = 1$]

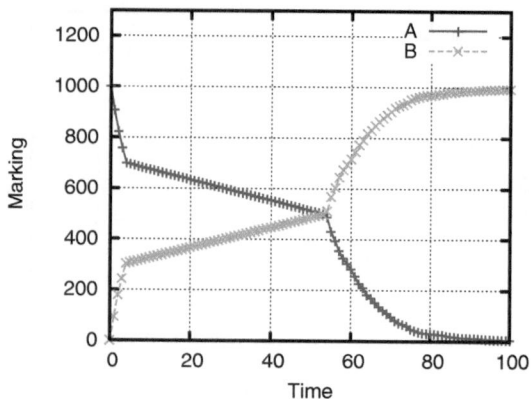

Fig. 13. Simulation result of the network given in Figure 12 (single run). There is a deterministic token flow from A to B for a token amount on A between 500 and 700, and stochastic flow else (EX6).

- Stochastic token flow from A to B from 0 up to time point 10, and starting at time 30 again.

  $\mathbf{P}_{=?}$ [$(time < 10 \lor 30 \leq time) \Rightarrow stochastic\_on = 1$]

- Deterministic token flow from A to B for a token amount on A between 500 and 700.

  $\mathbf{P}_{=?}$ [$(500 \leq A \land A < 700) \Rightarrow det\_on = 1$]

- Stochastic token flow from A to B for a token amount on A less than 500 or greater or equal 700.

  $\mathbf{P}_{=?}$ [$(A < 500 \lor 700 \leq A) \Rightarrow stochastic\_on = 1$]

All these properties are invariant properties, i.e. they yield probability 1.0, independently of the number and the length of considered simulation traces.

5   Lac Operon Model

We conclude by looking briefly at a classical example of prokaryotic gene regulation, the lac operon case. We follow the simplified version discussed in [Wil06] and specified there by a set of reaction equations and in an SBML-shorthand notation. We keep all naming conventions and the initial conditions, and translate the textual representation into a (qualitative) Petri net, reflecting explicitly the inherent structure of the regulatory network, compare Figure 14. Finally, we assign the rate equations as specified in the SBML code, and we get a stochastic Petri net.

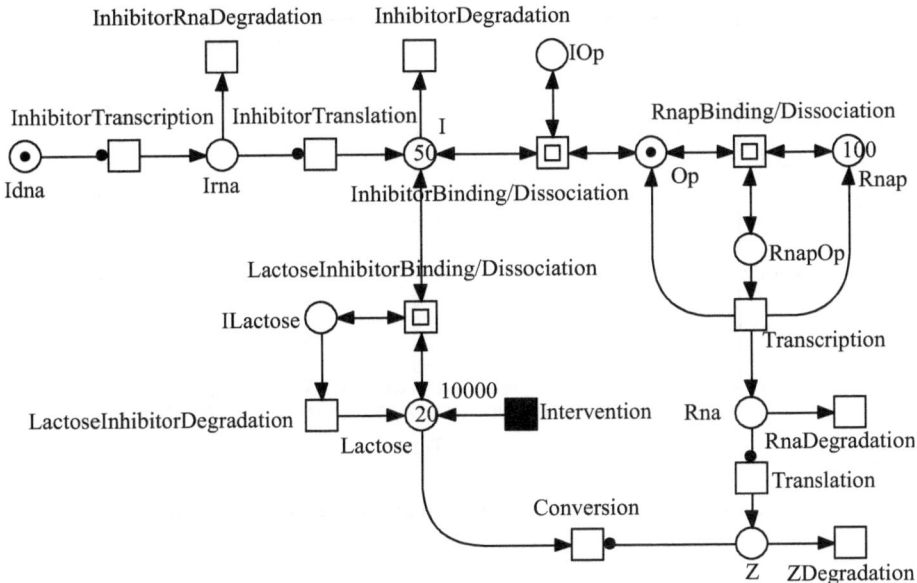

Fig. 14. Lac operon model according to [Wil06]. Macro transitions (drawn as two centric squares) indicate reversible reactions.

The core model of the network under consideration is extended by a special transition – an event in SBML terminology – modelling a timed intervention in a wetlab experiment. The transition *Intervention* (*FixedTimed-Firing_Periodic(50000,50000,_SimEnd)*[2]) introduces 10,000 molecules of Lactose every 50,000 time units, compare Figure 15.

To increase our confidence in the model we start with a preliminary structural analysis and compute the P-invariants and T-invariants[3]. There are input transitions, so the net can not be covered by P-invariants. However, there are three P-invariants, inducing mass-conserving subnetworks (modules) and enjoying obvious biological meaning. The preserved species is given first in the following short-hand notation:

- $pi_1 = \{Idna\}$,
- $pi_2 = \{Rnap, RnapOp\}$,
- $pi_3 = \{Op, IOp, RnapOp\}$.

Contrary, T-invariants do cover the net, which is a common consistency criteria for well-formed net structures, allowing e.g. a steady state behaviour. Each T-invariant induces a self-contained, state-repeating subnetwork (module). Besides the expected three trivial T-invariants for the three reversible reactions:

---

[2] Here we differ from the model given in [Wil06], where the modelled intervention occurs only once at a specified point of time.

[3] For all notions used in this section, but note introduced in this paper, see [HGD08].

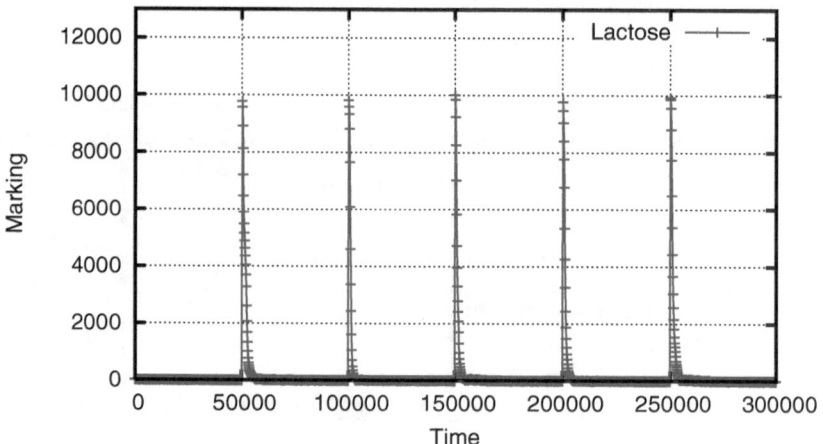

Fig. 15. Simulation result of the lac operon model: Lactose

- $ti_1 = \{LactoseInhibitorBinding, \underline{LactoseInhibitorDissociation}\}$,
- $ti_2 = \{InhibitorBinding, \underline{InhibitorDissociation}\}$,
- $ti_3 = \{RnapBinding, \underline{RnapDissociation}\}$,

we get the following six non-trivial T-invariants, each input/output behaviour is made of:

- $ti_4 = \{InhibitorTranscription, InhibitorRnaDegradation\}$,
- $ti_5 = \{InhibitorTranslation, InhibitorDegradation\}$,
- $ti_6 = \{InhibitorTranslation, LactoseInhibitorBinding,$
  $LactoseInhibitorDegradation\}$,
- $ti_7 = \{Intervention, Conversion\}$,
- $ti_8 = \{RnapBinding, Transcription, RnaDegradation\}$,
- $ti_9 = \{Translation, ZDegradation\}$.

There are four transitions (underlined), which are not involved in non-trivial T-invariants. However, they are crucial for the regulation mechanism between Z and Lactose. Please note, each T-invariant is given in a short-hand notation, enumerating the T-invariants' transitions in an order, which they may follow to reproduce a state, or what has to happen to get the system back in the steady state after some disturbences.

Remarkably, the net fulfills the Deadlock Trap Property (DTP), however is beyond the structural net class *extended simple*. In summary this allows the conclusion that there is no reachable dead state, in which any further system activities would be prevented. Actually, we expect the model to be live, which can not be proven with the analysis techniques available for (qualitatively) unbounded Petri nets.

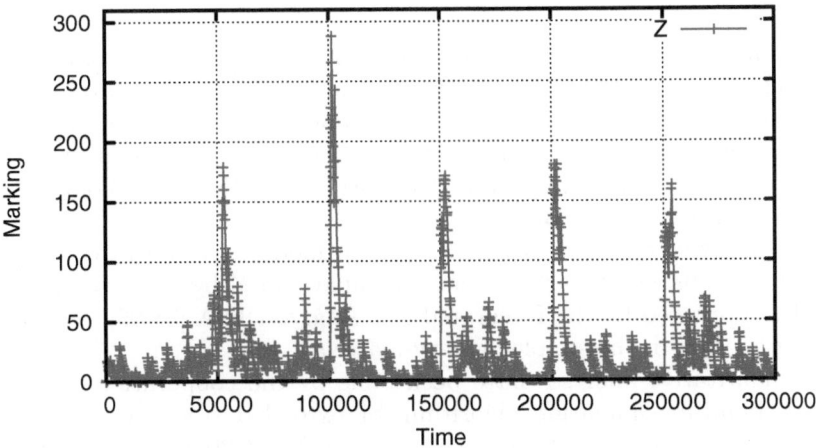

Fig. 16. Simulation result of the lac operon model: Z

However, there are property-preserving reduction rules downsizing the net structure, which are supported by the Integrated Net Analyser INA [SR99]. Applying these structural reduction rules, we get a smaller network, consisting of 2 places and 4 transitions. Liveness becomes obvious for this reduced network; see the supplementary material for more details.

The place Z models the enzyme β-Galactosidase; its reaction to the repeatedly sudden increase of Lactose molecules is shown in Figure 16. We analyse for the first intervention how a peak of *Lactose* triggers a peak of Z.

- The intervention causes *Lactose* to peak at time point 50,000 *(probabilities 1.0, 1.0, 0.65)*.
  $\mathbf{P}_{=?}$ [$(49,999 \leq time \wedge time < 50,000) \Rightarrow Lactose \leq 0.01 \cdot max(Lactose)$]
  $\mathbf{P}_{=?}$ [$time = 50,000 \Rightarrow Lactose \geq 0.99 \cdot max(Lactose)$]
  $\mathbf{P}_{=?}$ [$(52,000 \leq time \wedge time < 52,001) \Rightarrow Lactose \leq 0.1 \cdot max(Lactose)$]

- Z is highly likely to be at low concentration at time point 50,000 *(probability 0.9)*.
  $\mathbf{P}_{=?}$ [$time = 50,000 \Rightarrow Z \leq 0.1 \cdot max(Z)$]

- Z will rise to at least 80% of its maximal value within 2,000 time units *(probability 0.925)*.
  $\mathbf{P}_{=?}$ [$F((50,000 < time \wedge time < 52,000) \wedge Z \geq 0.6 \cdot max(Z))$]

- In summary, a peak of *Lactose* triggers a peak of Z within 2,000 time units *(probability 0.925)*.
  $\mathbf{P}_{=?}$ [$time = 50,000 \wedge Lactose \geq 0.99 \cdot max(Lactose) \wedge Z \leq 0.1 \cdot max(Z)$
  $\Rightarrow F(Z \geq 0.8 \cdot max(Z) \wedge time < 52,000)$]

6   Tools

The Petri net components and the lac operon model have been constructed using Snoopy [Sno08], [HRS08], a tool to design and animate or simulate hierarchical graphs, among them qualitative and continuous Petri nets, and the extended stochastic Petri nets as used in this paper. Snoopy provides export to various analysis tools as well as import and export of the Systems Biology Markup Language (SBML) [HFS$^+$03].

The qualitative analyses of the lac operon model have been made with the Petri net analysis tool Charlie [Cha08], complemented by the structural reduction rules supported by the Integrated Net Analyser INA [SR99]; see the corresponding log files in the supplementary material.

The quantitative analyses have been done by the cooperation of two tools: Snoopy's build-in simulation algorithm for extended stochastic Petri nets to generate the sets of simulation traces, and MC2 [MC208], a model checker by Monte Carlo sampling, for the simulative PLTL model checking. MC2 reads sets of simulation traces as, e.g., generated by Snoopy and expects additionally a file with the temporal-logical formulae.

As a proof of concept, we confined ourselves to rather small sets of 100 (1,000) runs only, allowing at the same time an affordable repetition of all computational experiments by the reader. A general recommendation is to start with smaller sets of simulation runs, just to check whether one got a formula right, before analysing larger sets, which could actually be done in parallel. None of the computational experiments for the typical components required more than 6 minutes per net example on a standard machine. Simulative model checking of the lac operon model is slightly more expensive. The traces have been generated on a workstation (2.83 GHz, 64 bit). The 100 exact traces (simulation time interval: 300,000) require about 5 GB. The model checking itself consumes less than 30 minutes on a standard machine.

Snoopy, Charlie as well as the data and analysis files of the discussed Petri net examples are available at

www-dssz.informatik.tu-cottbus.de/examples/xspn-components.

7   Summary

This paper extends the Markovian stochastic Petri nets $\mathcal{SPN}_{Bio}$ as introduced in [GHL07] to model and analyse biochemical networks. The extensions lead to the definition of Generalised Stochastic Petri nets $\mathcal{GSPN}_{Bio}$ and deterministic and stochastic Petri nets $\mathcal{DSPN}_{Bio}$. They include read and inhibitor arcs as well as several time-dependent transition types, which in summary preclude standard Markovian analysis approaches. Therefore we applied simulative model checking, approximating the probability of a given temporal logic formula by considering finite sets of finite paths through the state space. These paths are generated by stochastic simulation algorithms, adjusted to deal with the extended modelling features.

We discussed some typical net components demonstrating the usability of $\mathcal{DSPN}_{Bio}$ for the envisaged application scenario of model-based experiment design and evaluation. These components have been analysed by checking sets of stochastic simulation traces against PLTL properties. Invariant properties have been used to prove at the same time the plausibility of the applied simulation algorithm. We concluded with briefly looking at the lac operon case study, one of the classical examples of prokaryotic gene regulation.

Currently we consider some further extensions of our modelling formalism; among them are variable deterministic firing delays specified by an interval or an arbitrary marking-dependent function, reset and equal arcs as well as marking-dependent arc weights.

Simulative model checking is an extremely powerful tool. By way of introduction we have deliberately deployed some basic features of PLTL only. There is an interesting extension, PLTLc [DG08], supporting free variables, and thus allowing richer and more elegant properties, which however are also more complicated to write and to interpret. Thus, demonstrating these advanced features to more sophisticated users is beyond the scope and space limits of this paper.

Acknowledgements. The stochastic features of the Snoopy tool have been developed by Sebastian Lehrack, which cumulated in his Master thesis [Leh07]. This work has been financially supported by MPI Martinsried and MPI Madgeburg. Snoopy's quality improvements by Christian Rohr are crucial for the computational experiments presented in this paper. We would like to thank Robin Donaldson for his responsive assistance in MC2 issues.

References

[ASSB96]   Aziz, A., Sanwal, K., Singhal, V., Brayton, R.K.: Verifying Continuous-time Markov Chains. In: Alur, R., Henzinger, T.A. (eds.) CAV 1996. LNCS, vol. 1102, pp. 269–276. Springer, Heidelberg (1996)

[BGHO08]   Breitling, R., Gilbert, D., Heiner, M., Orton, R.: A structured approach for the engineering of biochemical network models, illustrated for signalling pathways. Briefings in Bioinformatics 9(5), 404–421 (2008)

[BK02]   Bause, F., Kritzinger, P.S.: Stochastic Petri Nets. Vieweg (2002)

[CGP01]   Clarke, E.M., Grumberg, O., Peled, D.A.: Model checking. MIT Press, Cambridge (2001) (third printing)

[Cha08]   Charlie Website. A Tool for the Analysis of Place/Transition Nets. BTU Cottbus (2008),
           http://www-dssz.informatik.tu-cottbus.de/software/
           charlie/charlie.html

[DG08]   Donaldson, R., Gilbert, D.: A model checking approach to the parameter estimation of biochemical pathways. In: Heiner, M., Uhrmacher, A.M. (eds.) CMSB 2008. LNCS (LNBI), vol. 5307, pp. 269–287. Springer, Heidelberg (2008)

[GBHD09]   Gilbert, D., Breitling, R., Heiner, M., Donaldson, R.: An introduction to biomodel engineering, illustrated for signal transduction pathways. In: Corne, D.W., Frisco, P., Paun, G., Rozenberg, G., Salomaa, A. (eds.) WMC 2008. LNCS, vol. 5391, pp. 13–28. Springer, Heidelberg (2009)

[Ger01]   German, R.: Performance analysis of communication systems with non-Markovian stochastic Petri nets. John Wiley and Sons Ltd., Chichester (2001)

[GHL07]   Gilbert, D., Heiner, M., Lehrack, S.: A unifying framework for modelling and analysing biochemical pathways using Petri nets. In: Calder, M., Gilmore, S. (eds.) CMSB 2007. LNCS (LNBI), vol. 4695, pp. 200–216. Springer, Heidelberg (2007)

[GHR+08]   Gilbert, D., Heiner, M., Rosser, S., Fulton, R., Gu, X., Trybiło, M.: A Case Study in Model-driven Synthetic Biology. In: Proc. 2nd IFIP Conference on Biologically Inspired Collaborative Computing (BICC), IFIP WCC 2008, Milano, pp. 163–175 (2008)

[Gil77]   Gillespie, D.T.: Exact stochastic simulation of coupled chemical reactions. The Journal of Physical Chemistry 81(25), 2340–2361 (1977)

[Haa03]   Haas, P.J.: Stochastic Petri nets: Modelling, Stability, Simulation. Springer, Heidelberg (2003)

[HDG10]   Heiner, M., Donaldson, R., Gilbert, D.: Petri Nets for Systems Biology. In: Iyengar, M.S. (ed.) Symbolic Systems Biology: Theory and Methods. Jones and Bartlett Publishers, Inc., USA (in Press, 2010)

[HFS+03]   Hucka, M., Finney, A., Sauro, H.M., Bolouri, H., Doyle, J.C., Kitano, H., et al.: The Systems Biology Markup Language (SBML): A Medium for Representation and Exchange of Biochemical Network Models. J. Bioinformatics 19, 524–531 (2003)

[HGD08]   Heiner, M., Gilbert, D., Donaldson, R.: Petri nets in systems and synthetic biology. In: Bernardo, M., Degano, P., Zavattaro, G. (eds.) SFM 2008. LNCS, vol. 5016, pp. 215–264. Springer, Heidelberg (2008)

[HJ94]   Hansson, H., Jonsson, B.: A logic for reasoning about time and reliability. Formal Aspects of Computing 6(5), 512–535 (1994)

[HLMP04]   Hérault, T., Lassaigne, R., Magniette, F., Peyronnet, S.: Approximate probabilistic model checking. In: Steffen, B., Levi, G. (eds.) VMCAI 2004. LNCS, vol. 2937, pp. 307–329. Springer, Heidelberg (2004)

[HRS08]   Heiner, M., Richter, R., Schwarick, M.: Snoopy - a tool to design and animate/simulate graph-based formalisms. In: Proc. PNTAP 2008, associated to SIMUTools 2008. ACM digital library (2008)

[Leh07]   Lehrack, S.: A tool to model and simulate stochastic Petri nets in the setting of biochemical networks (in German). Master thesis, BTU Cottbus, Dep. of CS (2007)

[MBC+95]   Ajmone Marsan, M., Balbo, G., Conte, G., Donatelli, S., Franceschinis, G.: Modelling with Generalized Stochastic Petri Nets, 2nd edn. Wiley Series in Parallel Computing. John Wiley and Sons, Chichester (1995)

[MC208]   MC2 Website. MC2 - PLTL model checker. University of Glasgow (2008), http://www.brc.dcs.gla.ac.uk/software/mc2/

[Mur89]   Murata, T.: Petri Nets: Properties, Analysis and Applications. Proc.of the IEEE 77(4), 541–580 (1989)

[Pnu81]   Pnueli, A.: The temporal semantics of concurrent programs. Theor. Comput. Sci. 13, 45–60 (1981)

[PW03]   Priese, L., Wimmel, H.: Theoretical Informatics - Petri Nets (in German).
          Springer, Heidelberg (2003)
[Rei82]   Reisig, W.: Petri nets; An introduction. Springer, Heidelberg (1982)
[Sno08]   Snoopy Website. A Tool to Design and Animate/Simulate Graphs. BTU
          Cottbus (2008),
          `http://www-dssz.informatik.tu-cottbus.de/software/snoopy.html`
[SR99]   Starke, P.H., Roch, S.: INA - The Intergrated Net Analyzer. Humboldt
          University Berlin (1999),
          `http://www.informatik.hu-berlin.de/~starke/ina.html`
[Wil06]   Wilkinson, D.J.: Stochastic Modelling for System Biology, 1st edn. CRC
          Press, New York (2006)

A Projective Brane Calculus with Activate, Bud and Mate as Primitive Actions

Maria Pamela C. David[1,*], Johnrob Y. Bantang[1,2,3,*], and Eduardo R. Mendoza[1,4]

[1] Faculty of Physics and Center for Nanoscience, Ludwig-Maximilians-Universität München, Geschwister-Scholl-Platz 1, D-80539 München, Germany
[2] Max-Planck-Institut für Dynamik komplexer technischer Systeme, Sandtorstraße 1, D-39106 Magdeburg, Germany
[3] National Institute of Physics, College of Science, University of the Philippines, Diliman, Quezon City 1101 Philippines
[4] Department of Computer Science, University of the Philippines, Diliman, Quezon City 1101 Philippines

Abstract. We modify and extend Cardelli's Brane Calculus and Danos and Pradalier's Projective Brane Calculus (PBC) to improve consistency with biological characteristics of membrane reactions. We propose a Projective Activate-Bud-Mate (PABM) calculus as an alternative to the Phago-Exo-Pino (PEP) basic calculus of L. Cardelli. PABM uses a generalized formalism for Action activation with receptor-ligand type channel construction that incorporates multiple association and affinity similar to Priami's beta binders. Calculus elements are finite. Volumes are associated with systems for more realistic compartment-based reaction probabilities. PABM also uses *Brane domains* that partition membranes into controllable, independent groupings of projective actions. Domains eliminate the need for parameters in Phago and Bud and allow lateral and cross-membrane interactions. We show that PABM can emulate bitonal membrane reactions. PABM also realizes the idea of L. Cardelli (Cardeli, 2004) on modeling molecules as systems.

1 Introduction

Cellular organization plays a key role in biological systems through the physical regulation of reactions. Enzymes, for instance, are typically sequestered in membrane-bound systems to which access is only made possible through cascades of equally regulated and timed signals. Most current formalisms for modeling, however, do not possess an explicit functionality for modeling compartmentalization. In deterministic models, compartmentalization is modeled with the use of additional variables that differentiate a species S that is within some compartment X from S that is within another compartment Y. While this has been used with some success, S in X is actually not different from S in Y, unless it has already reacted with other species in either compartment. It has only been in

---

* The first two authors contributed equally to this work.

C. Priami et al. (Eds.): Trans. on Comput. Syst. Biol. XI, LNBI 5750, pp. 164–186, 2009.

recent years that several calculi were developed so that membrane compartmentalization: (a) becomes an inherent part of computations and (b) is emphasized in simulating reactions[1, 2, 3, 4].

Brane calculus is a formalism that can be used to describe systems as mem-brane-bound compartments that may contain other systems[3]. These compartments can merge, split or be hierarchically reorganized through uptake (phagocytosis) or extrusion (exocytosis) mechanisms, based on the capabilities, known as *actions*, of the membranes that enclose them[3, 5]. An important aspect of these actions — directly adapted from pi calculus — is that they are triggered via highly specific channel-based communication. Nevertheless, the mapping between channels is not necessarily one-to-one, with some channels having more than one communication partner. Although the original concept of channels in pi calculus was for mobile telecommunication systems, it is compatible with the representation of biological interactions, from enzyme-substrate systems that interact to form a chemically distinct product to receptor-mediated intermembrane communication that leads to membrane reorganization.

Another formalism that includes compartments is Priami's beta binders[4]. Here, much emphasis is given to the promiscuousity of the channels ("beta binders") through which the compartments interact, as well as its effects on the dynamic evolution of the compartment contents, interactions, and interfaces. As in Brane calculus, compartments can merge and split as a result of binder-based communication. While inherent in Brane Calculus, beta binders needed an extension to include hierarchical construction of compartments. Recent extensions, however, only permit intuitive representation of static hierarchical structures, but still forbid the explicit nesting of compartments[6]. The main advantage of beta binders over Brane calculus is its natural representation of affinity to channel pairings, a concept that is adapted in the proposed extension in this paper.

The uniqueness of Cardelli's Brane calculus lies in the representation of all computations *on* membranes. This is important, particularly since it is actually the dynamic property of membranes that determines its capability to interact with other membranes and its general environment *in vivo*. Consequently, this property also determines how a membrane-bound system would evolve. Structure hierarchy can likewise be easily represented in Brane calculus, where nested systems are effectively organized in tree structures[3, 5].

Brane calculus has been previously extended by Danos and Pradalier to incorporate the idea that the inner and outer surfaces of a membrane are not identical[5]. *In vivo*, it could be frequently observed that the membrane protein domains exposed to the extracellular matrix are different from the domains exposed to the cytosol. It is even possible for membrane proteins to possess either an extracellular domain or a cytosolic domain. As a result, the definition of the inner and outer membranes are different. Additional physical restrictions are introduced on which reactions could take place, in particular only directed actions on membrane surfaces that could "see" each other are allowed to interact. This extension using directed actions is known as *projective Brane calculus* (PBC).

Nevertheless, there are a number of aspects in both calculi that involve concepts not observed in biological system. The purpose of this paper is to introduce further modifications and extensions combining the strength of both Brane calculi, with the aim of making it even more consistent with the biological characteristics of membrane reactions. Specifically, we introduce the following changes that result to the proposed extension, the Projective Activate-Bud-Mate calculus (PABM):

1. Use of the minimal set $\mathcal{S}_{min} \equiv \{bud, mate, !, 0\}$ instead of the set $\mathcal{S} \equiv \{phago, exo, pino, mate, bud, drip\}$ for the possible actions $a \in \mathcal{S}_{min}$ (see §2). All other actions in $\mathcal{S}$ are realized using only the actions in $\mathcal{S}_{min}$ together with directed Actions of PBC.
2. Abstraction of specific send-receive channel pairing into less specific channel name equality, eliminating the distinction between input and output channels. Together with the previous revision, it allows generalized representations in the form ax for Actions and Coactions, where x is a named channel.
3. Introduction of *Brane domains*, which are autonomous groups of directed Actions within a Brane. The use of Brane domains would also allow inter-domain interactions within and across the same membrane.
4. Removal of the parameters for Bud and Phago, allowing the dynamic nature of membranes to be reflected in the calculus.
5. Inclusion of volume information as a system attribute to reflect its effects on the probability at which collisions will occur inside a compartment.
6. Association of rates to channels emulating an affinity feature similar to beta binders.
7. Treatment of Brane constituents and contents as finite quantities.
8. Elaboration of molecules as systems, a concept previously introduced by L. Cardelli[3].

These modifications are also geared towards the development of a machine for Brane calculus that can handle large-scale biological models.

2   Modified Notations

Table 1 summarizes the proposed notation and conceptual changes to the current Brane calculus, provided as a quick reference to the detailed explanations for these changes in the succeeding sections.

2.1   Actions

Notations and terms. In the documentation for the design of a machine for Brane calculus[7], stochastic pi calculus notations for input and output channels are used to distinguish between actions and coactions. At this point, it is important to make a distinction between an action (small 'a'), a and an Action (capital 'A'), σ. An action is an element of the set, $a \in \mathcal{S}$ currently defined as:

$$\mathcal{S} \equiv \{phago, pino, exo, mate, bud, drip\}; \tag{1}$$

Table 1. Comparison of the currently-established Brane calculus (Cardelli's Brane calculus and PBC) with the proposed calculus. Note that $a, a_i \in \mathcal{S}$ $\sigma, \sigma_i, \tau \in \mathcal{A}$, $i = 1, 2$, with $\bar{a}$ as the coaction of a. Conventions for parallel composition from PBC [5] are used.

Definition	Brane/Pi Calculus	PABM
Channel ($x \in \mathcal{C}$)	$!x \longleftrightarrow ?x$	$x \longleftrightarrow x$
Set of actions ($a \in \mathcal{S}$)	$\mathcal{S} \equiv \{\text{phago, pino, exo,}$ $\text{mate, bud, drip}, \ldots\}$	$\mathcal{S}_{\min} \equiv \{0, \text{bud, mate}, !\}$
Action ($\sigma, \tau \in \mathcal{A}$)	$a!x \longleftrightarrow \bar{a}?x$	$!x \longleftrightarrow ax \ (a \neq !)$
Brane domain	undefined	$\rho \equiv \langle \sigma_1 \,;\, \sigma_2 \rangle$
Brane	$\langle \sigma_1 \,;\, \sigma_2 \rangle$	$[\,\rho\,]$
Directionality	σ_1 is outside, σ_2 is inside	σ_1 is outside, σ_2 is inside
System	$\langle \sigma_1 \,;\, \sigma_2 \rangle (\!(P)\!)$	$[\,\rho\,](\!(P)\!)$
Parameter (bud and phago)	$\sigma(\tau)$	$\tau.\sigma$ and ρ (see text for details)
Choice	$\sigma_1 + \sigma_2$	$\sigma_1 + \sigma_2$
Series	$\sigma_1.\sigma_2$ or $\sigma_1\sigma_2$	$\sigma_1.\sigma_2$ or $\sigma_1\sigma_2$
Parallel	$\sigma_1 \mid \sigma_2$ $\quad$ $P \circ Q$	σ_1, σ_2 $\rho_1 \mid \rho_2$ $P \circ Q$
Replication	$!\sigma \doteq \sigma, \sigma, \ldots$ (infinite) $!P \doteq P \circ P \circ \ldots$ (infinite)	$(n)\sigma \doteq \sigma, \sigma, \ldots, \sigma$; n parallel $(\sigma)^n \doteq \sigma.\sigma.\ldots.\sigma$; n series $(n)\rho \doteq \rho \mid \rho \mid \ldots \mid \rho$; n parallel $(n)P \doteq P \circ P \circ \ldots \circ P$; n parallel

while an Action is an element of the set, $\sigma \in \mathcal{A}$ currently defined as:

$$\mathcal{A} \equiv \{ax; a \in \mathcal{S}, x \in \mathcal{C}\}, \tag{2}$$

where the set $\mathcal{C}$ contains all possible channels. These notations are used throughout the text. In PABM, we use the following (minimal) set:

$$\mathcal{S}_{\min} \equiv \{m, b, 0, !\}; \tag{3}$$

where m is Mate, b is Bud, and two new actions, 0 and !, as the null and activate actions, respectively. The set $\mathcal{A}$ remains the same but with $\mathcal{S}$ replaced by $\mathcal{S}_{\min}$. We demonstrate that all other elements $a \in \mathcal{S}$ (Eq. 1) can be derived from a combination of these modifications with the directed Actions of PBC. Note that with the changes, the action becomes a passive entity (i.e. an action waits for an activation signal) by default.

Since the definitions of mate and bud were not changed and have been discussed elsewhere [3, 5, 8], we will only review these definitions briefly.

Activate action, ! Cardelli's Brane calculus requires two levels of matching before an Action could be executed/activated: (a) input and output channels; and (b) action and co-action. The use of an activation signal is expected to improve the symmetry of form for the Actions with respect to channel and activation pairings. Here, we introduce an activate action, '!', to approximate the input/output channel functionality of stochastic pi calculus, consequently precluding the need for an explicit distinction between actions and coactions within Brane. The Action in the form !x acts as an initiator of membrane interaction through channel x. This Action may be interpreted as a binding event, analogous to the required output signal from the intiating membrane or molecule before any non-activate Action can be executed/activated.

The use of an activation signal, instead of a more strictly-bound action-coaction pair with matching channels, is based on the fact that a single compound, modeled here as a communication channel, can interact with more than one substance, which may range from proteins to oligosaccharides, on the cell membrane. As discussed in §2.2, typical biological interactions involving receptors, logically corresponding to channels, are one-to-many relationships, rather than one-to-one pairs. Nevertheless, such cardinality does not imply that reaction specificity is lost.

Another biological characteristic taken into account is the dependence of the kind of reaction that occurs on the receptor type, rather than on the ligand (i.e. it is the receiver that determines which effect will occur). This characteristic is particularly marked in cells of the immune system, as well as antibodies, which have different effector functions associated with each class and subclass. The proposed form emphasizes that interaction specificity is conferred by the channel, but the receiver determines the type of action to execute.

The null action, 0. The null action, 0, blocks actions that precede it; an Action in the form $\sigma.0x_0$ can thus be used to model a blocked Action, σ. The null action can be deactivated with !x, making σ accessible. Biologically, blocking occurs in the event of temporary receptor internalization [9], binding-induced conformational changes [10, 11], and binding-induced physical blocking of other available binding sites. The use of 0 will be useful for modeling bind-and-release, molecular functionality switching, and other membrane-bound mechanisms.

Bud. Bud refers to the arbitrary splitting of a membrane, resulting in two membrane-bound compartments[3]. Cardelli makes a distinction between bud and *drip*; bud occurs when the split occurs with *one* internal membrane, while *drip* refers to the separation of *zero* internal membranes. In PABM, this distinction is not made.

Mate. Mate causes the irreversible mixing of actions of membranes that fuse either horizontally (i.e. membranes at the same level of nesting) or vertically through an exocytosis-type process[3].

2.2   Choice, Parallel, and Series

All discussions of choice, parallel and series compositions are made with reference to Actions, unless otherwise indicated. Parallel and series compositions are not valid for actions and channels.

Choice. The concepts of parallel composition, choice and prefix(series) are retained from pi calculus. The notation for choice will be retained ('+'). Choice could either be between actions a_1 and a_2 or channels x_1 and x_2. These are equivalent to having a choice between two (or more) Actions in the basic form, ax. In particular, the following choices within action-channel pairs (Actions) would be equivalent to their respective choices between Actions:

$$(a_1 + a_2)x \equiv a_1 x + a_2 x \tag{4}$$

$$a(x_1 + x_2) \equiv ax_1 + ax_2 \tag{5}$$

$$(a_1 + a_2)(x_1 + x_2) \equiv a_1 x_1 + a_2 x_1 + a_1 x_2 + a_2 x_2 \tag{6}$$

These equivalences remove the need of implementing Actions in their non-basic forms. Hence, implementing choice in actions and/or channels will be unnecessary since all cases can always be reduced to a choice between (at least) two Actions.

Aside from simplifying the implementation of Actions, Eqns. 4 and 5 reflect biological phenomena. For instance, Eqn. 5 is illustrated by membrane-bound receptors that have multiple ligands, with each ligand binding with a different affinity. At least three virus families, *Orthomyxoviradae*, *Paramyxoviradae* and *Reoviradae*, for example, use sialic acid in cell surfaces to enter via the endocytic pathway(s)[14]. A biological phenomenon that illustrates Eqn. 4, on the other hand, is the receptor for advanced glycation of end products (RAGE), expressed in a wide variety of cell types. RAGE is characterized by its ability to recognize numerous ligands, each of which result in different effects[15]. This is equivalent to having several actions associated with the same channel. Although the reactions of RAGE do not involve membrane structure deformations, a feature that would allow the direct modeling of such events may be of interest. Furthermore, Eqn. 5 implies that several receptors (or channels), can be used to initiate the same actions. Different receptors, for instance, are used by different viruses to enter the cell. Equation 6 is included for the purpose of completeness, but may not have any biological significance.

Parallel. Parallel pi processes and Actions, as indicated in Table 1, are represented following the notations in Danos and Pradelier and Cardelli.

Prefix/Series. The original notation will be maintained for the series. A recurring series of the same Action would be used instead of replication to indicate the finite reusability of an Action.

2.3   Rates

PABM incorporates rates by associating a real number, r_x, to the channel of each Action ax. When r_x is associated with $!x$, it corresponds to the rate with which $!x$ reacts on average with its receiver — the *basic rate*. When this real number is instead associated with ax, $a \neq !$, r_x is a factor of the basic rate which reflects the efficiency of the reaction. A value of 1.0 indicates that the specific reaction rate with a particular receiver is the same as the basic rate. Association of rates to channels is adapted from beta binders[4].

2.4   Affinity

Affinity describes the strength of non-covalent interactions between a ligand to its specific binding site on the receptor surface; this value is independent of the number of binding sites[12]. Higher affinities are associated with factors such as the exposure of large, interactive amino acid side-chains, highly electronegative groups, or the deformability of a surface; these characteristics generally enable a ligand to form more non-covalent bonds with the receptor[13]. Empirically, a value known as the affinity constant (K_a) is used to approximate affinity for ligand-receptor systems[1]. It is determined by measuring the concentration of free ligand required to fill half of the binding sites on the receptor. When half the sites are filled, [Ligand $\cdot$ Receptor] = [Ligand] and $K_a = 1/$[Receptor], where '$[X]$' is used to indicate the molar concentration of X; common K_a values range from as low as 5×10^4 to as high as 10^{11} liters/mole[12].

The use of affinity in process calculi for biology has been proposed by C. Priami and P. Quaglia as a feature for beta binders[4]. Affinity is incorporated as a probability $P(a, b)$ that an interaction between two different interfaces a and b can take place, effectively relaxing the requirement for an exact matching of interface[6] — a distinct digression from pi calculus, where interactions occur on syntactically identical ports (lock-and-key model).

In PABM, affinity is inherent with choice. Since channels in PABM represent receptor-ligand functionality, the execution of a single action a can be associated with its interaction through more than one channel, say $x_1, x_2, \ldots,$ and x_n, $n > 1$, resulting to the Action: $a(x_1+x_2+\ldots+x_n)$ that reduces to $ax_1+ax_2+\ldots+ax_n$ (Eqn. 5). When $a = !$, this results in multiple rates of execution, which depends on the Action $a'x_l$ that is activated ($a' \leq !$ and $l \leq n$). A similar situation occurs when $a \neq !$ and $a' = !$, albeit with a different biological implication. Table 2 summarizes the difference between this approach and Priami's implementation in beta binders.

Table 2. Comparison of affinity in beta binders and PABM

	Beta Binders	PABM
Association	each reduction	each channel
Implementation	reaction probability $P(a,b)$ between two non-identical interfaces a and b	multiple channels using choice

2.5   Branes and Systems

The definition for Systems as sets of nested Branes is retained, and notations for these are adapted from PBC[5]. Null Systems are represented as ⋄. Notations for parallel composition of Systems are also retained. The same replication rules are

---

[1] Notably for antibodies and antigens.

applied to both Actions and Systems (Table 1). Branes however, are redefined as a composition of Brane domains; a Brane consisting of a single domain reduces to the original definition (Table 1).

Directed Brane domains and directed Actions. In this section, the concept of directed actions in PBC is extended to *Brane domains*. A Brane domain, represented as a vector, ρ, is a grouping of directed Actions that approximates the occurrence of composition and functional non-homogeneity ("patchiness") observed in biological membranes. Consequently, a Brane is now defined as a parallel composition of Brane domains. A Brane defined using a single domain is homogenous, and reduces to a Brane in PBC.

Brane domains were introduced to facilitate greater control in processes like membrane budding. As opposed to Cardelli's calculus where a parameter is used to define the characteristics of the Brane that will be budded out, the proposed calculus makes these characteristics entirely dependent on the current, dynamic state of the parent membrane. Budding processes, however, are highly localized, and the derived system should *not* have all the characteristics of the parent membrane. In the proposed calculus, only specific Brane domains are transferred in budding processes, unless the parent membrane is homogenous.

Alternately, a Brane domain can be visualized as a set of directed Actions occurring proximally in a membrane. As an example, a system with Brane domains is subsequently represented as follows:

$$[\, \rho_1 | \rho_2 \,]\!(\, [\, \rho_3 \,]\!(\, P \,]\!) \circ Q \,]) \tag{7}$$

where ρ_n is of the form $\langle \sigma_1 ; \sigma_2 \rangle$ (Table 1). As in the original Cardelli calculus, both ρ_1 and ρ_2 are visible to ρ_3. Using the rules of PBC, only the "outside" Actions of ρ_3 can interact with the "outside" Actions of both ρ_1 and ρ_2. The advantage of this feature is relevant in modeling competition between parallel membrane processes (§3.6).

It is important to note that Brane domains represent active or functional sites on membranes or molecules and not the molecules themselves. Nevertheless, since at least one active site is associated with proteins, these can be represented as Brane domains. Brane domains can be used to model membrane proteins that function together such as lipid rafts[17] and SNARE complexes[18].

Lateral and cross-membrane interaction. Since interactions are now between two Brane domains, apart from the interaction of a domain from one membrane with another in a different membrane, domain-domain interactions within a membrane is now possible. Actions can now be activated by Activate Actions on a neighboring domain (lateral membrane interaction). Activations by Actions on opposite sides of a single membrane (cross-membrane) can also be facilitated, provided that one of the Actions is translocated to the other side of the membrane by a mechanism similar to diffusion or channel-mediated transport. This capability can be used to model ligands that interact with receptors on the same membrane surface or on the opposite side of the same membrane. Spontaneous membrane and in-membrane operations such as pinocytosis, drip, inversion, and fusion of proteins to form rafts and complexes[17, 18] can now be easily modeled.

The following equation shows the competition between a_1x and a_2x since lateral- and cross-membrane interactions are allowed.

$$[\langle !x\,;\ a_1x \rangle | \langle a_2x\,;\ - \rangle | \langle -\,;\ a_3x \rangle \,]\!(\cdots)\!) \Rightarrow \quad a_1 \quad \text{or} \quad a_2 \qquad (8)$$

Note that a_3 cannot be activated since cross-membrane interaction are allowed only within the same Brane domain. With PABM, the transport of functional particles (e.g. molecules) through the membrane without introduction of atonal reduction rules can also be modeled (see §3.7).

Volume information. A single enzyme-substrate experiment in a controlled nanoenvironment has shown that the frequency of collisions between two molecules is inversely proportional to the size of the vesicle where these molecules are contained[19]. Consequently, volume information will be associated with each System, representative of a compartment, allowing adjustments to be made in the probabilities at which the contained reactions will occur.

2.6   Replication

For the purpose of a calculus geared towards discrete biological system modeling, PABM uses a more controlled form of replication for Actions (also applicable to Brane domains and Systems), where the cardinality of replication is indicated (see Table 1). For instance, even if the initial counts of cellular components that are in the order of $\gtrsim 10^4$ to $\sim 10^{10}$ [16] are large enough to warrant the use of ∞, these are still finite quantities that may be critical determinants of biological system viability, especially in simulations that run for relatively prolonged periods of time (≥ 24 hours). Finite replication also reflects the finite lifetimes, masses and/or energies of both the components of biological systems and the systems themselves, appropriately manifested in the calculus in the form of finite Brane or Action usage. The numbers representing the finite number of replications can also be made stochastic to mimic the heterogeneity of membrane domains in terms of the absolute numbers of its constituents. Finite replication is conceptually similar to energy in beta binders [4], since the special entity E^j (with $j \in \Re^+$) can be mapped to the cardinality of replication in PABM.

2.7   Sample Notation: Mitogen-Induced Proliferation of Schwann Cells

Cell proliferation induced by an external signal is one of the simplest biological examples that is, nevertheless, difficult to express in an intuitive manner without the use of spatial information. Mitogens, which induce cell division, are typically associated with one or more cognate receptors through which it can enter a cell. In Schwann cells, which form the insulation for vertebrate neurons in the peripheral nervous system and which are critical for axon regeneration, proliferation is induced by the following mitogens in the neonatal stage: glial growth factor (GGF), platelet-derived growth-factor B (PDGF-BB) and basic fibroblast growth factor (bFGF) [32].

For purposes of illustration, a coarse-grained model of the system can be defined in the above notational changes as follows:

$$[\langle !x_G \,;\, !x_a \rangle \,]\!(\, X \,]\!) \circ [\, \langle !x_P \,;\, !x_a \rangle \,]\!(\, X \,]\!) \circ [\, \langle !x_{bF} \,;\, !x_a \rangle \,]\!(\, X \,]\!) \circ$$
$$[\, \langle mx_G, mx_P, mx_{bF} \,;\, bx_b \rangle \,]\!(\, SC \circ [\, \langle !x_b.0x_a \,;\, - \rangle \,]\!(\, R \,]\!)\,]\!)$$

where each X represents a growth factor associated with some channel $!x_s$, where s represents the part of X that binds to the GGF receptor (G), the PDGF-BB receptor (P) or the bFGF receptor (bF). The corresponding receptors, mx_G, mx_P, mx_{bF} are all associated with the Schwann cell (SC). R represents the inactive replication machinery of the cell. This can only be activated on the fusion of one of the growth factors with SC, removing $0x_a$, and making $!x_b$ available for interaction. The availability of $!x_b$ in R allows SC to bud through its interaction with Action bx_b.

3   Projective Activate, Bud, and Mate Calculus

In this section, we demonstrate that all Actions in $\mathcal{S}$ (Eq. 1) can be expressed as the actions in $\mathcal{S}_{\min}$ (Eq. 3) combined with the directed Actions of PBC. The use of $\mathcal{S}_{\min}$ as primitives has similarities to the basic Mate-Bud-Drip (MBD) calculus [8], which is one of two possible basic calculi for membrane interactions, together with the Phago-Exo-Pino (PEP) calculus. It has been shown [3, 8], however, that an encoding of MBD can be obtained with PEP, but not the opposite, because the maximum level of membranes (i.e. the membrane nesting) cannot grow during computation in MBD. Furthermore, the same articles prove that PEP calculus is Turing complete and Turing powerful, as opposed to MBD.

Given these limitations, the use of a Bud- and Mate-based basic calculus appears counterintuitive. However, events indicated in the derivation of the MBD primitives using PEP (Fig. 1A) are not observed in biological systems (Fig. 1B). Although it is partly superfluous to observe that the derivations of MBD were previously qualified as performed for computational purposes only, it is clear that *in vivo* membrane fusion is characterized by membrane perturbances rather than a series of phagocytosis and exocytosis events[20]. Specifically, the prevalent hypothesis regarding membrane fusion involves the reduction of the distance between the fusing membranes, followed by the local perturbation of the lipid structure and merger of proximal monolayers. Stalk formation and stalk expansion, and finally, pore formation are postulated to follow. Furthermore, there is a requirement that each of these steps has to be driven by an energy gradient towards lower energies. The stalk hypothesis is mainly based on the observation that the merger of proximal monolayers precedes the merger of distal monolayers. These events are followed by the intravesicular solvent exchange[20].

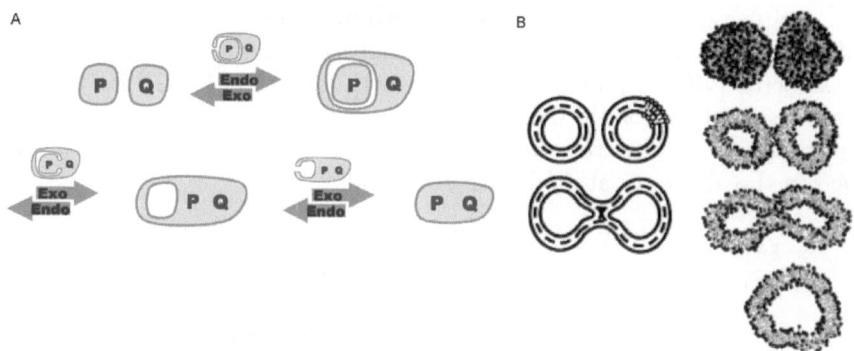

Fig. 1. PEP derivation of Mate [3] (A) and the latest model of how membrane fusion occurs [20] (B)

The succeeding discussions focus on the proposed PABM calculus as an encompassing calculus that conforms better with biologically observed phenomena.

3.1 Mate and Bud as Inverses of the Other

We consider Bud and Mate membrane actions as the primitives of this calculus, together with the Activate action, which controls their execution. Figure 2(top to bottom) shows a local deformation of the membrane separating the spaces labeled as P and $Q \circ R$ resulting from its interaction with Q. The increase in local curvature is then followed by the movement of Q towards the newly formed protrusion. On the fusion of the initial points of deformation, a new membrane-bound space containing Q is formed within P (Fig. 2, bottom), completing Bud. The reverse process, Mate, can be obtained using an opposite perspective. Here, the membrane separating Q from P merges with the membrane separating P and R (bottom to top). Colors are used to indicate tonality; in these processes, bitonality is conserved, as in PBC[5]. Since Bud and Mate are opposite operations, it would be possible to think of these as belonging to a single operation.

3.2 Projective Equivalence

Projective equivalence arose from the introduction of directed actions by Danos and Pradalier [5]. Briefly, projective equivalence refers to the idea that the nature of membrane interactions is such that one does not make a distinction between top and bottom, or in this case, outside and inside. Consequently, by using a simple point-of-view change (i.e. what one considers inside before, which is a bounded space, is now viewed as the outside, which is unbounded), one reverses the process. If one uses a pointed bitonal tree representation for the structure, the equivalence is simply a change in the *distinguished vertex* [5], which is a change in the root of the tree. One can then generalize phago and bud as a single budding action, and exo and mate as a single mate action.

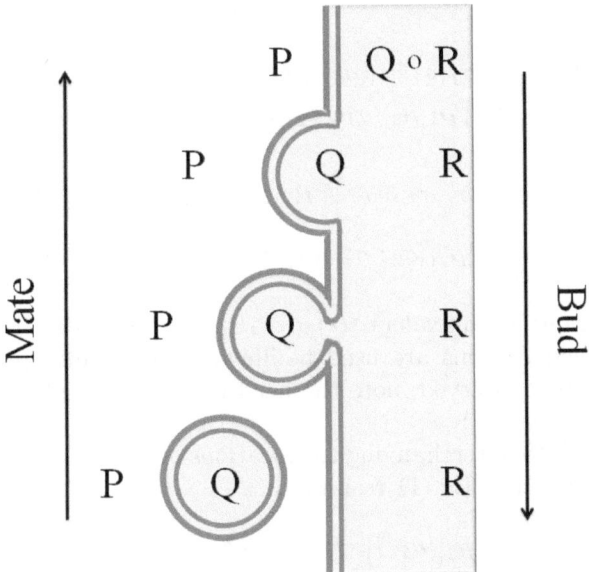

Fig. 2. Mate and Bud as inverse actions of the other. Bud is shown as a sequence from top to bottom while Mate as the reverse. Note that a distinction is not made between "inside" and "outside" spaces. A bilayer is used to illustrate directionality.

3.3 Basic Reduction Rules

The basic reduction rules of PABM are entirely based on Bud and Mate. Reduction rules are applied between interacting Brane domains, where the location of the activation signal with respect to the receiver (i.e. the directionality of the Action) determines if a Bud will be inward or outward, or if a Mate will be horizontal (i.e. membranes at the same level will merge) or vertical (i.e. the membrane of a content will merge with the membrane of its parent); this is conceptually similar to what has been done in PBC[5]. In the design of a Brane model, the directions at which Bud and Mate proceed are naturally integrated. The reduction rules of PABM are as follows, with "$\sim$" used for indicating projective equivalence[5].

– Bud:

$$P \circ [\, \boldsymbol{\rho}_1 | \langle \sigma_1 \,;\, \sigma_2, \tau_2.b x \rangle \,]\!(\, [\, \boldsymbol{\rho}_2 | \langle \sigma_4, \tau_4.!x \,;\, \sigma_3 \rangle \,]\!(\, Q' \,)\, \rangle \circ R \,)$$
$$\longrightarrow \;\; P \circ [\, \langle \sigma_1 \,;\, \sigma_2, \tau_2, \tau_4 \rangle \,]\!(\, [\, \boldsymbol{\rho}_2 | \langle \sigma_4 \,;\, \sigma_3 \rangle \,]\!(\, Q' \,)\,)\, \circ [\, \boldsymbol{\rho}_1 \,]\!(\, R \,) \qquad (9)$$

$$\sim$$

$$[\, \boldsymbol{\rho}_1^\dagger | \langle \sigma_2, \tau_2.b x \,;\, \sigma_1 \rangle \,]\!(\, P \,) \circ [\, \boldsymbol{\rho}_2 | \langle \sigma_4, \tau_4.!x \,;\, \sigma_3 \rangle \,]\!(\, Q' \,) \circ R$$
$$\longrightarrow \;\; [\, \boldsymbol{\rho}_1^\dagger \,]\!(\, P \circ [\, \langle \sigma_1 \,;\, \sigma_2, \tau_2, \tau_4 \rangle \,]\!(\, [\, \boldsymbol{\rho}_2 | \langle \sigma_4 \,;\, \sigma_3 \rangle \,]\!(\, Q' \,)\,)\,)\, \circ R \qquad (10)$$

– Mate:

$$P \circ [\, \rho_2 | \langle \sigma_1, \tau_1 .! x \,;\, \sigma_2 \rangle \,]\!(\, Q \,)\! \circ [\, \rho_1 | \langle \sigma_1', \tau_3 . mx \,;\, \sigma_2' \rangle \,]\!(\, R \,)$$

$$\longrightarrow \qquad P \circ [\, \rho_1 | \rho_2 | \langle \sigma_1 \,;\, \sigma_2 \rangle | \langle \sigma_1', \tau_1, \tau_3 \,;\, \sigma_2' \rangle \,]\!(\, Q \circ R \,) \qquad (11)$$

$$\sim$$

$$\left[\, \rho_1^{\dagger} | \langle \sigma_2' \,;\, \sigma_1', \tau_3 . mx \rangle \, \right]\!(\, P \circ [\, \rho_2 | \langle \sigma_1, \tau_1 .! x \,;\, \sigma_2 \rangle \,]\!(\, Q \,)\,)\,) \circ R$$

$$\longrightarrow \qquad \left[\, \rho_1^{\dagger} | \rho_2 | \langle \sigma_2 \,;\, \sigma_1 \rangle | \langle \sigma_1', \tau_1, \tau_3 \,;\, \sigma_2' \rangle \, \right]\!(\, P \,) \circ Q \circ R \qquad (12)$$

Note that Q in Mate is equivalent to $[\, \sigma_4, \tau_4 .! x \,]\!(\, \sigma_3 \,) \, Q'$ in Bud. Odd-even subscripts and primed Actions are used to illustrate bitonality preservation. For directionality to be conserved, note the need for the reversal of ρ_1 to $\rho_1^{\dagger}$ when the perspective is changed.

In the case of Mate, interchanging the locations of mx and $!x$ results in slightly different Brane domains. Eq. 11 results in:

$$P \circ [\, \rho_1 | \rho_2 | \langle \sigma_1, \tau_1, \tau_3 \,;\, \sigma_2 \rangle | \langle \sigma_1' \,;\, \sigma_2' \rangle \,]\!(\, Q \circ R \,); \qquad (13)$$

while Eq. 12 results in:

$$[\, \rho_1 | \rho_2 | \langle \sigma_1 \,;\, \sigma_2 \rangle | \langle \sigma_1', \tau_1, \tau_3 \,;\, \sigma_2' \rangle \,]\!(\, P \,) \circ Q \circ R. \qquad (14)$$

It is only in the absence of τ_1 and τ_3 that the location of mx and $!x$ does not result in different succeeding states.

3.4   Non-primitive Actions with Bud and Mate

As shown in Eqs. 9 to 12, congruence exists between an inward and outward Bud, and between a horizontal and vertical Mate. This is more clearly illustrated in Fig. 2, where one sees that a simple perspective shift makes the same Bud or Mate operation inward or outward, or vertical or horizontal. For instance, when one chooses P as the "inside", Q can be viewed as budding in towards P, or that the membrane containing Q is mating with the membrane separating P and R. As a result, PABM considers $a \in \mathcal{S}$, $a \notin \mathcal{S}_{min}$ as membrane operations congruent to either Bud or Mate operation or its specific cases.

Phago and Exo. Fig. 3(top) shows how Q is exocytosed from R or endocytosed into P via Mate and Bud, respectively. This is congruent to the Mate-Bud reactions in Fig. 2, with R as the outside and P as the inside. Phago is expressed as Bud in Eq. 10 and is congruent to the usual bud in Eq. 9.

Pino and Drip. Pino and Drip may be spontaneous or induced Bud actions (see Eq. 10), where a null System ($Q =$ null) is created inside or outside the bounded space P. The activation may also be induced by an appropriate Activate Action outside or inside P, or within-membrane activations (see §2.5). Pino and drip are obtained when $Q =$ null in Fig. 3.

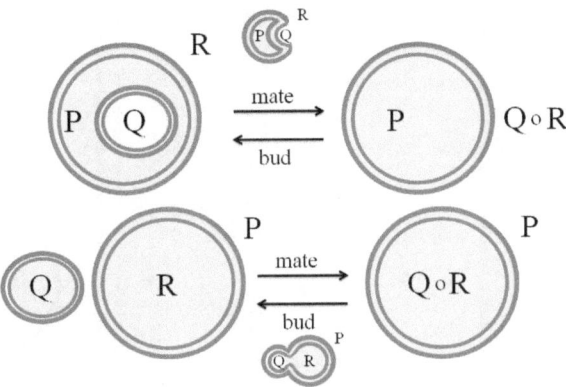

Fig. 3. Specific cases of Mate and Bud: (top, forward) Exo and (top, reverse) Phago; (bottom, forward) Cardelli's Mate and (bottom, reverse) Bud operations

3.5   Enhanced Membrane Dynamics

A fundamental difference of the proposed calculus from the Brane calculus of Cardelli is the dynamic nature of the reacting membranes. In Cardelli's version, the properties of budded membranes are specified as parameters to provide control; the same is true for the "endosomes" formed during Phago. In biological systems, however, the characteristics of the budded membrane are necessarily dependent on the *state* of the parent membrane at the time of budding or phagocytosis. Fig. 4 reflects this particular case of budding, when a sequential action is associated with the activation Action on an initiating membrane (σ_1, green). Note the incorporation of σ_1 in the budded membrane. Also note that only a portion of the membrane is budded out.

This dynamic property of the membrane generally implies that systems involved in a Mate followed by a Bud ($b.m$) will not evolve equivalently when Bud is performed before Mate ($m.b$). For instance, consider the following initial system:

$$Q_0 \equiv [\, \langle \sigma_4, \tau_4.mx_M \,;\, \sigma_3, \tau_3.bx_B \rangle \,]\!(\, [\, \langle \sigma_1, \tau_1.!x_B \,;\, \sigma_2 \rangle \,]\!(\, P_2 \,)\! \circ P_3 \,)$$
$$\circ [\, \langle \sigma_6, \tau_6.!x_M \,;\, \sigma_7 \rangle \,]\!(\, P_1 \,) \tag{15}$$

Depending on which operation occurs first, the system will evolve in two different ways. First, on performance of $b.m$, the system will evolve as:

$$Q_0 \;\xrightarrow{\;m\;}\; Q_0' \equiv [\langle \sigma_4, \tau_4, \tau_6 \,;\, \sigma_3, \tau_3.bx_B \rangle | \langle \sigma_6 \,;\, \sigma_7 \rangle]\!($$
$$P_1 \circ [\, \langle \sigma_1, \tau_1.!x_B \,;\, \sigma_2 \rangle \,]\!(\, P_2 \,)\! \circ P_3) \tag{16}$$

$$Q_0' \;\xrightarrow{\;b\;}\; Q_1' \equiv [\, \langle \sigma_6 \,;\, \sigma_7 \rangle \,]\!(\, P_1 \circ P_3 \,)$$
$$\circ [\, \langle \sigma_4, \tau_4, \tau_6 \,;\, \sigma_3, \tau_1, \tau_3 \rangle \,]\!(\, [\, \langle \sigma_1 \,;\, \sigma_2 \rangle \,]\!(\, P_2 \,)\,)\,) \tag{17}$$

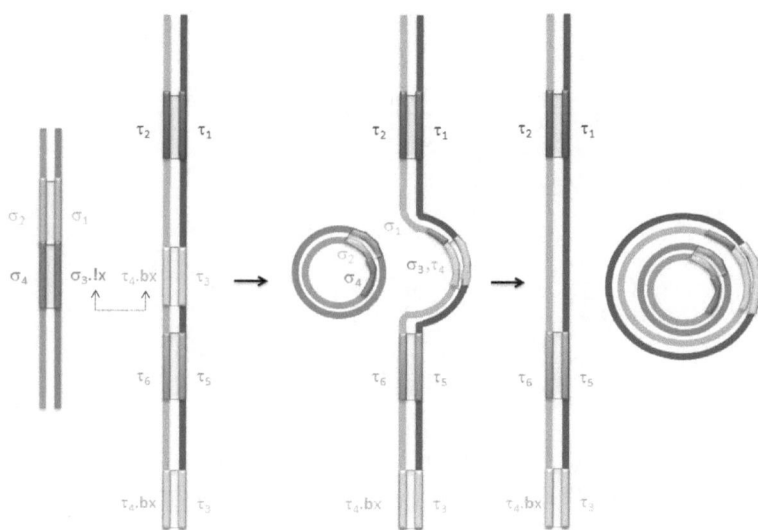

Fig. 4. Budding as a dynamic process. Note that a new action, σ_1, associated with the '!' is incorporated into the budded membrane. Domains are illustrated as line segments; only selected domains proximal to the activated action are budded out.

Second, with $m.b$, the system will evolve as:

$$Q_0 \xrightarrow{\quad b \quad} Q_0'' \equiv [\, - \,]\!(\, P_3 \,)) \circ [\, \langle \sigma_4, \tau_4.mx_{\mathrm{M}} \,;\, \sigma_3, \tau_3, \tau_1 \rangle \,]\!(\, [\, \langle \sigma_1 \,;\, \sigma_2 \rangle \,]\!(\, P_2 \,)) \,))$$
$$\circ [\, \langle \sigma_6, \tau_6.!x_{\mathrm{M}} \,;\, \sigma_7 \rangle \,]\!(\, P_1 \,)) \tag{18}$$

$$Q_0'' \xrightarrow{\quad m \quad} Q_1'' \equiv [\, - \,]\!(\, P_3 \,))$$
$$\circ [\, \langle \sigma_4, \tau_4, \tau_6 \,;\, \sigma_3, \tau_3, \tau_1 \rangle | \langle \sigma_6 \,;\, \sigma_7 \rangle \,]\!(\, P_1 \circ [\, \langle \sigma_1 \,;\, \sigma_2 \rangle \,]\!(\, P_2 \,)) \,)) \tag{19}$$

Clearly, $Q_1' \neq Q_1''$. This asymmetry example ($b.m \neq m.b$) is depicted in Fig. 5. The difference disappears when the Mate and Bud are placed in separate domains, $\langle \sigma_4, \tau_4.mx_{\mathrm{M}} \,;\, \sigma_3 \rangle$ and $\langle \sigma_4' \,;\, \sigma_3', \tau_3.bx_{\mathrm{B}} \rangle$.

3.6   Competition of Parallel Membrane Processes

Using the concept of Brane domains, competition of two or more parallel membrane processes can be easily modeled. Given the following system (longhand), the interactions of $!x$ is restricted to bx associated with σ_1 or that associated with σ_3. If it interacts with bx in $\langle \sigma_1 \,;\, bx \rangle$, then the system reduction will be in the form:

$$[\, \langle \sigma_1 \,;\, bx \rangle | \langle \sigma_3 \,;\, bx \rangle \,]\!(\, [\, \langle !x \,;\, \sigma_5 \rangle \,]\!(\, Q' \,)) \circ R \,))$$
$$\longrightarrow [\, \langle \sigma_1 \,;\, - \rangle \,]\!(\, [\, \langle - \,;\, \sigma_5 \rangle \,]\!(\, Q' \,)) \,)) \circ [\, \langle \sigma_3 \,;\, bx \rangle \,]\!(\, R \,)) \tag{20}$$

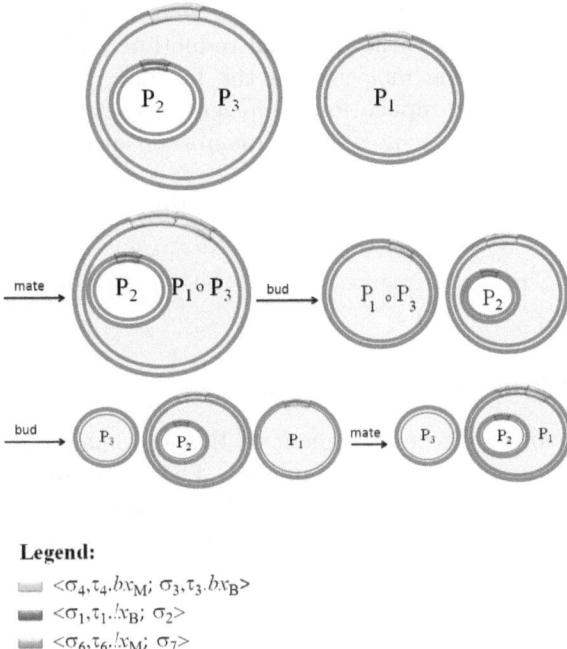

Legend:

$\langle\sigma_4,\tau_4.bx_M; \sigma_3,\tau_3.bx_B\rangle$

$\langle\sigma_1,\tau_1.!x_B; \sigma_2\rangle$

$\langle\sigma_6,\tau_6.!x_M; \sigma_7\rangle$

Fig. 5. Differences in final system states based on the order at which reactions occur. (top) Initial configuration; (middle) Mate then Bud; (bottom) Bud then Mate.

On the other hand, if it interacts with bx in $\langle\sigma_3 ; bx\rangle$ instead, then the system reduction will be as follows:

$$[\langle\sigma_1 ; bx\rangle|\langle\sigma_3 ; bx\rangle] \!| [[\langle !x ; \sigma_5\rangle]\!| Q') \circ R)$$
$$\longrightarrow [\langle\sigma_3 ; -\rangle]\!| [[\langle - ; \sigma_5\rangle]\!| Q')) \circ [\langle\sigma_1 ; bx\rangle]\!| R) \tag{21}$$

Competition can also be realized in lateral and cross-membrane processes (see Eq. 8).

3.7 Molecules as Systems

Molecules can either serve as ligands or receptors. In this proposed modification, Molecules can be modeled as null Systems containing Activators associated with Actions or other Activators. It may also be in the form of blocking functions, $\sigma.0x$, which could only be activated by $!x$. A molecule can be modeled as:

$$\text{Molecule} : [\langle\sigma_1, (n)(!x_{\text{name}}+!x_{\text{generic}}) ; \sigma_2, (m)(!x_{\text{name}}+!x_{\text{generic}})\rangle]\!| \diamond) \tag{22}$$

where τ may be a null Brane and the "name" could be the name of the molecule making the channel unique for the molecule and "generic" refers to the generic channel of the molecule. For example, "RNA" can be a generic channel having the name of the protein that it encodes for its specific name.

Together with cross-membrane interactions, molecule diffusion through a membrane can be modeled without atonal reduction. This is illustrated in the reduction below, where the molecule on the left-hand side enters the system P. Note the change in the replication coefficient, reflecting the reduction of the active sites in both the molecule and the membrane surrounding P.

$$\left[\langle (!x)^{(n_1)} \,;\, (bx.!y)^{(n_1)} \rangle \right]\!(\!\lozenge\!)\circ \left[(n_2)\langle my\,;\, -\rangle \right]\!(\!\left.P\right)\!$$
$$\longrightarrow \left[(n_2-1)\langle my\,;\,-\rangle | \langle (bx.!y)^{(n_1-1)}.bx\,;\,(!x)^{(n_1-1)}.!x\rangle \right]\!(\!\left.P\right)\!$$
$$\longrightarrow \left[(n_2-1)\langle my\,;\,-\rangle \right]\!(\!\left. P\circ \left[\langle (!x)^{(n_1-1)}\,;\,(bx.!y)^{(n_1-1)}\rangle \right]\!(\!\lozenge\!) \right) \quad (23)$$

3.8   Mass and Energy Conservation

Since budding involves direct movement of Brane domains, mass (represented by an Action) conservation is also simulated. The "consumption" of an Action after reduction can be seen as the usage of the available energy used for and/or transfer of mass during the process, i.e. transformation of the structure into new ones.

4   PABM as an Extension of Existing Brane Calculi

Equivalent expressions for the multiple association of the activation action using Cardelli's original notations can be derived. Suppose there are three systems that can interact via a generic action a, with coaction $\bar{a}$. Multiple association can be realized with:

$$a!x \,(\!\left. Q_0 \right)\circ \bar{a}?x \,(\!\left. P_1 \right)\circ \bar{a}?x \,(\!\left. P_2 \right), \quad (24)$$

where system Q_0 can proceed with the action $(a \leftrightarrow \bar{a})$ on both systems P_1 and P_2 through the same channel $!x \rightarrow ?x$. It is possible to eliminate the use of coactions $(\bar{a})$ through the following representation:

$$!x \,(\!\left. Q_0 \right)\circ a?x \,(\!\left. P_1 \right)\circ a?x \,(\!\left. P_2 \right) \quad (25)$$

with $!x$ possibly activating either P_1 or P_2 via $a?x$. This minor notation change is immediately compatible with Pi calculus, and would require a minor code translation for recognition by the Stochastic Pi Machine (SPiM) [21].

However, the proposed calculus also involves the removal of the sender-receiver pairing (viz. $!x \rightarrow ?x$), apart from the action-coaction pairing. Moreover, in Eq. 25, the notation is asymmetric, with the activator $!x$ having a different form from $a?x$. It is possible to use $\alpha!x$ as a universal activator to conserve symmetry, but α will be underutilized. The use of a single "sender" is proposed for all the other actions in the form $!x$, while *simultaneously* making the notation symmetric with the use of the same form (ax, see Table 1). Both conceptually and implementationwise, these major differences could be seen as improvements over the current representation.

Hence, the multiple association expressed as (24) would be written in PABM as:

$$[\,\langle !x\,;\,-\rangle\,]\!(\!(\,Q_0\,)\!)\circ[\,\langle ax\,;\,-\rangle\,]\!(\!(\,P_1\,)\!)\circ[\,\langle ax\,;\,-\rangle\,]\!(\!(\,P_2\,)\!) \tag{26}$$

with x as the channel; "!" now belongs to the same class as $a \neq$! (Table 1). The choice of the symbol "!" for the activate action is directly inspired by the Pi calculus notation.

As indicated previously, the other major departure from Cardelli's Brane calculus is the utilization of Brane domains in dynamic membranes to eliminate the use of parameters in phagocytosis and budding. With these domains, a Brane in PBC becomes a special case when a Brane in PABM is homogenous (i.e. is comprised of a single Brane domain). For purposes of comparison, the reduction rules of the original calculus [taken from [8]] are shown in Fig. 6.

(phago)	$\mho_n.\sigma\|\sigma_0(\!(P)\!) \circ \mho_n^{\perp}(\rho).\tau\|\tau_0(\!(Q)\!) \rightarrow$ $\tau\|\tau_0(\!(\rho(\!(\sigma\|\sigma_0(\!(P)\!))\!)) \circ Q)\!)$
(exo)	$\mho_n^{\perp}.\tau\|\tau_0(\!(\mho_n.\sigma\|\sigma_0(\!(P)\!) \circ Q)\!) \rightarrow$ $P \circ \sigma\|\sigma_0\|\tau\|\tau_0(\!(Q)\!)$
(pino)	$\circledcirc(\rho).\sigma\|\sigma_0(\!(P)\!) \rightarrow \sigma\|\sigma_0(\!(\rho(\!(\,)\!) \circ P)\!)$
(mate)	$mate_n.\sigma\|\sigma_0(\!(P)\!) \circ mate_n^{\perp}.\tau\|\tau_0(\!(Q)\!) \rightarrow$ $\sigma\|\sigma_0\|\tau\|\tau_0(\!(P \circ Q)\!)$
(bud)	$bud_n^{\perp}(\rho).\tau\|\tau_0(\!(bud_n.\sigma\|\sigma_0(\!(P)\!) \circ Q)\!) \rightarrow$ $\rho(\!(\sigma\|\sigma_0(\!(P)\!))\!) \circ \tau\|\tau_0(\!(Q)\!)$
(drip)	$drip(\rho).\sigma\|\sigma_0(\!(P)\!) \rightarrow \rho(\!(\,)\!) \circ \sigma\|\sigma_0(\!(P)\!)$

Fig. 6. Cardelli's Brane calculus reduction rules taken from [8]

The realization of non-primitive actions that were illustrated utilizes the same concepts as in the projective equivalence of Danos and Pradalier [5], with the exception that no arguments are explicitly used for the Bud action. Furthermore, the simplicity of the current basis and reduction makes the calculus closer to actual biological membrane operations. Finally, PBC becomes a subset of the proposed calculus since PBC Branes can be simulated using homogenous PABM Branes.

5   Summary and Outlook

We end this paper with brief discussions on a potential application of the calculus, as well as a strategy for its possible implementation.

5.1   Application Example: Viral Infection

Influenza A causes highly contagious respiratory infections in humans that range in severity from acute to lethal. New strains arise annually, which lead to 250,000 to 500,000 deaths worldwide[22, 23, 24]. It is particularly interesting for biologists because of its ability to evolve very quickly, a trait that makes the development of an efficient vaccine against it particularly challenging[25, 26]. To date, a number of qualitative studies have been performed to investigate its life cycle, but most involved separate analyses of steps in the infection process [24, 26].

One of the recent most extensive quantitative models of influenza A in cell culture is that by Sidorenko and Reichl[24]. It consists of 49 ordinary differential equations (ODEs) that involve the use of additional parameters to approximate the movement of viruses and its components across cell compartments. The main results obtained from the model include the identification of factors that limit the growth rate of viral progeny; these results are particularly useful in molecular engineering, where engineered viruses are created for vaccine production[24]. Nevertheless, it is clear that much is still not known about the influenza A life cycle, primarily owing to the complexity of the virus. Some details, for instance, that have not been included in the Sidorenko-Reichl model include the following:

1. Distinction between each of the eight strands of genetic material (vRNA), complexed with three proteins (collectively known as vRNP), throughout the replication cycle
2. Distribution of 11 protein-coding genes across the eight vRNAs
3. Indirect genetic material replication (vRNA → cRNA → vRNA), with the intermediate cRNA being able to interact with the same proteins that vRNA interacts with
4. Requirement for precise viral assembly

Accordingly, several key issues remain unanswered:

1. time it takes to assemble vRNPs
2. ratio of infective to non-infective viruses
3. instances of 'infectivity recovery' in the event that two complementary non-infective viruses enter a cell

Since compartments can be naturally represented in Brane, its use for modeling the influenza A life cycle is probably an elegant, quantitative alternative that would allow the inclusion of details such as those enumerated previously. Fig. 7 is a general illustration for the possible usage of PABM to model the influenza infection cycle. This particular model is an interesting application for Brane calculus on account of its scale. Note that all operations used for modeling the system are restricted to budding and mating, including the simulation of the bind-and-release action in the nucleus. The position of activation signals are not explicitly indicated in the figure, but could be deduced from the illustrated processes.

In addition to these, it would also be possible to include details that are known to affect influenza infectivity, as well as efficiency[27]:

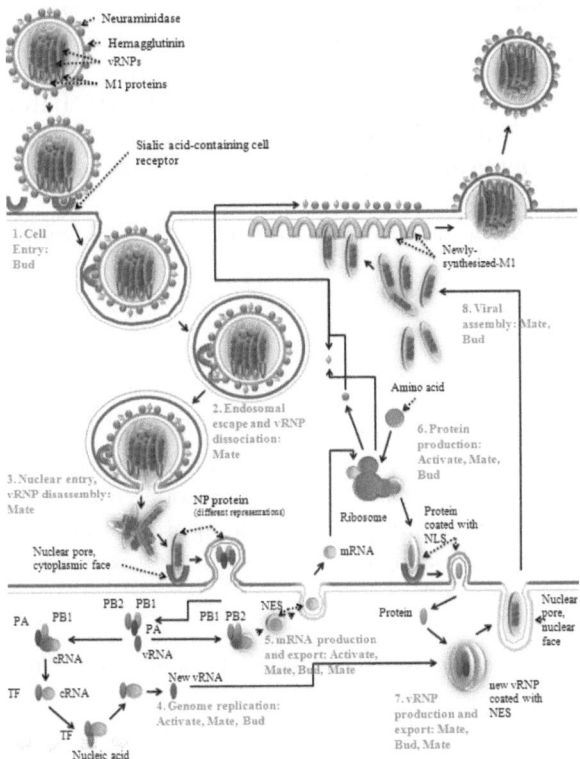

Fig. 7. Influenza A infection cycle model being implemented using PABM. All operations are performed with Bud and Mate, including nuclear import and export through the use of the NP protein, nuclear localization sequences (NLS) or nuclear export sequences (NES). All processes are conformant with the actual events in influenza infection.

1. cleavage efficiency of HA
2. distinction between transcriptionally active and inactive vRNPs

5.2   Implementation and Compatibility with SPiM

Previous efforts have been made to implement Brane calculus[28, 29]. These implement calculi based on the set $\mathcal{S}$ (Eq. 1) and were found useful for studying events having the same scale as the Semliki forest virus life cycle, which was used as the illustrative example in [3]. Nevertheless, these are not powerful enough to handle models having the scale of the influenza A life cycle. It is consequently of interest to develop an implementation that is both scalable and robust.

The stochastic pi machine (SPiM) was developed by Andrew Phillips, and uses a simulation algorithm for stochastic pi calculus that is particularly suited for simulating biological systems involving a large number of molecules. This simulation algorithm makes the execution cost dependent on the number of species, rather than the actual number of molecules, unlike in direct implementations of the Gillespie algorithm [21]. SPiM has been used on a number of occasions for a variety of biological problems [21, 30, 31]. Lately, the algorithm in SPiM has been extended to include compartment-based computation, using the Bioambients formalism [7]. SPiM also has a graphical interface, which significantly improves its ease of use.

PABM should be compatible with SPiM using the following equivalences:

$$!x \equiv (m!x + b!x + 0!x) \tag{27}$$

$$ax \equiv \bar{a}?x \tag{28}$$

$$\sigma.!x \equiv a!x(\sigma) \tag{29}$$

where $a = m, b, 0$ and $\bar{a}$ is the corresponding coaction. Encoding more specific stochastic pi calculus constructs would only require the use of very specific channel names. For PABM to be implemented on top of SPiM, compartments, Brane domains, and action directionality have to be appropriately represented. A separate implementation approach that focuses on the rewriting rules of PABM is also currently being explored.

Acknowledgments. We wish to thank Luca Cardelli and Andrew Phillips for helpful discussions.

References

[1] Păun, G.: Introduction to membrane computing. In: Applications of Membrane Computing, pp. 1–42 (2006)

[2] Regev, A., Shapiro, E.: The π-calculus as an abstraction for biomolecular systems. In: Modelling in Molecular Biology (2004)

[3] Cardelli, L.: Brane calculi: interactions of biological membranes. In: Danos, V., Schachter, V. (eds.) CMSB 2004. LNCS (LNBI), vol. 3082, pp. 257–278. Springer, Heidelberg (2005)

[4] Priami, C., Quaglia, P.: Beta binders for biological interactions. In: Danos, V., Schachter, V. (eds.) CMSB 2004. LNCS (LNBI), vol. 3082, pp. 20–33. Springer, Heidelberg (2005)

[5] Danos, V., Pradalier, S.: Projective brane calculus. In: Danos, V., Schachter, V. (eds.) CMSB 2004. LNCS (LNBI), vol. 3082, pp. 134–148. Springer, Heidelberg (2005)

[6] Guerriero, M.L., Priami, C., Romanel, A.: Modeling Static Biological Compartments with Beta-binders. In: Anai, H., Horimoto, K., Kutsia, T. (eds.) Ab 2007. LNCS, vol. 4545, pp. 247–261. Springer, Heidelberg (2007)

[7] Phillips, A., Cardelli, L.: Efficient, correct simulation of biological processes in the stochastic pi-calculus. In: Calder, M., Gilmore, S. (eds.) CMSB 2007. LNCS (LNBI), vol. 4695, pp. 184–199. Springer, Heidelberg (2007)

[8] Busi, N., Zandron, C.: Modelling and analysis of biological processes by (mem)brane calculi and systems. In: Proceedings of the 2006 Winter Simulation Conference, pp. 1646–1655 (2006)

 [9] Guglielmo, G.D., Drake, P., Baass, P., Authier, F., Posner, B., Bergeron, J.: Insulin receptor internalization and signalling. Mol. Cell. Biochem. 182, 59–63 (1998)

[10] Hsu, S., Bonvin, A.: Atomic insight into the CD4 binding-induced conformational changes in HIV-1 gp120. Proteins: structure, function and bioinformatics 3, 582–593 (2004)

[11] Keskin, O.: Binding induced conformational changes of proteins correlate with their intrinsic fluctuations: a case study of antibodies. BMC Structural Biology 7, 31 (2007)

[12] Alberts, B., Bray, D., Lewis, J., Raff, M., Roberts, K., Watson, J.: Molecular Biology of the Cell, New York (2002)

[13] David, M., Asprer, J., Ibana, J., Concepcion, G., Padlan, E.: A study of the structural correlates of affinity maturation: antibody affinity as a function of chemical interactions, structural plasticity and stability. Mol. Immunol. 44, 1342–1351 (2006)

[14] Dimitrov, D.: Virus entry: molecular mechanisms and biomedical applications. Nature Reviews Microbiology 2, 109–122 (2004)

[15] Kim, W., Hudson, B., Moser, B., Guo, J., Rong, L., Yu, L., Qu, W., Lalla, E., Lerner, S., Chen, Y., Yan, S.D., D'Agati, V., Naka, Y., Ramasamy, R., Herold, K., Yan, S., Schmidt, A.: Receptor for advanced glycation end products and its ligands: A journey from the complications of diabetes to its pathogenesis. Annals of the New York Academy of Sciences 1043, 553–561 (2006)

[16] Thulke, S., Radonic, A., Nitsche, A., Siegert, W.: Quantitative expression analysis of HHV-6 cell receptor CD46 on cells of human cord blood, peripheral blood and G-CSF mobilised leukapheresis cells. Virology Journal 3, 77–81 (2006)

[17] Simons, K., Vaz, W.L.C.: Model Systems, Lipid Rafts, and Cell Membranes. Annual Review of Biophysics and Biomolecular Structure 33(1) (June 2004)

[18] Sutton, R.B., Fasshauer, D., Jahn, R., Brunger, A.T.: Crystal structure of a SNARE complex involved in synaptic exocytosis at 2.4 Å resolution. Nature 395(6700), 347–353 (1998)

[19] Chiu, D., Wilson, C., Karlsson, A., Danielsson, A., Lunqvist, A., Stroemberg, A., Ryttsen, F., Davidson, M., Nordholm, S., Orwar, O., Zare, R.: Manipulating the biochemical nanoenvironment around single molecules contained within vesicles. Chem. Phys. 247, 133–139 (1999)

[20] Jahn, R., Grubmüller, H.: Membrane fusion. Current Opinion in Cell Biology 14, 488–495 (2002)

[21] Phillips, A., Cardelli, L.: A Correct Abstract Machine for the Stochastic Picalculus. In: Concurrent Models in Molecular Biology (2004)

[22] Poland, G.A., Tosh, P., Jacobson, R.M.: Requiring influenza vaccination for health care workers: seven truths we must accept. Vaccine 23(17-18), 2251–2255 (2005); Vaccines and Immunisation. Based on the Fourth World Congress on Vaccines and Immunisation

[23] Baccam, P., Beauchemin, C., Macken, C.A., Hayden, F.G., Perelson, A.S.: Kinetics of Influenza A Virus Infection in Humans. J. Virol. 80(15), 7590–7599 (2006)

[24] Sidorenko, Y., Reichl, U.: Structured model of influenza virus replication in mdck cells. Biotechnology and bioengineering 88, 1–14 (2004)

[25] Bardiya, N., Bae, J.: Influenza vaccines: recent advances in production technologies. Applied Microbiology and Biotechnology 67(3), 299–305 (2005)

[26] Genzel, Y., Schulze-Horsel, J., Möhler, L., Sidorenko, Y., Reichl, U.: Influenza vaccines –challenges in mammalian cell culture technology. Cell Technology for Cell Products, 503–508 (2007)

[27] Nayak, D.P., Hui, E.K.-W., Barman, S.: Assembly and budding of influenza virus. Virus Research 106, 147–165 (2004)

[28] de Ronde, J.J., Ndjehan, C.P.: Modelling Networks and Pathways in Systems Biology. Technical report, CA545 Practicum, School of Computing, Dublin City University (2005/2006)

[29] David, M.P.C.: BCD: Design and implementation of a stochastic brane machine. Master's thesis, Department of Computer Science, University of the Philippines, Diliman, Quezon City (2008)

[30] Segata, N., Blanzieri, E., Priami, C.: Stochastic π-calculus modelling of multisite phosphorylation based signaling: in silico analysis of the Pho4 transcription factor and the PHO pathway in Saccharomyces cerevisiae. Technical report, Center for Computational and Systems Biology, The Microsoft Research – University of Trento (2007)

[31] Yap, J.M.: A Pi-Calculus Model of the CD95 Receptor Medicated Pathway of Apoptosis. Philippine Information Techonology Journal 1(1) (2008)

[32] Zhang, B.T., Hikawa, N., Horie, H., Takenaka, T.: Mitogen induced proliferation of isolated adult mouse Schwann cells. J. Neurosci. Res., 648–654 (1995)

Accepting Networks of Non-inserting Evolutionary Processors

Jürgen Dassow[1] and Victor Mitrana[2],[*]

[1] Faculty of Computer Science, University of Magdeburg
P.O.Box 4120, 39016 Magdeburg, Germany
`dassow@iws.cs.uni-magdeburg.de`
[2] Faculty of Mathematics, University of Bucharest
Str. Academiei 14, 70109 Bucharest, Romania
Department of Information Systems and Computation
Technical University of Valencia,
Camino de Vera s/n. 46022 Valencia, Spain
`mitrana@fmi.unibuc.ro`

Abstract. In this paper we consider four variants of accepting networks of evolutionary processors with in-place computations, that is the length of every word in every node at any step in the computation is bounded by the length of the input word. These devices are called here accepting networks of non-inserting evolutionary processors (ANNIEP shortly). The variants differ in two respects: filters that are used to control the exchange of information, i.e., we use random context conditions and regular languages as filters, and the way of accepting the input word, i.e., at least one output node or all output nodes are nonempty at some moment in the computation. The computational power of these devices is investigated. In the case of filters defined by regular languages, both variants lead to the class of context-sensitive languages. If random context conditions are used for defining filters, all linear context-free languages and some non-semilinear (even over the one-letter alphabet) can be accepted with both variants. Moreover, some closure properties of the classes of languages ANNIEPs with random context filters are also given.

1 Introduction

The origin of networks of evolutionary processors (NEP for short) is a basic architecture for parallel and distributed symbolic processing, related to the Connection Machine [9] as well as the Logic Flow paradigm [7], which consists of several processors, each of them being placed in a node of a virtual complete graph, which are able to handle data associated with the respective node. All the nodes send simultaneously their data and the receiving nodes handle also simultaneously all the arriving messages, according to some strategies, see, e.g., [8,9]. Similar ideas may be met in other bio-inspired models like *membrane systems* [16], *evolutionary systems* [4], or models from Distributed Computing area

---

[*] Work supported by the Alexander von Humboldt Foundation.

C. Priami et al. (Eds.): Trans. on Comput. Syst. Biol. XI, LNBI 5750, pp. 187–199, 2009.

like *parallel communicating grammar systems* [15], *networks of parallel language processors* [3].

In a series of papers (see [14] for an early survey) one considers that each node may be viewed as a cell having genetic information encoded in DNA sequences which may evolve by local evolutionary events, that is point mutations. Each node is specialized just for one of these evolutionary operations. Furthermore, the data in each node are organized in the form of multisets of words (each word appears in an arbitrarily large number of copies), and all the copies are processed in parallel such that all the possible events that can take place do actually take place. Obviously, the computational process just described is not exactly an evolutionary process in the Darwinian sense. But the rewriting operations we have considered might be interpreted as mutations and the filtering process might be viewed as a selection process. Recombination is missing but it was asserted that evolutionary and functional relationships between genes can be captured by taking only local mutations into consideration [17].

In [13] one presents a characterization of the complexity class **NP** based on accepting networks of evolutionary processors (ANEP for short). This characterization is extended in [12] to **PSPACE** and **P**. The work [10] discusses how ANEPs can be considered as problem solvers. In [11], one shows that every recursively enumerable language can be accepted by an ANEP with 24 nodes. More precisely, one proposes a method for constructing, for every **NP**-language, an ANEP of size 24 deciding that language in polynomial time. While the number of nodes of this ANEP does not depend on the language, the other parameters of the network (rules, symbols, filters) depend on it.

From a computational point of view it is of interest to consider ANEPs with in-place computations, that is the length of every word in every node at any step in the computation is bounded by the length of the input word. This is our main reason to consider here some variants of networks of evolutionary processors without insertion nodes, called here accepting networks of non-inserting evolutionary processors, ANNIEP shortly. The differences between the variants of ANNIEPs consist in the filters and in the way of accepting the input word.

Besides accepting networks of evolutionary processors, generating networks of such processors have been investigated (see [2], [5], [14]). In the paper [6], the generative power of networks where only two types of point mutations are allowed for the nodes have been investigated. In case of non-inserting processors one only gets the set of all finite languages. This paper presents the counterpart for accepting networks, where the situation is completely different.

We study the computational power of accepting networks of non-inserting processors. In the case of filters defined by regular languages, both variants of accepting lead to the same class of languages, namely the class of context-sensitive languages. If random context conditions are used for defining filters, all linear context-free languages and some non-semilinear (even over the one-letter alphabet) can be accepted with both variants. Therefore the power of accepting networks is much greater than that of generating networks (both with non-inserting processors). Moreover, some closure properties of the classes of languages accepted by ANNIEPs with filters defined by random context conditions are also discussed.

2   Some Notations and Definitions

Throughout the paper we assume that the reader is familiar with the basic notions of the theory of formal languages. We here only recall some notation and notions as they are used in the paper.

An *alphabet* is a finite and nonempty set of symbols. The cardinality of a finite set A is written $card(A)$. Any sequence of symbols from an alphabet V is called *word* over V. The set of all words over V is denoted by V^* and the empty word is denoted by ε. A language over V is a subset of V^*.

The length of a word x is denoted by $|x|$ while $alph(x)$ denotes the (with respect to inclusion) minimal alphabet W such that $x \in W^*$. A morphism $h : V^* \longrightarrow U^*$ is said to be *literal* if $|h(a)| = 1$ for all $a \in V$; it is *weak literal* if $|h(a)| \leq 1$ for all $a \in V$. In other words a (weak) literal morphism is called (weak) coding.

Let V be an alphabet. We say that a rule $a \to b$, with $a, b \in V \cup \{\varepsilon\}$ is a *substitution rule* if both a and b are not ε; it is a *deletion rule* if $a \neq \varepsilon$ and $b = \varepsilon$. The set of all substitution and deletion rules over an alphabet V are denoted by Sub_V and Del_V, respectively. Given a rule σ as above and a word $w \in V^*$, we define the following *actions* of σ on w:

- If $\sigma \equiv a \to b \in Sub_V$, then $\sigma^*(w) = \begin{cases} \{ubv : \exists u, v \in V^* \ (w = uav)\}, \\ \{w\}, \ \text{otherwise} \end{cases}$

$$\sigma^r(w) = \begin{cases} \{ub : w = ua\}, \\ \{w\}, \ \text{otherwise} \end{cases} \qquad \sigma^l(w) = \begin{cases} \{bv : w = av\}, \\ \{w\}, \ \text{otherwise} \end{cases}$$

- If $\sigma \equiv a \to \varepsilon \in Del_V$, then $\sigma^*(w) = \begin{cases} \{uv : \exists u, v \in V^* \ (w = uav)\}, \\ \{w\}, \ \text{otherwise} \end{cases}$

$$\sigma^r(w) = \begin{cases} \{u : w = ua\}, \\ \{w\}, \ \text{otherwise} \end{cases} \qquad \sigma^l(w) = \begin{cases} \{v : w = av\}, \\ \{w\}, \ \text{otherwise} \end{cases}$$

The action $\alpha \in \{*, l, r\}$ expresses the way of applying a substitution or deletion rule to a word, namely at any position ($\alpha = *$), in the left ($\alpha = l$), or in the right ($\alpha = r$) end of the word, respectively. For every rule σ, any action $\alpha \in \{*, l, r\}$, and any $L \subseteq V^*$, we define the α-*action of σ on* L by $\sigma^\alpha(L) = \bigcup_{w \in L} \sigma^\alpha(w)$.

Given a finite set of rules M, we define the α-*action of M on* the word w and the language L by:

$$M^\alpha(w) = \bigcup_{\sigma \in M} \sigma^\alpha(w) \ \text{ and } \ M^\alpha(L) = \bigcup_{w \in L} M^\alpha(w),$$

respectively.

If $\theta V^* \longrightarrow \{0,1\}$ is a predicate and $L \subseteq V^*$, we write:

$$\theta(L) = L \cap \theta^{-1}(1).$$

We are interested in some special predicates. For two disjoint subsets P and F of an alphabet V, a regular set R over V, and a word x over V, we define the predicates

$$\theta^{s,P,F}(x) = 1 \text{ if and only if } P \subseteq alph(x) \text{ and } F \cap alph(x) = \emptyset,$$
$$\theta^{w,P,F}(x) = 1 \text{ if and only if } alph(x) \cap P \neq \emptyset \text{ and } F \cap alph(x) = \emptyset,$$
$$\theta^R(x) = 1 \text{ if and only if } x \in R.$$

The first two predicates are based on *random context conditions* defined by the two sets P (*permitting contexts/symbols*) and F (*forbidding contexts/symbols*). Informally, the first condition requires (s stands for strong) that all permitting symbols are and no forbidding symbol is present in x, while the second (w stands for weak) is a weaker variant such that at least one permitting symbol appears in x but still no forbidding symbol is present in x. We call these two predicates random context predicates. The third predicate asks for membership in a regular set, and is called a regular predicate.

A *non-inserting evolutionary processor over V* is a tuple (M, φ, ψ), where:

- M is a set of either substitution or deletion rules over the alphabet V; formally, $M \subseteq Sub_V$ or $M \subseteq Del_V$. The set M represents the set of evolutionary rules of the processor. As one can see, a processor is "specialized" in one evolutionary operation, only.
- φ is the *input predicate*, while ψ is the *output predicate* of the processor. Informally, these two predicates work as filters. A word w can enter or leave the processor, if it satisfies the predicate φ or ψ, respectively.

We are interested in two types of processors, random context non-inserting evolutionary processor over V (or short rcNIEP$_V$) and regular non-inserting evolutionary processor over V (or short regNIEP$_V$). These processors are defined by the requirement that,

- for an rcNIEP$_V$, both predicates are of the form $\theta^{s,P,F}$ or of the form $\theta^{w,P,F}$ for certain subsets P and F of V,
- for an regNIEP$_V$, both predicates are of the form θ^R for some regular set $R \subseteq V^*$.

We want to stress from the very beginning that the evolutionary processor we discuss here is a mathematical object only and the biological hints presented in the introduction are intended to explain in an informal way how some biological phenomena are *sources of inspiration* for our mathematical computing model. We denote the set of non-inserting evolutionary processors over V by $NIEP_V$.

An *accepting network of non-inserting evolutionary processors* (ANNIEP for short) is a 8-tuple $\Gamma = (V, U, G, N, \alpha, x_{In}, Out)$, where:

- V and U are the input and network alphabet, respectively, satisfying $V \subseteq U$.
- $G = (X_G, E_G)$ is an undirected graph without loops with the set of vertices X_G and the set of edges E_G. G is called the *underlying graph* of the network.
- $N : X_G \longrightarrow NIEP_V$ is a mapping which associates with each node $x \in X_G$ the evolutionary processor $N(x) = (M_x, \varphi_x, \psi_x)$.
- $\alpha : X_G \longrightarrow \{*, l, r\}$ is a mapping which associates with each node a type of action; $\alpha(x)$ gives the action mode of the rules of node x on the words existing in that node.
- $x_{In} \in X_G$ is the *input* node of Γ.
- $Out \subset X_G$ is the set of *output* nodes of Γ.

An ANNIEP is a random context ANNIEP or regular ANNIEP if all its non-inserting evolutionary processors are random context or regular non-inserting evolutionary processors, respectively.

We say that $card(X_G)$ is the size of Γ. A *configuration* of an ANNIEP Γ as above is a mapping $C : X_G \longrightarrow 2_f^{V^*}$ which associates a finite set of words with every node of the graph. A configuration may be understood as the sets of words which are present in any node (or in the associated prozessor) at a given moment. Given a word $z \in V^*$, the initial configuration of Γ on z is defined by $C_0^{(z)}(x_{In}) = \{z\}$ and $C_0^{(z)}(x) = \emptyset$ for all $x \in X_G \setminus \{x_{In}\}$.

A configuration can change either by an *evolutionary step* or by a *communication step*. When changing by an evolutionary step, each component $C(x)$ of the configuration C is changed in accordance with the set of evolutionary rules M_x associated with the node x and the way of applying these rules $\alpha(x)$. Formally, we say that the configuration C' is obtained in *one evolutionary step* from the configuration C, written as $C \Longrightarrow C'$, iff

$$C'(x) = M_x^{\alpha(x)}(C(x)) \text{ for all } x \in X_G.$$

When changing by a communication step, each node processor $x \in X_G$ sends one copy of each word it has, which is able to pass the output filter of x, to all the node processors connected to x and receives all the words sent by any node processor connected with x provided that they can pass its input filter.

Formally, we say that the configuration C' is obtained in *one communication step* from configuration C, written as $C \vdash C'$, iff

$$C'(x) = (C(x) - \psi_x(C(x))) \cup \bigcup_{\{x,y\} \in E_G} (\psi_y(C(y)) \cap \varphi_x(C(y))) \text{ for all } x \in X_G.$$

Note that words that cannot pass the output filter of a node remain in that node and can be further modified in the subsequent evolutionary steps, while words that can pass the output filter of a node but cannot pass the input filter of any node are lost.

Let Γ be an ANNIEP, the computation of Γ on the input word $z \in V^*$ is a sequence of configurations $C_0^{(z)}, C_1^{(z)}, C_2^{(z)}, \ldots$, where $C_0^{(z)}$ is the initial configuration of Γ on z, $C_{2i}^{(z)} \Longrightarrow C_{2i+1}^{(z)}$ and $C_{2i+1}^{(z)} \vdash C_{2i+2}^{(z)}$, for all $i \geq 0$. Note that the configurations are changed by alternative steps. By the previous definitions, each configuration $C_i^{(z)}$ is uniquely determined by the configuration $C_{i-1}^{(z)}$. A computation *halts* (and it is said to be *weak (strong) halting*) if one of the following two conditions holds:

(i) There exists a configuration in which the set of words existing in at least one output node (all output nodes) is non-empty. In this case, the computation is said to be a weak (strong) *accepting computation*.

(ii) There exist two identical configurations obtained either in consecutive evolutionary steps or in consecutive communication steps.

The *language weakly (strongly) accepted* by Γ are defined as:

$$L_{wa}(\Gamma) = \{z \in V^* \mid \text{the computation of } \Gamma \text{ on } z \text{ is a weak accepting one}\}$$
$$L_{sa}(\Gamma) = \{z \in V^* \mid \text{the computation of } \Gamma \text{ on } z \text{ is a strong accepting one}\}.$$

In the theory of networks some types of underlying graphs are common like *rings, stars, grids*, etc. Networks of evolutionary words processors, seen as language generating or accepting devices, with underlying graphs having these special forms have been considered in several papers, see, e.g., [14] for an early survey. We focus here on *complete* ANNIEPs i.e., ANNIEPs having a complete underlying graph. Therefore, in what follows we replace the graph G in the definition of an ANNIEP by the set of its nodes usually denoted by χ.

Moreover, we present an evolutionary network by its nodes x and the parameters corresponding to x, where instead of $\varphi^{\beta,PI_x,FI_x}$ and ψ^{β,PO_x,FO_x}, in case of random context processors, and instead of φ^{R_x} and $\varphi^{R'_x}$ for regular processors, we only mention PI_x, FI_x, PO_x, FO_x, β and R_x, R'_x, β, respectively.

For $x \in \{wa, sa\}$ and $y \in \{rc, reg\}$, by $\mathcal{L}_x(yANNIEP)$ we denote the set of all languages which can be accepted by yANNIEPS.

The following two notions will be very useful in the sequel. If h is a one-to-one mapping from U to W and $\Gamma = (V, U, \chi, N, \alpha, x_{In}, Out)$ is an ANNIEP, then we denote by Γ_h the ANNIEP $\Gamma_h = (h(V), h(U), \chi, h(N), \alpha, x_{In}, Out)$, where by $h(N)$ we mean $h(N)(x) = (h(M_x), \varphi^{\beta,h(PI_x),h(FI_x)}, \psi^{\beta,h(PO_x),h(FO_x)})$ for every $x \in \chi$, provided that $N(x) = (M_x, \varphi^{\beta,PI_x,FI_x}, \psi^{\beta,PO_x,FO_x})$. Further, $h(a \to b) = h(a) \to h(b)$ for any evolutionary rule $a \to b$. Now, given two ANNIEPs $\Gamma_i = (V_i, U_i, \chi_i, N_i, \alpha_i, x_{In}^i, Out_i)$, $i = 1, 2$, $\chi_1 \cap \chi_2 = \emptyset$, we denote by $\Gamma_1 \sqcup \Gamma_2 = (V_1, U_1 \cup U_2, \chi_1 \cup \chi_2, N, \alpha, x_{In}^1, Out_2)$, where $\circ \mid_{\chi_i} = \circ_i$ for all $\circ \in \{N, \alpha\}$ and $i = 1, 2$.

3   Computational Power of Regular ANNIEPs

We start with a relation between the strong and weak acceptance modes.

Theorem 1. $\mathcal{L}_{wa}(regANNIEP) \subseteq \mathcal{L}_{sa}(regANNIEP)$.

Proof. Let $L \in \mathcal{L}_{wa}(regANNIEP)$. Then $L = L_{wa}(\Gamma)$ for some regular ANNIEP $\Gamma = (V, U, \chi, N, \alpha, x_{In}, Out)$. Let $N(x) = (M_x, \varphi^{R_x}, \psi^{R'_x})$ for a node x of

χ. Without loss of generality we may assume that $M_x = \emptyset$ for all $x \in Out$. We now construct the regular ANNIEP

$$\Gamma' = (V, U \cup \{Z\}, \chi \cup \{x_{Out}\}, N', \alpha', x_{In}, \{x_{Out}\}),$$

where $N'(x) = N(x)$ for $x \in \chi \setminus Out$, and

$$y : \{a \to Z \mid a \in U\}, \ R_y, \ Z^*, \alpha'(y) = * \text{ for } y \in Out,$$
$$x_{Out} : \emptyset, \ Z^*, \ \emptyset, \ \alpha'(x_{Out}) = *.$$

Obviously, if there is a non-empty node y of Out in some configuration of Γ, then y contains some word in some configuration of Γ', too. If this word is ε, then ε is not changed and sent to x_{Out}. If the word in y is non-empty, then all its letters are replaced by Z (note that it cannot leave the node as long as it still contains letters different than Z) and it is send to x_{Out}. Conversely, if a word eventually arrives in x_{Out}, then it contains only Z's which means that it was in a node from Out at some previous step. Thus Γ' accepts the same language as Γ does. Moreover, since the set of output nodes of Γ' is a singleton, we have $L_{wa}(\Gamma) = L_{wa}(\Gamma') = L_{sa}(\Gamma')$. $\qquad\square$

Note that we have shown a stronger result than given in Theorem 1 because we have shown that the number of output nodes of an ANNIEP accepting in the weak mode can be decreased to one only.

We now compare the families of languages generated by ANNIEPs with the family of context-sensitive languages denoted here by $\mathcal{L}(CS)$.

Theorem 2. $\mathcal{L}(CS) \subseteq \mathcal{L}_{wa}(regANNIEP)$.

Proof. Let L be a context-sensitive language. Then $L = L(G)$ for some context-sensitive grammar $G = (N, T, P, S)$ in Kuroda normal form, i.e., all its rules are of the form $A \to a$, $A \to BC$ and $AD \to BC$ with $A, B, C, D \in N$ and $a \in T$. Let P' be the set of rules of the form $A \to BC$ and $AD \to BC$. For every $p \in P'$ with its right-hand side BC we set

$$R_p = (N \cup T)^* \{B_p\} (N \cup T)^*,$$
$$R'_p = (N \cup T)^* \{B_p C_p\} (N \cup T)^*,$$
$$R''_p = (N \cup T)^* \{C_p\} (N \cup T)^*$$

and $R = \bigcup_{p \in P'} R_p$. We construct the ANNIEP $\Gamma = (T, U, \chi, H, \alpha, x_{In}, \{x_{Out}\})$ with

$$U = N \cup T \cup \{B_p, C_p \mid p = AD \to BC \text{ or } p = A \to BC\},$$
$$\chi = \{x_{In}, x_{Out}\} \cup \{p, p', p'' \mid p \in P'\},$$
$$x_{In} : M_{x_{In}}, \ (N \cup T)^*, \ R, \ \alpha = *$$
$$M_{x_{In}} = \{a \to A \mid A \to a \in P\} \cup \{B \to B_p \mid p = AD \to BC \text{ or } p = A \to BC\},$$
$$p : \{C \to C_p\}, \ R_p, \ R'_p, \alpha = * \text{ for } p = AD \to BC \text{ or } p = A \to BC,$$

$$p' : \{B_p \to A\}, \ R'_p, \ R''_p, \ \alpha = * \text{ for } p = AD \to BC \text{ or } p = A \to BC,$$

$$p'' : \begin{cases} \{C_p \to D\}, \ R''_p, \ (N \cup T)^*, \ \alpha = * \text{ for } p = AD \to BC, \\ \{C_p \to \varepsilon\}, \ R''_p, \ (N \cup T)^*, \ \alpha = * \text{ for } p = A \to BC, \end{cases}$$

$$x_{Out} : \emptyset, \ \{S\}, \ \{S\}, \ \alpha = *.$$

The network simulates a derivation in G backwards. Let w be the input word; we claim that for any word $z \in (N \cup T)^+$ in x_{In} at any computation step we have that $z \Longrightarrow^* w$ in G. Initially, this assertion is true as w lies in x_{In}. Assume that a word $z \in (N \cup T)^+$ is in the node x_{In} at some step. If we apply a rule $a \to A$ to z, the new word remains in x_{In} and the assertion holds for this new word.

Now assume that we apply $B \to B_p$ to z for a rule $p = AD \to BC$. Then the obtained word $z' = z_1 B_p z_2$, where $z = z_1 B z_2$, is sent to the node p, where some C is replaced by C_p. If $B_p C_p$ is not a subword, then the word cannot go out from this node; moreover any word further obtained from this word can never go out from the node p. If $B_p C_p$ is a subword, the word is sent out to the node p', where B_p is replaced by A. This new word is sent out to p''. There C_p is either replaced by D, provided that $p = AD \to BC$, or deleted provided that $p = A \to BC$. Finally, the obtained word, say z', is sent to x_{In}. Altogether, we started with $z = vBCu$ and obtained $z' = vADu$, which implies that $z' \Longrightarrow z \Longrightarrow^* w$.

Moreover, since a word only reaches x_{Out}, if it is S, we infer that a word is weakly accepted by Γ if and only if it is generated by G. Thus $L_{wa}(\Gamma) = L(G)$.
□

Theorem 3. $\mathcal{L}_{sa}(regANNIEP) \subseteq \mathcal{L}(CS).$

Proof. For an ANNIEP $\Gamma = (V, U, \chi, N, \alpha, x_{In}, Out)$, we construct a linearly bounded automaton, which accepts $L_{sa}(\Gamma)$. We do not give a complete formal construction; we only give an informal description of the automaton and leave the details of the construction to the reader.

Let $r = card(Out)$. The automaton has r tapes, and on each tape it nondeterministically follows the itinerary of a copy of the input word. The states are vectors of size $2r$, each ith entry, $1 \le i \le r$, being associated with the node containing the word on the tape i, and each ith entry, $r + 1 \le i \le 2r$, being 0 or 1 that indicates whether the node associated with the $(i - r)$th entry has finished its task on the word on tape i (in this case the entry is 1) or not. Initially, all tapes contain the input word w, the first r entries of the initial states are associated with the input node x_{In}, and the last r entries are 0.

Let us now consider an arbitrary configuration of the automaton: the first r elements of the current state state are associated with the nodes $x_1, x_2, \ldots x_r$, the last r elements are 0, and on the i-th tape, $1 \le i \le r$, the word w_i stands. Now the automaton performs on each tape i the following actions:

- Changes the word w_i according to an application of a rule in M_{x_i}; let v_i be the result.
- Checks whether v_i can pass the output filter of x_i. In the non-affirmative case the automaton blocks the computation. In the affirmative case, the

automaton changes the ith entry of the state into an entry associated with the node y_i, which is a nondeterministically chosen node among the nodes of $\chi \setminus \{x_i\}$.
- Check whether v_i can pass the input filter of y_i. In the non-affirmative case the automaton blocks the computation. In the affirmative case, the $i+r$ entry becomes 1. From now on, no move is observed on the ith tape and no change is made for the entries i and $i+r$, until all the entries $r+1, r+2, \ldots, 2r$ are 1.
- Checks whether the state with the last r entries 1 has its first r entries associated with all output nodes of Γ. In the affirmative case the automaton accepts the input; otherwise it changes the last r entries into 0 and resumes the actions explained above.

It is rather plain that the automaton accepts $L_{sa}(\Gamma)$. Since in any evolutionary step one deletes or substitutes one letter, the length of the words on any tape is bounded by the length of the input word. Thus the workspace of this automaton is linearly bounded. □

By the Theorems 1, 2 and 3, we get immediately the following two statements.

Corollary 1

1. $\mathcal{L}_{wa}(regANNIEP) = \mathcal{L}_{sa}(regANNIEP) = \mathcal{L}(CS)$.
2. Every language in $\mathcal{L}_X(regANNIEP)$, $X \in \{wa, sa\}$, can be weakly/strongly accepted by a regANNIEP Γ such that the action mode of every node of Γ is $*$. □

4   Computational Power of Random Context ANNIEPs

We start with two statements that immediately follows from Theorems 1 and 3.

Theorem 4

1. $\mathcal{L}_{wa}(rcANNIEP) \subseteq \mathcal{L}_{sa}(rcANNIEP)$.
2. $\mathcal{L}_{sa}(rcANNIEP) \subseteq \mathcal{L}(CS)$. □

We do not know whether the second inclusion is proper or equality holds. Thus we give some further relations to other known language families inside $\mathcal{L}(CS)$ and some closure properties which give some more information about the classes $\mathcal{L}_{wa}(rcANNIEP)$ and $\mathcal{L}_{sa}(rcANNIEP)$.

Theorem 5

1. $\mathcal{L}_{wa}(rcANNIEP)$ includes the class of linear context-free languages.
2. $\mathcal{L}_{wa}(rcANNIEP)$ contains non-semilinear languages.

Proof. 1. Let $G = (N, T, S, P)$ be a linear context-free grammar; without loss of generality we may assume that the following conditions hold:

– Every rule in P is of one of the following three forms: $A \to aB$, $A \to Ba$, $A \to a$, where $A, B \in N$ and $a \in T$,
– If both rules $A \to aC$ and $B \to Db$ belong to P, then $A \neq B$,
– The set of nonterminals N of G is $\{A_1, A_2, \ldots, A_n\}$ for some $n \geq 1$ and $S = A_1$,
– There is no rule $A \to aA$ or $A \to Aa$ for any $A \in N$ and $a \in T$.

We construct the following ANNIEP with the input alphabet T, the working alphabet $U = T \cup \{a_i, a'_i \mid 1 \leq i \leq n\} \cup \{Z\}$, and only one output node x_{Out}.

$$
x_{In} : \begin{cases} M = \{a \to a_1 \mid a \in T\}, \\ PI = T, FI = \{a_i \mid a \in T, 1 \leq i \leq n\}, \\ PO = \emptyset, FO = T, \\ \alpha = *, \ \beta = w, \end{cases}
\qquad
x_{Out} : \begin{cases} M = \emptyset, \\ PI = \{Z\}, FI = U \setminus \{Z\}, \\ PO = U, FO = \emptyset, \\ \alpha = *, \ \beta = s, \end{cases}
$$

If there exists $A_i \to aA_j \in P$ for some $a \in T$ and $1 \leq j \neq i \leq n$, then the node x_i is defined by

$$
x_i : \begin{cases} M = \{a_i \to a'_j \mid A_i \to aA_j \in P\}, \\ PI = \{a_i \mid a \in T\}, FI = U \setminus \{a_i \mid a \in T\}, \\ PO = \{a'_j \mid a \in T, 1 \leq j \neq i \leq n\}, FO = \emptyset, \\ \alpha = l, \ \beta = w, \end{cases}
$$

If there exists $A_i \to A_j a \in P$ for some $a \in T$ and $1 \leq j \neq i \leq n$, then the node x_i is defined by

$$
x_i : \begin{cases} M = \{a_i \to a'_j \mid A_i \to aA_j \in P\}, \\ PI = \{a_i \mid a \in T\}, FI = U \setminus \{a_i \mid a \in T\}, \\ PO = \{a'_j \mid a \in T, 1 \leq j \neq i \leq n\}, FO = \emptyset, \\ \alpha = r, \ \beta = w, \end{cases}
$$

Moreover, we set

$$
x'_i : \begin{cases} M = \{a_j \to a_i \mid a \in T, 1 \leq j \neq i \leq n\}, \\ PI = \{a'_i \mid a \in T\}, FI = \emptyset, \\ PO = \{a_i \mid a \in T\}, FO = \{a_j \mid a \in T, 1 \leq j \neq i \leq n\}, \\ \alpha = *, \ \beta = w, \end{cases} \quad \text{for } 1 \leq i \leq n,
$$

$$
\bar{x}_i : \begin{cases} M = \{a'_i \to \varepsilon \mid a \in T\}, \\ PI = \{a'_i \mid a \in T\}, FI = \{a_j \mid a \in T, 1 \leq j \neq i \leq n\}, \\ PO = \{a_i \mid a \in T\}, FO = \{a'_i \mid a \in T\}, \\ \alpha = l, \ \beta = w, \end{cases} \quad \text{for } 1 \leq i \leq n,
$$

$$
\tilde{x}_i : \begin{cases} M = \{a'_i \to \varepsilon \mid a \in T\}, \\ PI = \{a'_i \mid a \in T\}, FI = \{a_j \mid a \in T, 1 \leq j \neq i \leq n\}, \\ PO = \{a_i \mid a \in T\}, FO = \{a'_i \mid a \in T\}, \\ \alpha = r, \ \beta = w, \end{cases} \quad \text{for } 1 \leq i \leq n,
$$

$$
y : \begin{cases} M = \{a_i \to Z \mid A_i \to a \in P, a \in T, 1 \leq i \leq n\}, \\ PI = \bigcup_{i=1}^{n} \{a_i \mid a \in T\}, FI = U \setminus (\bigcup_{i=1}^{n} \{a_i \mid a \in T\}), \\ PO = \{Z\}, FO = \emptyset, \\ \alpha = r, \ \beta = w, \end{cases}
$$

The general idea of this construction is that for every $1 \le i \le n$, the following statement holds:

Fact: *If* $S \Longrightarrow^t uA_iv \Longrightarrow^+ uwv = z$ *for some* $t \ge 0$, *with* $|z| = m$, *then* $h_i(w) \in (C^{(z)}_{2m(t+1)+2t}(x_i) \cap C^{(z)}_{2m(t+1)+2t}(y))$, *where* h_i *is a literal morphism from* T *to* $\{a_i \mid a \in T\}$ *defined by* $h(a) = a_i$ *for any* $a \in T$.

This fact can be proved by a standard induction argument on t. Now, if $t = m - 1$, then w is reduced to a letter from T, say a, therefore after the word a_i is transformed into Z in the node y, it arrives in x_{Out} and the computation halts successfully. This means that z is accepted by the network.

On the other hand, if $C^{(z)}_0, C^{(z)}_1, C^{(z)}_2, \ldots, C^{(z)}_p$ is an accepting computation on z and $h_i(w) \in (C^{(z)}_t(x_i) \cap C^{(z)}_t(y))$ for some $t < p$, then the derivation $S \Longrightarrow^* uA_iv \Longrightarrow^+ uwv = z$ holds in G, which concludes the proof of the first statement of the theorem.

2. The network with the nodes defined by:

$$x_{In} : \begin{cases} M = \{a \to \bar{a}\}, \\ PI = \{a\}, FI = \{\bar{a}, \tilde{a}\}, \\ PO = \{\bar{a}\}, FO = \emptyset, \\ \alpha = *, \ \beta = s, \end{cases} \qquad x_1 : \begin{cases} M = \{a \to \tilde{a}\}, \\ PI = \{\bar{a}\}, FI = \{\tilde{a}\}, \\ PO = \{\tilde{a}\}, FO = \emptyset, \\ \alpha = *, \ \beta = s, \end{cases}$$

$$x_2 : \begin{cases} M = \{\bar{a} \to \varepsilon\}, \\ PI = \{\bar{a}, \tilde{a}\}, FI = \emptyset, \\ PO = \{\tilde{a}\}, FO = \emptyset, \\ \alpha = *, \ \beta = s, \end{cases} \qquad x_3 : \begin{cases} M = \{\tilde{a} \to a'\}, \\ PI = \{\tilde{a}\}, FI = \{\bar{a}\}, \\ PO = \{a'\}, FO = \{\tilde{a}\}, \\ \alpha = *, \ \beta = s, \end{cases}$$

$$x_4 : \begin{cases} M = \{a' \to a\}, \\ PI = \{a'\}, FI = \{a, \bar{a}, \tilde{a}\}, \\ PO = \{a\}, FO = \{a'\}, \\ \alpha = *, \ \beta = s, \end{cases} \qquad x_{Out} : \begin{cases} M = \emptyset, \\ PI = \{\bar{a}\}, FI = \{a, \bar{a}, \tilde{a}\}, \\ PO = \{\bar{a}\}, FO = \emptyset, \\ \alpha = *, \ \beta = s, \end{cases}$$

weakly accepts the non-semilinear language $\{a^{2^n} \mid n \ge 0\}$. Indeed, the computation of this netwok on every input is divided in two phases. In the first phase, the input word looses one occurrence of a and changes another one to a' by visiting the nodes x_{In}, x_1, x_2, x_3. This process resumes until no occurrence of a is observed in the current word. There are three possiblities: (1) it contains only a's, (2) it contains only a's excepting an occurrence of $\bar{a}$, (3) it equals $\bar{a}$. Now the second phase of the computation starts. In the first case, the word enters x_4 where all a's are transformed into original a's and the first phase resumes from x_{In} with a word that is exactly twice shorter than the word present in the input node in the beginning of the previous first phase. In this case, we have checked whether the length of that word was an even number. In the second case listed above, the computation cannot continue anymore, hence the network will eventually halt without accepting. In the third case, the computation halts accepting the input word. This means that the length of the input word could be divided iteratively by 2 until the result was one, hence the length of the input word was a power of 2.   □

Theorem 6

1. *The class $\mathcal{L}_{wa}(rcANNIEP)$ is closed under boolean union, literal morphism, inverse weak literal morphism, mirror image.*
2. *The class $\mathcal{L}_{sa}(rcANNIEP)$ is closed under boolean intersection, literal morphism, inverse weak literal morphism, concatenation, mirror image.*

Proof. 1. We give an informal proof for union that can be easily formalized by the reader. Let Γ_1 and Γ_2 be to ANNIEPs; we construct a new ANNIEP Γ that contains three subnetworks. In the input node of the first subnetwork, an arbitrary symbol of the input word is substituted by either its primed copy or its barred copy. All words containing a primed symbol are received by a specific node while those containing a barred symbol are received by another specific node. All symbols of the words arrived in these two nodes are replaced by their primed and barred copies, respectively. When this process is finished, each of the two nodes contains only one word. The word containing primed symbols only is given as an input word to the subnetwork formed from Γ_1 modified accordingly. The other word is processed analogously by the subnetwork formed from Γ_2 modified accordingly. The set of output nodes of Γ is the union of the sets of output nodes of Γ_1 and Γ_2 modified accordingly. Clearly, $L_{wa}(\Gamma) = L_{wa}(\Gamma_1) \cup L_{wa}(\Gamma_2)$.

If $h : V \longrightarrow U$ is a literal morphism and Γ is an ANNIEP with the input alphabet V, then let Γ' be the ANNIEP with the input alphabet U formed by two subnetworks as follows. In the input node of the first subnetwork, each symbol b of the input word is substituted by a symbol a' such that a' is a copy of $a \in V$ that does not appear in $V \cup U$ and $h(a) = b$. When all symbols of the input word were substituted, all the words obtained are sent to the input node of the subnetwork formed from Γ modified accordingly. It is plain that $h(L_{wa}(\Gamma)) = L_{wa}(\Gamma')$. The construction for the closure under inverse weak literal morphism is pretty similar and left to the reader.

The closure under mirror image follows pretty simple; it suffices to interchange all the action modes l and r of the nodes.

2. The closure under intersection, literal morphism and inverse literal morphism follows similarly to the previous case. Note the fundamental role played by the strong acceptance in the case of intersection.   □

It is known that every recursively enumerable language can be written as the image of the intersection of two linear languages through a weak literal morphism. Therefore, the following statement is a consequence of the second statement of Theorem 4 and Theorem 6:

Corollary 2

1. *Every recursively enumerable language is the weak literal morphic image of a language in $\mathcal{L}_{sa}(ANNIEP)$.*
2. *$\mathcal{L}_{sa}(ANNIEP)$ is not closed under weak literal morphism.*   □

5   Final Remarks

As we showed in this note, the computational power of ANNIEPs is very different than that of generating networks of non-inserting processors. The role of evolutionary operations in generating networks of evolutionary processors, that is generating networks with nodes specialized in all three evolutionary operations, in two operations out of these three and in only one operation, was considered in [1]. A similar investigation on ANEPs has already started.

References

1. Alhazov, A., Dassow, J., Rogozhin, Y., Truthe, B.: Personal communication
2. Castellanos, J., Martín-Vide, C., Mitrana, V., Sempere, J.: Networks of evolutionary processors. Acta Informatica 38, 517–529 (2003)
3. Csuhaj-Varj, E., Salomaa, A.: Networks of parallel language processors. In: Păun, G., Salomaa, A. (eds.) New Trends in Formal Languages. LNCS, vol. 1218, pp. 299–318. Springer, Heidelberg (1997)
4. Csuhaj-Varj, E., Mitrana, V.: Evolutionary systems: a language generating device inspired by evolving communities of cells. Acta Informatica 36, 913–926 (2000)
5. Csuhaj-Varjú, E., Martín-Vide, C., Mitrana, V.: Hybrid NEPs are computationally complete. Acta Informatica 41, 257–272 (2005)
6. Dassow, J., Truthe, B.: On the power of networks of evolutionary processors. In: Durand-Lose, J., Margenstern, M. (eds.) MCU 2007. LNCS, vol. 4664, pp. 158–169. Springer, Heidelberg (2007)
7. Errico, L., Jesshope, C.: Towards a new architecture for symbolic processing. In: Artificial Intelligence and Information-Control Systems of Robots, vol. 94, pp. 31–40. World Scientific, Singapore (1994)
8. Fahlman, S., Hinton, G., Seijnowski, T.: Massively parallel architectures for AI: NETL, THISTLE and Boltzmann Machines. In: Proc. AAAI National Conf. on AI, pp. 109–113. William Kaufman, Los Altos (1983)
9. Hillis, W.: The Connection Machine. MIT Press, Cambridge (1985)
10. Manea, F., Martín-Vide, C., Mitrana, V.: On the size complexity of universal accepting hybrid networks of evolutionary processors. Mathematical Structures in Computer Science 17, 753–771 (2007)
11. Manea, F., Mitrana, V.: All NP-problems can be solved in polynomial time by accepting hybrid networks of evolutionary processors of constant size. Information Processing Letters 103, 112–118 (2007)
12. Manea, F., Margenstern, M., Mitrana, V., Perez-Jimenez, M.: A new characterization of **NP**, **P**, and **PSPACE** with accepting hybrid networks of evolutionary processors (submitted)
13. Margenstern, M., Mitrana, V., Perez-Jimenez, M.: Accepting hybrid networks of evolutionary systems. In: Ferretti, C., Mauri, G., Zandron, C. (eds.) DNA 2004. LNCS, vol. 3384, pp. 235–246. Springer, Heidelberg (2005)
14. Martín-Vide, C., Mitrana, V.: Networks of evolutionary processors: results and perspectives. In: Molecular Computational Models: Unconventional Approaches, pp. 78–114. Idea Group Publishing, Hershey (2005)
15. Păun, G., Sntean, L.: Parallel communicating grammar systems: the regular case. Annals of University of Bucharest, Ser. Matematica-Informatica 38, 55–63 (1989)
16. Păun, G.: Computing with membranes. Journal of Computer and System Sciences 61, 108–143 (2000)
17. Sankoff, D., et al.: Gene order comparisons for phylogenetic inference: evolution of the mitochondrial genome. In: Proceedings of the National Academy of Sciences of the United States of America, vol. 89, pp. 6575–6579 (1992)

Discrete Modeling of Biochemical Signaling with Memory Enhancement

John Jack[1] and Andrei Păun[1,2,3]

[1] Department of Computer Science/IfM
Louisiana Tech University, P.O. Box 10348, Ruston, LA 71272, USA
{johnjack,apaun}@latech.edu
http://www.latech.edu
[2] Departamento de Inteligencia Artificial, Facultad de Informática
Universidad Politécnica de Madrid,
Campus de Montegancedo S/N, Boadilla del Monte, 28660 Madrid, Spain
http://www.upm.es
[3] Bioinformatics Department, National Institute of Research and
Development for Biological Sciences,
Splaiul Independenţei, Nr. 296, Sector 6, Bucharest, Romania

Abstract. We present an enhancement of the Nondeterministic Waiting Time algorithm. This work is a continuation of our group's previous modeling efforts. We have improved our algorithm with a "memory enhancement". Previously, we have used our algorithm to explore the Fas-mediated apoptotic pathway in cells with a particular focus on cancerous or HIV-1-infected T cells. In this paper, we will describe the memory enhancement and give a simple three reaction model to illustrate the differences between our technique and a continuous, concentration-based approach using a system of ordinary differential equations. Furthermore, we provide our results from the modeling of two well-known models: the Lotka-Volterra predator-prey and a circadian rhythm model. For these models, we provide the results of our simulation technique in comparison to results from ordinary differential equations and the Gillespie Algorithm. We show that our algorithm, while being faster than Gillespie's approach, is capable of generating oscillatory behavior where ordinary differential equations do not.

Keywords: Discrete modeling, Lotka-Volterra, predator - prey, circadian rhythm, Gillespie, ordinary differential equations.

1   Introduction

Systems biology, the systematic study of biological systems through a combined effort between computational and experimental results, has received a great deal of attention in recent years [17,18]. There has been an expansive effort from mathematicians and computer scientists to use models to unravel the mysteries behind biochemical/biological systems – e.g., signal transduction, viral dynamics, gene transcription. With the ever-increasing wealth of information

C. Priami et al. (Eds.): Trans. on Comput. Syst. Biol. XI, LNBI 5750, pp. 200–215, 2009.

flowing in from biological labs around the world on protein dynamics, the challenge remains for computer scientists to develop/refine efficient algorithms for modeling molecular signaling cascades. Computational tools are being applied to interpret biological results and make predictions into the underlying molecular mechanisms involved in cancer, autoimmune disorders, and neurological disorders.

We see two important efforts being undertaken by computer scientists with respect to modeling molecular signaling cascades. First, the algorithms developed to interpret molecular interactions need visibility to biochemists and non-computer scientists. Notably, the authors of [10,12] have made great strides in developing software, designed for biochemists with little to no knowledge on the modeling algorithm, to design, develop, and implement biochemical network simulations based on their experimental observations.

The second major effort is the development of more efficient algorithms to drive biochemical simulation software. With many labs focusing on the modeling of individual pathways – e.g., Fas-mediated apoptosis [11], p53 network [19,22] and the EGF-receptor system [24] – the concept of a realistic whole-cell simulation remains a very distant goal. There are too many unknowns biochemical aspects to build an accurate and reliable model. However, while the biochemical questions surrounding whole-cell simulation are being answered in experimental labs, there is still work to be done in modifying (and developing new) algorithms for simulating biochemical systems.

1.1 Motivation Behind the Paper

Many signal transduction models in the literature contain as many (or more) reactions as proteins. Although the human genome contains over three billion base-pairs, it only encodes approximately 20,000-25,000 genes. The proteins encoded by these genes are entangled in an intricate and diverse web of interactions. The dynamics of these proteins – e.g., expression levels and reactions – define the complexity and physiological characteristics of the human cell.

Some reactions in signaling cascades can sometimes share common reactants and compete for resources. These competing reactions typically have different kinetic rates – i.e., some of the competing reactions utilize a given reactant faster than other reactions use the same reactant. Hence, when the numbers of some molecules are very small, stochastic (or nondeterministic) methods for biochemical modeling can play an important role in interpreting the results of lab experiments, and offer insight into unknown aspects of the system.

When modeling biochemical networks via systems of ordinary differential equations (ODEs), the data are considered in terms of concentrations instead of numbers of molecules, and the reactions are deterministically applied. While the ODEs are satisfactory for predicting the average behavior of a biochemical system, they are not ideal for extrapolating the different cellular responses resulting from molecular signaling cascades – especially ones involving low numbers of molecules. The Gillespie Algorithm [7,8], which is a numerical simulation algorithm for the chemical master equation, has been extensively employed to address these low molecular multiplicity situations. However, even though it has been modified to run more efficiently [9], it does not scale well with respect to the number of reactions. Hence, an algorithm capable of realistic and efficient whole-cell simulation, a significant goal for systems biology, is still being explored. Our algorithm is designed to be faster than Gillespie's algorithm and its derivatives – such as, Gibson's Next Reaction Method – yet more sensitive than ODE-based simulations.

Our group has previously argued in [3,13,14] that an approach involving the Membrane Systems paradigm of computing offers a unique perspective on biochemical network simulation. Specifically, in [13,14] we discuss the advantages of our simulation technique: the *Nondeterministic Waiting Time (NWT)* algorithm. Our algorithm is distinct from the Gillespie Algorithm. Yet, it is a discrete, nondeterministic technique which can offer a different perspective than systems of ordinary differential equations on the biochemistry of a cell.

In this paper, we will describe a modification to our algorithm. In order to improve the deterministic aspects of reaction competition for our simulation technique, we have added a *memory enhancement* to the NWT. This enhances the sensitivity of our algorithm with respect to reaction competition over limited resources. Since our algorithm relies on the law of mass action to drive the population dynamics, fast reactions may be allowed to use up all the resources of a slow reaction. With the modified algorithm, a slow reaction will *remember* how long it has waited when a fast reaction uses all available reactants. The memory can be factored into the equation for calculating the next time the slow reaction will occur, once reactants become available again.

In Section 2 we provide the necessary background on the NWT algorithm, a simulation technique based on Membrane Systems. We will discuss the specifics on the memory enhancement in Section 3, as well as results for a simple biochemical model involving fast-slow reaction competition. In Section 4 we show the results of the NWT algorithm for simulating two popular models: The Lotka-Volterra predator-prey model and a circadian rhythm model [1]. For both models, we compare the results of the NWT algorithm with an ODE-based simulation and a simulation based on the Gillespie Algorithm. Section 5 contains our final remarks and a discussion on the future research interests of our modeling group.

2   The Nondeterministic Waiting Time Algorithm

The NWT algorithm is a discrete, nondeterministic biochemical simulation algorithm. We track the evolution of a Membrane System where the rules (or reactions) occur over discrete time intervals in an asynchronous manner. Before we give a step-by-step description of the algorithm, it is important to discuss the concept of reaction Waiting Times.

Our NWT algorithm is driven by the law of mass action – the time a reaction takes to occur is directly proportional to the number of its reactant molecules.

When dealing with concentration-based kinetics we need to calculate a discrete kinetic constant (for molecules instead of nMs, μMs, etc). We initialize the discrete kinetic constants with Equation 1.

$$const_R = \frac{k_R}{V^{i-1} \times N^{i-1}} \tag{1}$$

where V is the volume of the system, N is Avogadro's constant (6.0221415×10^{23}) and i is the number of reactants involved in the reaction.

Once the kinetic constants are initialized, for every reaction in the system, we calculate the initial Waiting Time – the amount of time required for one instance of a reaction – using Equation 2.

$$WT_{R_1} = \frac{1}{const_{R_1} * |A|} \tag{2}$$

where A is the reactant, $const_{R_1}$ is the discrete kinetic constant, and $|A|$ represents the number of molecules present in the system at the moment of WT calculation.

Equation 2 represents the calculation for a first order reaction (involving only one reactant). For second and third order reactions (two and three reactants, resp.), we need to use Equations 3 and 4.

$$WT_{R_2} = \frac{1}{const_{R_2} * |A| * |B|} \tag{3}$$

and

$$WT_{R_3} = \frac{1}{const_{R_3} * |A| * |B| * |C|} \tag{4}$$

where A, B, and C are the reactants, $const_{R_2}$ and $const_{R_3}$ are the discrete kinetic constants, and $|A|$, $|B|$ and $|C|$ represent the numbers of molecules present in the system at the moment of WT calculation.

With the calculation of reaction Waiting Time, we have the amount of time it will take for each reaction to occur. If there are insufficient reactants for a reaction, then we set the Waiting Time equal to infinity; this is easily done in the C programming language. We can now provide the following description of the NWT algorithm (n.b., Step 7 is the new memory enhancement step, which will be explained in Section 3):

1. **Build Membrane System:** Import model information (alphabet, rules, etc.). For every reaction, R_i, calculate the initial Waiting Time, WT_{R_i}. Choose simulation end-time τ_{fin}. Set current simulation time to zero ($\tau = 0$).
2. **Build Heap:** Using the reaction Waiting Times, we build a min-heap of all reactions in the system.
3. **Select Rule:** Choose the reaction with the lowest Waiting Time – the top of the min-heap. Upon selecting the top node, recursively check to see if there are any children nodes sharing the minimum Waiting Time. If such a tie for minimum Waiting Time exists, proceed to Step 4. If no tie exists, then proceed to Step 5.

4. **Handle Tie:** Check the multiplicities of the reactant species for all tied reactions. If there are enough reactants to satisfy all of the reactions with the minimum Waiting Time, implement all tied reactions. If there are not enough reactants to accommodate all the reactions, then nondeterministically apply as many reactions as possible.

5. **Apply Rule:** Update the multiplicities of the reactant(s) and product(s) for the reaction(s) from Step 3. Aggregate the simulation time ($\tau = \tau + WT_{applied}$).

6. **Update Rules:** Recalculate the Waiting Time for all reactions whose reactants include the products or reactants of the applied reaction(s). That is, we need to see how the multiplicity changes from the applied reaction(s) have affected the Waiting Times for all rules dependent on those proteins with changed multiplicity. For each such reaction compare the new Waiting Time with the existing Waiting Time and keep the smallest of the two (unless the new time is infinity).

7. **Memory Enhancement:** If the recalculation of a reaction's Waiting Time results in a value of infinity, then we must store the amount of time waited as a percentage (Mem_{perc}). If the recalculation of a reaction's Waiting Time results in a real value and the previous value was infinite, then the Waiting Time will need to be adjusted according to the stored memory percentage.

8. **Heap Maintenance:** Adjust the min-heap, bubbling reaction nodes up or down in order to satisfy the min-heap property, once reaction Waiting Times have been recalculated according to the multiplicity changes. N.B., to accommodate the multiple changes in Waiting Times, we employ nonstandard heap maintenance methods.

9. **Termination:** If $\tau = \tau_{fin}$, then terminate the simulation. Output the multiplicity information for entire simulation. Otherwise, go back to Step 3.

For a deeper explanation of the algorithm, we refer the interested reader to [13,14]. In the next section, we will clarify Step 7 of the algorithm – the *memory enhancement*.

3   Memory Enhancement

As we discussed in Section 1.1, there are often situations in biochemical networks, where one protein is a reactant for two or more reactions of different kinetic rates (fast vs. slow). In order to explain our *memory enhancement*, we will consider an example system (see Table 1).

The biochemical system in Table 1 involves three reactions (R_1, R_2 and R_3) acting on four proteins (A, B, C, D). We can mathematically describe the model as a system of ordinary differential equations (Equation 5)

$$\frac{d[A]}{dt} = -(k_1 + k_2)[A] + k_3[D]$$
$$\frac{d[B]}{dt} = k_2 * [A]$$

Table 1. An example biochemical system

Reaction	Rate Constant	Initial Molecules
R_1: $A \rightarrow C$	k_1 (**slow**)	$A = 1$
R_2: $A \rightarrow B$	k_2 (**fast**)	$B = 0$
R_3: $D \rightarrow D + A$	k_3	$C = 0$
		$D = 1$

$$\frac{d[C]}{dt} = k_1 * [A]$$

$$\frac{d[D]}{dt} = 0 \tag{5}$$

The system of ordinary differential equations in Equation 5 was specifically designed to illustrate the effects of the memory enhancement. We will compare the enhanced NWT algorithm with solutions to the systems of ODEs, providing two differences cases based on variable kinetics. By selecting different kinetic rates, we will show how the memory enhancement leads to agreement between the NWT and ODEs for strictly deterministic runs, but with nondeterministic decisions it can lead to distinct results and a divergence in overall behavior of the biochemical network. For the sake of simplicity, we will assume the rate constants (k_1, k_2 and k_3) are already in discrete form. Therefore, when refer to k_i in R_i above, we have $const_{R_i}$.

A model similar to the one described in Table 1 could be used to investigate the dynamics of HIV-1 Tat protein, since it is initially transcribed at very low numbers [16]. Once Tat is assembled in the cytosol, it can be exocytised or translocated to the nucleus [20]. When Tat is translocated to the nucleus it can begin upregulating HIV-1 proteins (including itself). Since the downstream effects of Tat translocation to the nucleus has profound impacts on the cell (causing upregulation of the HIV-1 proteins), a discrete and nondeterministic approach is beneficial to the study of the dynamics of the low levels of Tat [23].

In the system, molecules of A are formed from molecules of D. This reaction can basically be viewed as a combined transcription and translation rule with D being the gene and A being the protein encoded by the gene. Once a molecule of A is formed, it has the option of turning into a molecule of B at rate k_2 or it can turn into a molecule of C at a rate k_1. If we consider the species A to be analogous to HIV-1 Tat protein, then reaction R_1 ($A \rightarrow B$) could be the translocation of Tat from the cytosol to the nucleus, and reaction R_2 ($A \rightarrow C$) could be the translocation of Tat from the cytosol to the extracellular environment.

We will next consider two cases for the model described in Table 1 and discuss the implementation of the *memory enhancement*. The cases are determined by the values for the discrete kinetic rates. The first case shows that the memory enhancement can produce similar results between the ODEs and the NWT when no nondeterministic decisions are made. For the second case, we will show how the technique can produce different results, illustrating the NWT algorithm's ability to explore the stochastic nature of molecular signaling cascades. Note, the NWT algorithm remains the same in both cases, the only difference lies in the initialization of the discrete kinetic constants.

3.1   Case 1: Deterministic Memory Enhancement

If we let $k_1 = 4$, $k_2 = 10$, and $k_3 = 5$, then we can see the results of a simulation using the NWT algorithm plotted against the solution of the system of ordinary differential equations (Figure 1).

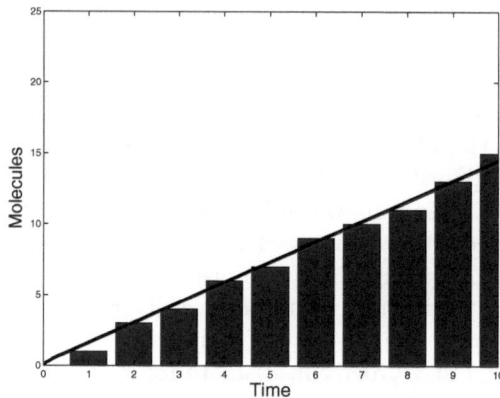

Fig. 1. The graph shows the accumulation of C molecules throughout a 10 second run. The bars are the discrete results from the NWT algorithm and the black line is the solution of the system of ordinary differential equations. With the kinetic values ($k_1 = 4$, $k_2 = 10$, $k_3 = 5$), there are no nondeterministic decisions for the entire length of the NWT simulation. Therefore, we are pleased to see the NWT algorithm results are similar to the solution to the system of ordinary differential equations.

At initialization ($t = 0$), there is exactly one molecule of D and one molecule of A. Therefore, from Equation 2, we see that all three reactions have real (finite) Waiting Times when the simulation begins. Moreover, we have $WT_{R_1} = 0.25$, $WT_{R_2} = 0.1$ and $WT_{R_3} = 0.2$. According to the reaction Waiting Times, the first reaction to occur is R_2, which immediately exhausts the system's supply of A molecules, yields one molecule of B and a simulation time of $t = 0.1$. The Waiting Times for the rules affected by the applied rule must be recalculated; since there are no molecules of A in the system, we have $WT_{R_1} = WT_{R_2} = \infty$. R_3 does not use any proteins involved in the applied reaction (A or C), so WT_{R_3} is left unchanged after the first reaction is executed. After the heap maintenance, R_3 will be at the top.

The next reaction to be applied is R_3, which gives us a new molecule of A and a simulation time of $t = 0.2$. Now the *memory enhancement* plays a role. In the first step, reaction R_2 exhausts the supply of A molecules. When this occurs,

R_1 has *waited* for 0.1 seconds of its total WT. The *memory enhancement* allows the simulator to keep track of the percentage of time waited. In other words, R_1 waited for 0.1 seconds out of its required 0.25 seconds, which means it has waited 40% of its Waiting Time. We store the percentage of time left to wait (60%). So, when a new molecule of A is formed (for instance, at $\tau = 0.2$), we can recalculate the WT_{R_1} using the percentage to adjust its Waiting Time.

Continuing after R_3 is applied at time $\tau = 0.2$, we have a new molecule of A. Since R_1 and R_2 both use A as a reactant, we must recalculate the Waiting Times of both reactions. The Waiting Time of R_2 and R_1 are calculated using Equation 2. However, the memory of R_1 allows us to take 60% of its recalculated WT. Therefore, the Waiting Time of R_2 is calculated as 0.1, but the Waiting Time of R_1 is recalculated as 0.15. This number stems from the equation

$$WT_{R_1} = Mem \frac{1}{k_1 * |C|} \tag{6}$$

where Mem is the percentage of time left to wait (60% in the example above). The second and third order reactions follow similarly.

Using this implementation, our NWT algorithm agrees with the ODEs for a strictly deterministic run. In the next subsection, we will explore the implications of the *memory enhancement* in a system requiring reaction competition over low numbers of molecules. Although we agree with ODEs in a deterministic run, we want to explore the nondeterministic effects of the *memory enhancement*.

3.2   Case 2: Nondeterministic Memory Enhancement

We will now assume different kinetic constants to highlight the effects of the nondeterministic component of the NWT algorithm in conjunction with the memory enhancement. Although the kinetics of our sample system are chosen in a deliberate manner in order to illustrate the nondeterministic effects of the algorithm, we will later show in Section 4.2 how our nondeterministic logic can have similar implications in a model reported in the literature.

For our next simulations, we assume $k_1 = 0.1$, $k_2 = 1.0$, and $k_3 = 0.5$. The initial Waiting Times are calculated as $WT_{R_1} = 10$, $WT_{R_2} = 1$, and $WT_{R_3} = 2$. In Figure 2, we see the accumulation of B and C molecules. The results of the ODE-based simulation are visibly different than the results of the simulation involving the NWT algorithm. The reasons for the differences are the nondeterministic decisions on reaction competition for A molecules.

Based on the initialized WTs, the first reaction to be applied is R_2. After R_2 is applied, the simulation time is aggregated ($t = 1$) and there are no more molecules of A present in the system. Hence, the WTs for R_2 and R_1 are both infinite after recalculation. We store the percentage of time the slow reaction, R_1, had left to wait when the WT changed to infinity – $Mem_{R_1} = 90\%$. The next rule to be applied is R_3, since it was unaffected by the application of R_2. The simulation time is adjusted ($t = 2$), and we now have a new molecule of A. With our new A molecule available, we must recalculate the WTs for R_1 and R_2.

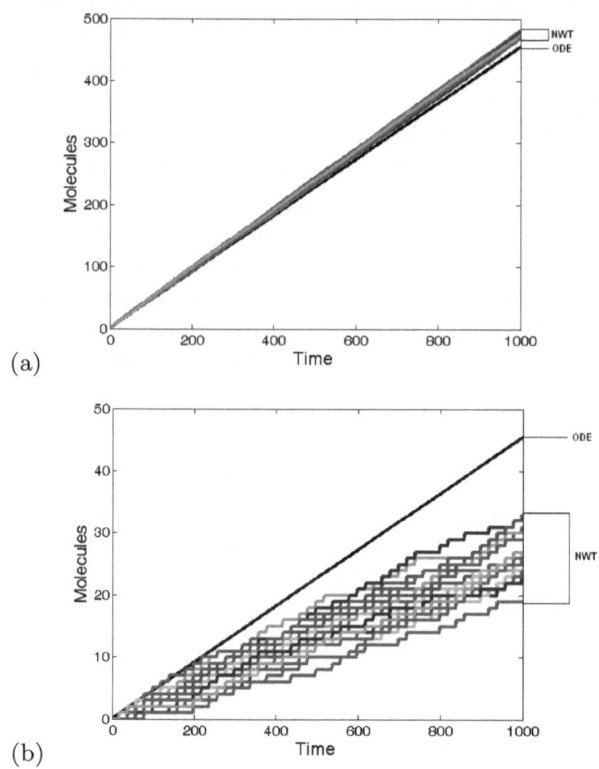

(a)

(b)

Fig. 2. In both graphs we see the results of the ODE-based simulation (straight black line) and the results from the modified NWT algorithm. (a) The accumulation of molecules of B and (b) the accumulation of C are shown. Molecules of B and C both come from A molecules. However, the reaction for B is faster than the reaction for C. In the ODE models, a molecule of A can be used to partially satisfy B and C. Since our NWT algorithm is discrete, the molecules are nondeterministically chosen to satisfy one or the other. The reaction changing A into C 'remembers' how long it has waited, and uses this information the next time a molecule of A is ready.

Using Mem_{R_1}, we calculate the WT for reaction R_1, using the fact that it need wait only 90% of its new Waiting Time. Therefore, when a new molecule of A is formed two seconds into the run, we recalculate WT_{R_1} using Equation 6. In our case, we have $WT_{R_1} = 9$ and $WT_{R_2} = 1$.

Continuing the calculations for the simulation, we skip ahead to a future event ($t = 18$). Up until this point, we have been creating molecules of A, and every single one of them has been deterministically chosen to change into molecule B via reaction R_2. But, at $t = 18$, a molecule of A has been created, and the Waiting Times of reaction R_1 and R_2 are equal $WT_{R_1} = WT_{R_2} = 1$, since we have $Mem_{R_1} = 10\%$. In other words, R_1 and R_2 are each attempting to use the A molecule to form a C and B, resp. The ODE-based simulation has no problems at this time-point, since it is continuously sending a fraction of each A to form a fraction of B and C, whereas our algorithm represents A discretely and only satisfies one reaction per molecule.

Our algorithm faces the question: at $t = 18$ should the A molecule be allowed to satisfy R_1 or R_2? The algorithm answers the question by making a nondeterministic choice between R_1 and R_2. If R_1 is chosen, then it is applied, and our results stay with the ordinary differential equations results (up to $t = 19$). Remember, the ordinary differential equations have been slowly and continuously aggregating fractions of C molecules throughout to reach one full molecule of C by $t = 19$. However, if R_2 is chosen, then our solution diverges from the previous solution. When the effects of the nondeterministic decisions are aggregated over 1000 seconds, we see the different results obtained from the NWT algorithm (Figure 2).

4 Other Models

4.1 Lotka-Volterra Predator-Prey

The Lotka-Volterra predator-prey model depicts the interactions of two species. There is a prey population and a predator population, where $P_1(t)$ and $P_2(t)$ represent the number of each species respectively at time t. The model can be written as the following pair of first-order, nonlinear, differential equations

$$\frac{dP_1}{dt} = P_1 * (a - b * P_2)$$
$$\frac{dP_2}{dt} = -P_2 * (c - d * P_1) \tag{7}$$

where prey species are born at a rate of a and consumed at a rate of b. Predator species are born at a rate of d and die at a rate of c.

In Figure 3, we see a picture of the predator-prey model. The picture (as well as the SBML code for the model) was generated with CellDesigner [5,6], which we also used to generate the SBML code to initialize our simulator. The way the system is designed, an increase in prey leads to an increase in predator, and an increase in predator leads to a decrease in prey. Total annihilation of prey leads to total extinction of predator, since the food supply of the predator will be exhausted.

We used three different simulation techniques to model the reactions described in the Lotka-Volterra model: solution to the system of ordinary differential equations, the Gillespie Algorithm, and our NWT algorithm. The ODEs were solved in MATLAB, while the other two algorithms were both coded in C. The results are given in Figure 4 and Figure 5.

The solution to the ordinary differential equations in Equation 7 shows consistent oscillations throughout the entire simulation run. The NWT shows dampened oscillations over time. The Gillespie Algorithm has difficulties producing the oscillations, due to the stochasticity of the algorithm. In this case, our NWT

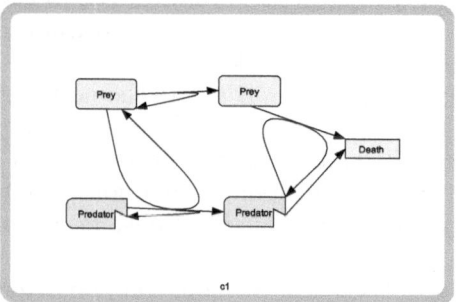

Fig. 3. The Lotka-Volterra model involves two interacting species. Prey species are born at a rate a and are consumed at a rate b by the predator species. The predator species are born at a rate of d if there is available food (prey). The way the system is designed, an increase in prey leads to an increase in predator. Total annihilation of prey leads to total extinction of predator, since there is no longer any food. The model deterministically leads to oscillatory behavior.

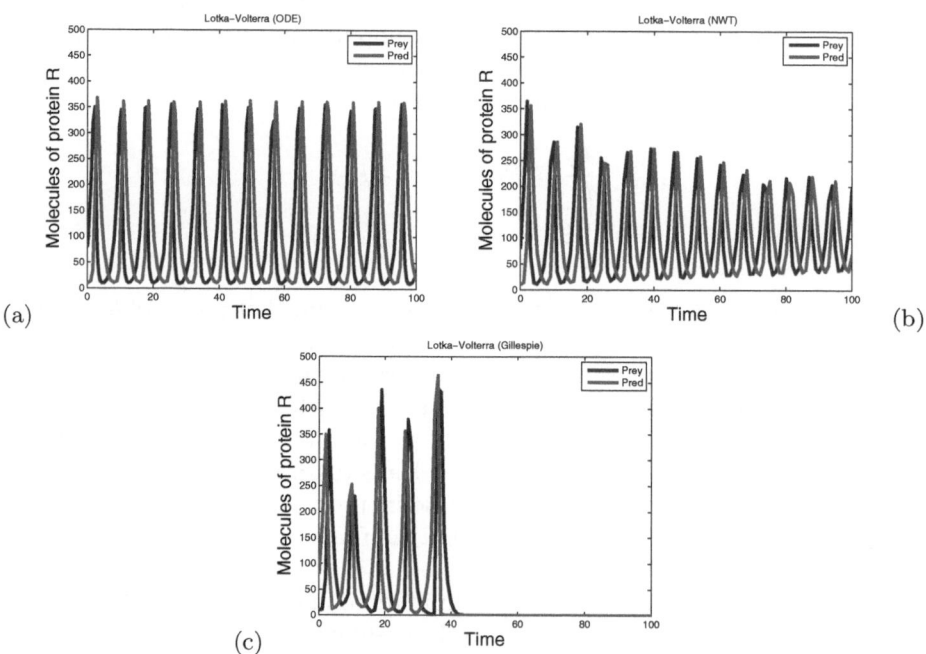

Fig. 4. Results of three simulation techniques for the Lotka-Volterra model (up to 100 seconds). (a) solution to ordinary differential equations, (b) the NWT algorithm, and (c) the Gillespie Algorithm.

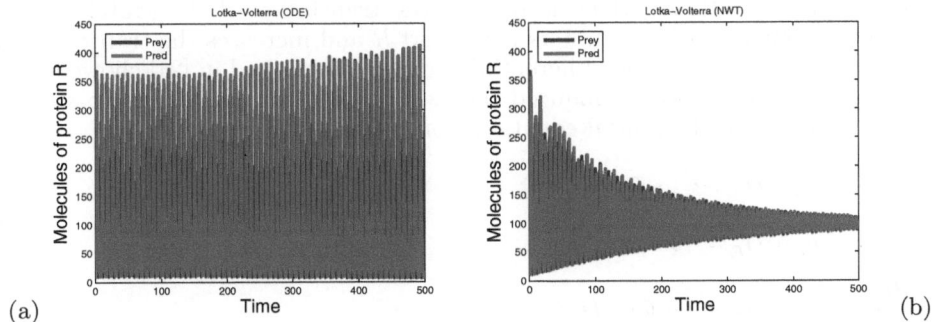

Fig. 5. Results of the two simulation techniques for the Lotka-Volterra model (up to 500 seconds). (a) solution to ordinary differential equations and (b) the NWT algorithm.

algorithm runs deterministically. The system is small enough and the dynamics are such that the NWT makes no nondeterministic decisions due to reaction competition. If we expand the results of the solution to system of ordinary differential equations and the NWT algorithm, we see further decline in the amplitude for the NWT algorithm. In Figure 5, we expand the simulation run for a total of five hundred seconds. The results for the solution to the system of ordinary differential equations and the NWT algorithm simulation are provided.

We modeled this classic system to illustrate the differences in the results of our simulation technique compared to the solution of the system of ordinary differential equations, the NWT algorithm, and the Gillespie Algorithm simulations. Our system was able to exhibit oscillatory behavior, albeit the oscillations are damped. However, as you can see in Figure 5, the oscillations persist with the NWT algorithm (and the ordinary differential equations). Yet, the Gillespie Algorithm will always reach a steady state, whereby the predator and prey species will eventually completely disappear. Since there are no nondeterministic decisions made during the run, we can only attribute the dampened oscillations to the fact that the system is discrete. We will next discuss a circadian rhythm model, which will illustrate how our algorithm can produce Gillespie-like results, even though we have a reduced complexity.

4.2   Circadian Rhythm

Circadian rhythm models are often explored in nature. These act as internal clocks which allow organisms to anticipate daily changes in the environment [1] – for instance, when to hunt for food, when to rest, etc. Yet, at the level of cellular biochemistry, circadian rhythms have also been reported [4]. Biological systems run by internal clocks – that is, certain proteins are created at certain parts of the day. Therefore, simulating circadian rhythm models is important in understanding the way DNA is interpreted and pre-existing proteins waiting to be activated are used by the body for daily survival [1].

We have chosen to model the circadian rhythm model described in [21]. The system describes an activator and a repressor gene (A and R). These genes are transcribed into mRNA, which leads into the translation of the proteins. The activator A binds to the promoters for A and R and increases the transcription rate. The system of ordinary differential equations described in [21] showed that intrinsic biochemical noise enhanced the oscillations. In Equation 8, we see the system of ordinary differential equations for the model.

$$\frac{dD_A}{dt} = \theta_A * D'_A - \gamma_A * D_A * A$$

$$\frac{dD_R}{dt} = \theta_R * D'_R - \gamma_R * D_R * A$$

$$\frac{dD'_A}{dt} = \gamma_A * D'_R * A - \theta_A * D'_A$$

$$\frac{dD'_R}{dt} = \gamma_R * D_R * A - \theta_R * D'_R$$

$$\frac{dM_A}{dt} = \alpha'_A * D'_A + \alpha_A * D_A - \delta_{M_a} * M_A$$

$$\frac{dA}{dt} = \beta_A * M_A + \theta_A * D'_A + \theta_R * D'_R - A * (\gamma_A * D_A + \gamma_R * D_R + \gamma_C * R + \delta_A)$$

$$\frac{dM_R}{dt} = \alpha'_R * D'_R + \alpha_R * D_R - \delta_{M_R} * M_R$$

$$\frac{dR}{dt} = \beta_R * M_R - \gamma_C * A * R + \delta_A * C - \delta_R * R$$

$$\frac{dC}{dt} = \gamma_C * A * R - \delta_A * C \tag{8}$$

where A and R represent the number of activator and repressor proteins, D'_A and D_A represent the number of activator genes with or without binding to A, D'_R and D_R represent the number of repressor genes with or without binding to R, M_A and M_R represent mRNA molecules of A and R, and C represent the corresponding inactivated complex formed by A and R.

Deterministic modeling techniques, like the solution to the systems of ordinary differential equations, for biochemical interactions fail to produce the oscillations of a circadian rhythm model. However, the stochastic noise from a Gillespie-based approach leads to repeated oscillations throughout an entire run. Our NWT algorithm can produce results similar to the Gillespie algorithm – genetic oscillations – but at a considerably reduced computational cost.

The results for the simulation of the circadian rhythm model are shown in Figure 7. We present the results from Gillespie's Algorithm, the solution of the system of ordinary differential equations (Equation 8), and our NWT algorithm. The NWT algorithm is able to reproduce the oscillations for the perturbed model, as is the case with the Gillespie approach [21]. Similar to Gillespie, the NWT shows some variability in both the amplitude – numbers of molecules – and the periodicity of oscillations.

The authors in [21] showed that parameter values can have a profound impact on oscillations. By reduction of the kinetic rate governing R degradation, the deterministic results produce a single peak followed by a steady state, while a stochastic simulation remains oscillating. Our NWT algorithm also produces oscillations instead of a steady state, but at a reduced computational cost from

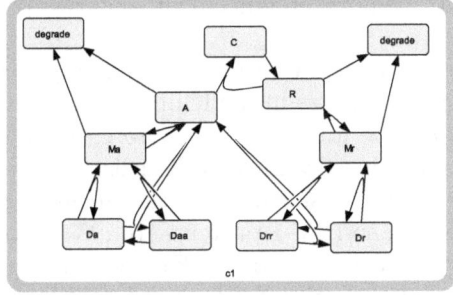

Fig. 6. The picture of this system was generated using CellDesigner. The system was described in [21]. We have modeled this system with the NWT algorithm, Gillespie's Algorithm, and as a system of ordinary differential equations.

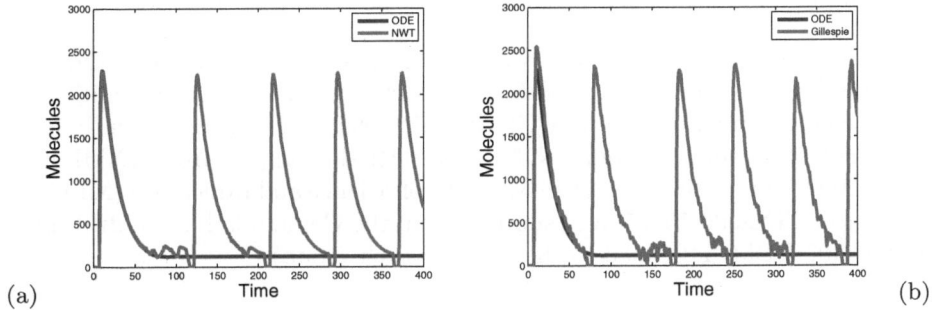

(a)   (b)

Fig. 7. The results for the circadian rhythm model: (a) the NWT algorithm and (b) the Gillespie Algorithm. Both algorithms are plotted against the solution to the system of ordinary differential equations.

the Gillespie Algorithm approach. This is the benefit of modeling with the NWT algorithm instead of the Gillespie Algorithm.

For our simulation to produce oscillations comparable to the Gillespie Algorithm, we require only 50 random numbers to be generated. This stems from the fact that the NWT algorithm relies on deterministic kinetics for the majority of reactions, but when reactants are limited and competition for reactants exists, nondeterministic decisions drive a variable response from the competing reactions.

5   Final Remarks

We have improved the sensitivity of our NWT algorithm through the addition of *memory* to Waiting Time calculation. We argue that this gives us an edge over ordinary differential equations in modeling reactions of low molecular multiplicity. The improvements were illustrated with multiple examples, one designed to specifically discuss the memory enhancement and two other models from the literature.

In the field of systems biology, there is a strong emphasis on using nondeterministic (or stochastic) techniques in modeling biochemical networks where low numbers of molecules can be found. We are interested in exploring these types of situations. For instance, in HIV-infected T cells, there are initially low levels of Tat protein, which after translocation to the nucleus, bind to receptor sites and cause upregulation of the HIV-1 proteins [16]. We have already published a paper on the effects of HIV-1 proteins on Fas-mediated apoptosis, and will be looking to use our refined algorithm for future development in this pathway.

Also, in regards to T cells, it seems that low levels of cytochrome C released from the mitochondria bind to IP$_3$R. This receptor binding leads to release of Ca+ form the mitochondria, and [2] showed that this was necessary for both the extrinsic (Fas-mediated) and the intrinisic apoptotic pathways. We are exploring this direction via wetlab experimentation, and we will be using the NWT algorithm to elucidate new aspects to Fas-mediated apoptotic events.

Acknowledgments. We gratefully acknowledge support in part from the LONI Institute: fellowship for J.J. and state-of-the-art parallel computing facilities, National Science Foundation Grant CCF-0523572, INBRE Program of the NCRR (a division of NIH), support from CNCSIS grant RP-13, support from CNMP grant 11-56 /2007, support from Spanish Ministry of Science and Education (MEC) under project TIN2006-15595, and support from the Comunidad de Madrid (grant No. CCG07-UPM/TIC-0386 to the LIA research group).

References

1. Barkai, N., Leibler, S.: Biological rhythms: Circadian clocks limited by noise. Nature 403, 267–268 (2000)
2. Boehning, D., van Rossem, D.B., Patterson, R.L., Snyder, S.H.: A peptide inhibitor of cytochrome c/inositol 1,4,5-triphosphate receptor binding blocks intrinsic and extrinisc cell death pathways. PNAS 102(5), 1466–1471 (2005)
3. Cheruku, S., Păun, A., Romero-Campero, F., Pérez-Jiménez, M., Ibarra, O.: Simulating FAS-Induced Apoptosis by Using P Systems. In: Proceedings of Bio-inspired computing: theory and applications (BIC-TA), Wuhan, China, September 18-22 (2006); also extended version published as Progress in Natural Science 17(4), 424–431 (2006)
4. Dunlap, J.: Circadian Rhythms: An End in the Beginning. Science 280(5369), 1548–1549 (1998)
5. Funahashi, A., Morohashi, M., Kitano, H.: CellDesigner: a process diagram editor for gene-regulatory and biochemical networks. BIOSILICO 1(5), 159–162 (2003)
6. Funahashi, A., Matsuoka, Y., Jouraku, A., Morohashi, M., Kikuchi, N., Kitano, H.: CellDesigner 3.5: A Versatile Modeling Tool for Biochemical Networks. Proceedings of the IEEE 96(8), 1254–1265 (2008)
7. Gillespie, D.T.: A General Method for Numerically Simulating the Stochastic Time Evolution of Coupled Chemical Reactions. Journal of Computational Physics 22, 403–434 (1976)

8. Gillespie, D.T.: Exact Stochastic Simulation of Coupled Chemical Reactions. Journal of Physical Chemistry 81(25), 2340–2361 (1977)
9. Gibson, M.A., Bruck, J.: Efficient Exact Stochastic Simulation of Chemical Systems with Many Species and Many Channels. Journal of Physical Chemistry A 104, 1876–1889 (2000)
10. Hoops, S., et al.: COPASI – a Complex Pathway Simulator. Bioinformatics 22(24), 3067–3074 (2006)
11. Hua, F., Cornejo, M., Cardone, M., Stokes, C., Lauffenburger, D.: Effects of bcl-2 levels on fas signaling-induced caspase-3 activation: molecular genetic tests of computational model predictions. The Journal of Immunology 175(2), 985–995 (2005); Correction 175(9), 6235–6237 (2005)
12. Hucka, M., et al.: The systems biology markup language (SBML): a medium for representation and exchange of biochemical network models. Bioinformatics 19(4), 524–531 (2003)
13. Jack, J., Romero-Campero, F.J., Perez-Jimenez, M.J., Ibarra, O.H., Păun, A.: Simulating Apoptosis Using Discrete Methods: A Membrane System and a Stochastic Approach. Language Theory in Biocomputing (2007)
14. Jack, J., Rodriguez-Paton, A., Ibarra, O.H., Păun, A.: Discrete Nondeterministic Modeling of the FAS Pathway. Int. J. Found. Comput. Sci. 19(5), 1147–1162 (2008)
15. Jack, J., Păun, A., Rodriguez-Paton, A.: Effects of HIV-1 Proteins on the Fas-mediated Apoptotic Signaling Cascade: A Computational Study of T cell Latency. In: Proceedings of WMC9: 2008. LNCS, vol. 5391, pp. 246–259 (2009)
16. Karn, J.: Tackling Tat. Journal of Molecular Biology 2(22), 235–254 (1999)
17. Kitano, H.: Computational Systems Biology. Nature 420 (2002)
18. Kitano, H.: Systems Biology: A Brief Overview. Science 295, 55–60 (2002)
19. Ma, L., Rice, J.J., Hu, W., Levine, A.J., Stolovitzky, G.A.: A plausible model for the digital response of p53 to DNA damage. PNAS 102(40), 14266–14271 (2005)
20. Selliah, N., Finkel, T.: Biochemical mechanisms of HIV induced T cell apoptosis. Cell Death and Differentiation 8, 127–136 (2001)
21. Vilar, J.M.G., et al.: Mechanisms of noise-resistance in general oscillations. PNAS 99(9), 5988–5992 (2002)
22. Wagner, J., Ma, L., Rice, J.J., Hu, W., Levine, A.J., Stolovitzky, G.A.: p53-Mdm2 loop controlled by a balance of its feedback strength and effective dampening using ATM and delayed feedback. IEE Proc.-Syst. Biol. 152(3), 109–118 (2005)
23. Weinberger, L., Burnett, J., Toettcher, J., Arkin, A., Schaffer, D.: Stochastic Gene Expression in a Lentivrial Positive-Feedback Loop: HIV-1 Tat Fluctuations Drive Phenotypic Diversity. Cell 122(2), 169–182 (2005)
24. Wiley, H.S., Shvartsman, S.Y., Lauffenburger, D.A.: Computational modeling of EGF-receptor system: a paradigm for systems biology. TRENDS in Cell Biology 13(1), 43–50 (2003)

Dynamical Systems and Stochastic Programming: To Ordinary Differential Equations and Back

Luca Bortolussi[1] and Alberto Policriti[2,3]

[1] Dept. of Mathematics and Computer Science, University of Trieste, Italy
luca@dmi.units.it
[2] Dept. of Mathematics and Computer Science, University of Udine, Italy
policriti@dimi.uniud.it
[3] Applied Genomics Institute (IGA), Udine, Italy
policriti@appliedgenomics.org

Abstract. In this paper we focus on the relation between models of biological systems consisting of ordinary differential equations (ODE) and models written in a stochastic and concurrent paradigm (sCCP stochastic Concurrent Constraint Programming). In particular, we define a method to associate a set of ODE's to an sCCP program and a method converting ODE's into sCCP programs. Then we study the properties of these two translations. Specifically, we show that the mapping from sCCP to ODE's preserves rate semantics for the class of biochemical models (i.e. chemical kinetics is maintained) and we investigate the invertibility properties of the two mappings. Finally, we concentrate on the question of behavioral preservation, i.e if the models obtained applying the mappings have the same dynamics. We give a convergence theorem in the direction from ODE's to sCCP and we provide several well-known examples in which this property fails in the inverse direction, discussing them in detail.

1 Introduction

The systemic approach to biology is nowadays a fertile and growing research area, considered by many as a promising track to the understanding of life [41,1]. A key ingredient of systems biology resides in coupling wet lab experiments with mathematical modeling and analysis of bio-systems [33]. Many mathematical instruments have been used for this purpose, some concerned with qualitative analysis, others encapsulating also quantitative data [32]. Quantitative modeling is essentially dominated by two main mathematical tools: (ordinary) differential equations on one side and stochastic processes on the other [32]. Both these methods are concerned with the study of dynamical evolution of systems; however, they differ in the description of the quantities of interest: differential equations represent them as continuous variables, stochastic processes operate, instead, on discrete quantities. Modeling formalisms mixing discrete and continuous ingredients, like hybrid automata [29], have also been used in modeling bio-systems [2].

C. Priami et al. (Eds.): Trans. on Comput. Syst. Biol. XI, LNBI 5750, pp. 216–267, 2009.

We will focus here mainly on the two former approaches, although commenting on the theme of the use of hybrid systems for restoring behavioral equivalence throughout the paper.

The theory of dynamical systems and differential equations (ODE's) is very attractive, being it a mature research area equipped with a huge set of analysis tools, ranging from static analysis of phase space topology to fast simulation via numerical integration [51,43]. However, writing ODE's for a given system is generally a difficult task, requiring a considerable expertise. In addition, the representation of biological entities as continuous variables is an approximation that can sometimes be too rough, especially for low populations [26].

Stochastic processes like Continuous Time Markov Chains (CTMC) [39], on the other hand, do not suffer from these approximation limits, as they represent biological entities as discrete quantities, thus being more adherent to reality. On the other hand, analyzing a stochastic model is much more difficult, both from an analytical and from a computational point of view [53]. Regarding the description of stochastic models, recently we have seen the application of stochastic process algebras (SPA) [48,45], a class of formal languages developed in theoretical computer science as formal tools to analyze (quantitatively) the performances of computing networks. These languages allow to build CTMC-based models following a simple, paradigmatic, identification of biological entities with (computing) processes. Moreover, they are *compositional*, allowing to build models by composing together sub-models.

Ideally, one would like to have a modeling technique that collects the advantages both of stochastic process algebras and differential equations, or, at least, to switch automatically between the two formalisms, depending on the particular task to be performed. In this direction, there are two related problems that must be faced: (a) studying the (mathematical) relation between the two modeling techniques and (b) finding automatic methods for converting one formalism into the other.

More specifically, we suggest the following workflow: first defining translation methods (for a specific process algebra), thus tackling (b), and then studying the mathematical relations intervening between the models obtained applying these translations. In this way we should be able to evaluate the appropriateness of conversion procedures between SPA and ODE's and to restrict the focus of the analysis required by (a).

There are two directions in the conversion between SPA and ODE: the first one associating a set of differential equations to a stochastic process algebra model, and the inverse one, mapping differential equations to stochastic process algebra programs. The first direction can be helpful for the analysis of SPA models, as ODE's can be solved and analyzed more efficiently. Associating SPA to ODE, instead, can help to clarify the logical pattern of interactions that are hidden in the mathematical structure of differential equations. Generally, as process algebra models can be written much more easily than differential equations, even by non-experts (possibly via a graphical interface), the first direction, from

SPA to ODE, looks potentially more fruitful, though having both mappings helps the study of the relationship between the two formalisms.

Supposing to have such transformations at our disposal, a crucial problem is to single out criteria to evaluate and validate them. The first possibility is to inspect the relationship intervening between a SPA program and the associated ODE's only from a mathematical point of view, forgetting any information about the system modeled. As both stochastic processes and differential equations are dynamical systems, this approach essentially corresponds to require that both models exhibit the same behavior, i.e. the same dynamical evolution. Of course, we may require agreement only from a qualitative point of view (so that the qualitative features of the dynamics are the same) or even from a quantitative one (numerical values agree). The difficulty with this approach is that stochastic processes have a noisy evolution, in contrast with the determinism characterizing differential equations. Hence, we need to remove the noise. One possibility is to look only at qualitative features of the dynamics, defining them in a precise way; we will go back to this problem in Section 3.2 below. Otherwise, we may average out noise from the stochastic models, thus considering the expected evolution of the system and requiring it to be described precisely (i.e. quantitatively) by the ODE's. Unfortunately, noise cannot be eliminated so easily, as sometimes it is the driving force of the dynamics [26,52]. Therefore, this second form of equivalence is not completely justified; we will comment more on this point while discussing some examples in the following.

A different approach in comparing stochastic and differential models can be defined if we consider some additional information, which is external to the mathematics of the two models. The idea is to validate the translation w.r.t. this additional information. We explain this point with an example. Consider a model of a set of biochemical reactions; there are different chemical kinetic theories that can be used to describe such system, the most famous one being the principle of mass action. Using such a kinetic theory, we can build (in a canonical way) both a model based on differential equations and a model based on stochastic process algebras. If we are concerned with the principle of mass action more than with dynamical behavior, we may ask that our translation procedures preserves the former, meaning that the ODE's associated to a mass action SPA program are exactly the ODE's built according to mass action principle, and viceversa. Essentially, this corresponds to requiring that the translation procedures defined are *coherent with (some) principles of the system* modeled. For instance, in the case of mass action, coherency corresponds to preserve the meaning of rates (the so called *rate semantics* in [17]). Notice that in this case we are not requiring anything about dynamics, so coherent models may exhibit a divergent behavior, and this is indeed a well known issue, see, for instance, [26] or Section 3.2 below. Therefore, this comparison is essentially different from the behavioral-based one, and it is essentially syntactic, in the sense that it is concerned only with how models are written, not with their time evolution.

The operation of associating ODE's to SPA can be seen also as the definition of an ODE-based semantic for the stochastic processes, as opposed to the

CTMC-based one. Consequently, the comparison of the stochastic model with the derived ODE's can also be seen as an attempt to discover the mathematical relationship between these two semantics.

The problem of associating ODE's to stochastic process algebras has been tackled only recently in literature. The forefather is the work of Hillston [31], associating ODE's to models written in PEPA [30], a stochastic process algebra originally designed for performance modeling. Successively, similar methods have been developed for stochastic π-calculus [16,11,44] and for stochastic Concurrent Constraint Programming [7,12]. All these methods build the ODE's performing a syntactic inspection and manipulation of the set of agents defining the SPA model. In fact, they all satisfy the coherency condition staten above, at least for mass action principle (a proof for stochastic π-calculus can be found in [17]). The inverse problem of associating SPA models to ODE's has received much less attention, the only example being [12], where we use stochastic Concurrent Constraint Programming as target SPA.

In this paper, we will retake the work previously done for sCCP in [12], presenting it in a more detailed and formal way. Basically, we will define two translation procedures: from sCCP to differential equations and viceversa. sCCP plays here a central role, thanks to some ingredients giving a noteworthy flexibility to it, the presence of functional rates above all. Therefore, in Section 2, we will recall the basics of sCCP and its application as a modeling language for biological systems, as presented in [15]. Further details on the language can be found in [8]. The translation procedure from sCCP to differential equations is presented formally in Section 3, while Section 4 is devoted to the presentation of the inverse mapping from general ODE's into sCCP. In Section 3, we will also show coherency conditions for a class of chemical kinetics and we will comment in detail the problem of behavioral equivalence in the conversion from sCCP to ODE's. This will be done mainly via examples, exhibiting biological systems for which the translation preservers also the behavior and other systems whose stochastic models show a different behavior than ODE's. The problem of behavioral equivalence is not new, and in fact some examples that we will give are famous ones [26]. However, the syntactic structure of process algebras in general, and sCCP specifically, give a new flavor to these classical examples, and brings the attention into new ones.

The issue of preservation of dynamic behavior in the mapping from ODE's to sCCP is tackled in Section 4. In this case we are able to exploit the structure of the mapping and thus to give a convergence theorem.

Throughout the paper, we will encounter several situations in which discreteness is a crucial ingredient for the dynamics of the system. This points to a third class of dynamical systems that is in the middle between SPA and ODE's and that can be used to approximate them, namely hybrid automata [29]. In [13,14] we deal with the problem of mapping sCCP programs into hybrid automata, showing that in this case we are able to deal correctly from a behavioral viewpoint with a broader class of sCCP systems. The idea of such mapping is that of translating to ODE's *locally* while retaining some level of discreteness in the finite control of the hybrid automata. By the way, the method of [13] can be extended into a general framework encompassing also the mapping presented in this paper as a particular case.

2   Preliminaries

In this section we briefly recall the basics of stochastic Concurrent Constraint Programming (sCCP, Section 2.1) and its application as a modeling language for biological systems (Section 2.2). The interested reader is referred to [8] for further details. In Section 2.3, we introduce some restrictions on the language that greatly simplify the mapping from and to ODE's.

2.1   Stochastic Concurrent Constraint Programming

Concurrent Constraint Programming (CCP, [49]) is a process algebra having two distinct entities: *agents* and *constraints*. Constraints are interpreted first-order logical formulae, stating relationships among variables (e.g. $X = 10$ or $X+Y < 7$). Agents in CCP, instead, have the capability of adding constraints (`tell`) into a sort of global memory (the *constraint store*) and checking if certain relations are entailed by the current configuration of the constraint store (`ask`). The communication mechanism among agents is therefore asynchronous, as information is exchanged through global variables. In addition to `ask` and `tell`, the language has all the basic constructs of process algebras: non-deterministic choice, parallel composition, procedure call, plus the declaration of local variables.

The stochastic version of CCP (sCCP [7,15]) is obtained by adding a stochastic duration to all instructions interacting with the constraint store $\mathcal{C}$, i.e. `ask` and `tell`. Each instruction has an associated random variable representing time (thus taking values in the positive reals), exponentially distributed with *rate given by a function associating a real number to each configuration of the constraint store*: $\lambda : \mathcal{C} \to \mathbb{R}^+$. This is a unusual feature in traditional stochastic process algebras like PEPA [30] or stochastic π-calculus [44] (although recently introduced in BioPEPA [18]), and it will be crucially used in the translation mechanisms. The syntax of sCCP can be found in Table 1.

Two different kind of actions are present in such table: stochastic actions, having a rate attached to them, and instantaneous actions, having an infinite rate.

Table 1. Syntax of sCCP

$$
\begin{array}{c}
Program = D.A \\
D = \varepsilon \mid D.D \mid p(\mathbf{x}) : -A \\
\pi = \text{tell}_\lambda(c) \mid \text{ask}_\lambda(c) \\
M = \pi.G \mid M + M \\
G = \mathbf{0} \mid \text{tell}_\infty(c).G \mid p(\mathbf{y}) \mid M \mid \exists_x G \mid G \parallel G \\
A = \mathbf{0} \mid \text{tell}_\infty(c).A \mid M \mid \exists_x A \mid A \parallel A
\end{array}
$$

This second class of actions can be used to model the happening of complex atomic events, like a sequence of store updates happening instantaneously. However, only `tell` actions can happen instantaneously, and moreover they are always guarded by a stochastic action. The same restriction applies to recursive calls.

Operational Semantics. The definition of the operational semantics is given specifying two different kinds of transitions: one dealing with instantaneous actions and the other with stochastically timed ones. The basic idea of this operational semantics is to apply the two transitions in an interleaved fashion: first we apply the transitive closure of the instantaneous transition, then we do one step of the timed stochastic transition. To identify a state of the system, we need to take into account both the agents that are to be executed and the current state of the store. Therefore, a configuration will be a point in the space $\mathcal{P} \times \mathcal{C}$, where $\mathcal{P}$ is the space of agents and $\mathcal{C}$ is the space of all possible configurations of the constraint store.

The instantaneous transition $\longrightarrow \subseteq (\mathcal{P} \times \mathcal{C}) \times (\mathcal{P} \times \mathcal{C})$ and the stochastic transition $\Longrightarrow \subseteq (\mathcal{P} \times \mathcal{C}) \times [0,1] \times \mathbb{R}^{+} \times (\mathcal{P} \times \mathcal{C})$ are defined according to the structural rules of Tables 2 and 3, respectively.

The fact that instantaneous actions and recursive calls are guarded by stochastic actions guarantees that $\longrightarrow$ can be applied only for a finite number of steps. Moreover, it can be proven to be confluent, see [8]. With the notation $\overrightarrow{\langle A, d \rangle}$ of Table 3, we denote by the configuration obtained by applying the transitions $\longrightarrow$ as long as it is possible (i.e., by applying the transitive closure of $\longrightarrow$). The confluence property of $\longrightarrow$ implies that $\overrightarrow{\langle A, d \rangle}$ is well defined.

The stochastic transition $\Longrightarrow$, instead, is labeled by two numbers: intuitively, the first one is the probability of the transition, while the second one is its global rate. Note that, after performing one step of the transition $\Longrightarrow$, we apply the transitive closure of $\longrightarrow$. This guarantees that all actions enabled after one $\Longrightarrow$ step are timed.

Using relation $\Longrightarrow$, we can build a labeled transition system, whose nodes are configurations of the system and whose labeled edges correspond to derivable

Table 2. Instantaneous transition for stochastic CCP

$(IR1)$ $\langle \text{tell}_{\infty}(c).A, d \rangle \longrightarrow \langle A, d \sqcup c \rangle$

$(IR2)$ $\langle p(\mathbf{x}), d \rangle \longrightarrow \langle A[\mathbf{x}/\mathbf{y}], d \rangle$ if $p(\mathbf{y}) : -A$

$(IR3)$ $\langle \exists_x A, d \rangle \longrightarrow \langle A[y/x], d \rangle$ with y fresh

$(IR4)$ $\dfrac{\langle A_1, d \rangle \longrightarrow \langle A_1', d' \rangle}{\langle A_1 \parallel A_2, d \rangle \longrightarrow \langle A_1' \parallel A_2, d' \rangle}$

Table 3. Stochastic transition relation for stochastic CCP. The function rate : $\mathcal{P} \times \mathcal{C} \to \mathbb{R}$ assigns to each agent its global rate. Its effect is to recursively traverse the syntactic tree of agents, adding up the rates of active stochastic actions. Its formal definition can be found in [15].

$$(SR1) \qquad \langle \mathrm{tell}_\lambda(c).A, d \rangle \Longrightarrow_{(1,\lambda(d))} \overrightarrow{\langle A, d \sqcup c \rangle}$$

$$(SR2) \qquad \langle \mathrm{ask}_\lambda(c).A, d \rangle \Longrightarrow_{(1,\lambda(d))} \overrightarrow{\langle A, d \rangle} \qquad \text{if } d \vdash c$$

$$(SR3) \qquad \frac{\langle M_1, d \rangle \Longrightarrow_{(p,\lambda)} \overrightarrow{\langle A_1', d' \rangle}}{\langle M_1 + M_2, d \rangle \Longrightarrow_{(p',\lambda')} \overrightarrow{\langle A_1', d' \rangle}}$$
$$\text{with } p' = \frac{p\lambda}{\lambda + \mathrm{rate}(M_2, d)} \text{ and } \lambda' = \lambda + \mathrm{rate}(M_2, d)$$

$$(SR4) \qquad \frac{\langle A_1, d \rangle \Longrightarrow_{(p,\lambda)} \overrightarrow{\langle A_1', d' \rangle}}{\langle A_1 \parallel A_2, d \rangle \Longrightarrow_{(p',\lambda')} \langle A_1' \parallel A_2, d' \rangle}$$
$$\text{with } p' = \frac{p\lambda}{\lambda + \mathrm{rate}(A_2, d)} \text{ and } \lambda' = \lambda + \mathrm{rate}(A_2, d)$$

steps of $\Longrightarrow$. As a matter of fact, this is a multi-graph, as we can derive more than one transition connecting two nodes. Starting from this labeled graph, we can build a Continuous Time Markov Chain (cf. [39] and brlow) as follows: substitute each label (p, λ) with the real number $p\lambda$ and add up the numbers labeling edges connecting the same nodes.

Continuous Time Markov Chains. A Continuous Time Markov Chain (CTMC for short) [39] is a continuous-time stochastic process $(X_t)_{t \geq 0}$ taking values in a discrete set of states S and satisfying the memoryless property:

$$P\{X_{t_n} = s_n \mid X_{t_{n-1}} = s_{n-1}, \ldots, X_{t_1} = s_1\} = P\{X_{t_n} = s_n \mid X_{t_{n-1}} = s_{n-1}\}, \tag{1}$$

for each $n, t_1, \ldots, t_n, s_1, \ldots, s_n$.

A CTMC can be represented as a directed graph whose nodes correspond to the states of S and whose edges are labeled by real numbers, which are the rates of exponentially distributed random variables. In each state there are usually several exiting edges, competing in a race condition in such a way that the fastest one is executed. The time employed by each transition is drawn from the random variable associated to it. When the system changes state, it forgets its past activity and starts a new race condition (this is the *memoryless property*). Therefore, the traces of a CTMC are made by a sequence of states interleaved by variable time delays, needed to move from one state to another.

The time evolution of a CTMC can be characterized equivalently by computing, in each state, the normalized rates of the exit transitions and their sum (called the *exit rate*). The next state is then chosen according to the probability distribution defined by the normalized rates, while the time spent for the transition is drawn from an exponentially distributed random variable with parameter equal to the exit rate. This second characterization is at the basis of several stochastic simulation algorithms for CTMC, like the well-known Gillespie's one [26].

Stream Variables and Implementation. Some variables of the system, like those used in the definition of rate functions, need to store a single number that may vary over time. Such variables, for technical reasons, are conveniently modeled as variables of the constraint store, which, however, must be rigid (over time). To deal with this problem we store time varying parameters as growing lists with an unbounded tail variable. In order to avoid heavy symbolism, we will use a natural notation where $X' = X + 1$ has the intended meaning of[1]: "extract the last ground element n in the list X, consider its successor $n + 1$ and add it to the list (instantiating the old tail variable as a list containing the new ground element and a new tail variable)". We refer to such variables as *stream variables.*

An interpreter for the language is available and can be used for running simulations. This interpreter is written in Prolog and uses standard constraint solver on finite domains as manager for the constraint store. All simulations of sCCP shown in the paper are performed with it.

2.2   Modeling Biological Systems in sCCP

In [8,15] we argued that sCCP can be conveniently used for modeling biological systems. In fact, while maintaining the compositionality of process algebras, the presence of a customizable constraint store and of variable rates gives a great flexibility to the modeler, so that different kinds of biological systems can be easily described within this framework. In [15], we showed that biochemical reactions and genetic regulatory networks are easily handled by sCCP. In [8] we added to this list also formation of protein complexes and the process of folding of a protein, whose description requires knowledge about spatial position of amino acids constituting the protein (a kind of information easily added building on expressive potential of the constraint store).

Finally, in [10] we showed how sCCP can be used to encode Kohn maps [34], a graphical formalism capable of describing implicitly biochemical networks subject to combinatorial explosion of the number of different kinds of protein complexes. In this case, the power of the constraint store is used to maintain a graph-based representation of complexes, allowing a *linear description* of Kohn Maps (i.e., the encoding requires a linear number of characters w.r.t. the ones needed to describe a Kohn map).

We recall now the modeling in sCCP of biochemical reactions. A general biochemical reaction has the form

$$R_1 + \ldots + R_n \rightarrow_{f(\mathbf{R},\mathbf{X};\mathbf{k})} P_1 + \ldots + P_m, \tag{2}$$

---

[1] The use of primed variables to denote values taken at the next time step is typical of model checking and is not to be confused with first derivatives (for which we will used dotted variables, as time is the only independent variable).

where $R_1, \ldots, R_n$ are the *reactants* and $P_1, \ldots, P_m$ are the *products*. The real-valued *kinetic function* of the reaction is $f(\mathbf{R}, \mathbf{X}; \mathbf{k})$, depending on the reactants $\mathbf{R}$, on other molecules $\mathbf{X}$ acting as modifiers, and on some parameters $\mathbf{k}$. This function can be one of the many used in biochemistry (cf. [20,50]) and it is required to satisfy the following *boundary condition*: it must be zero whenever one reactant is less than its amount consumed by the reaction. For instance, if a reactant R appears two times in the left hand size of (2), then f must be zero for $R = 0, 1$.[2]

Biochemical networks can be easily modeled in sCCP taking a *reaction-centric* approach, where each reaction (or action capability) is associated to a process, while molecules, whose concentration varies over time, are represented by integer variables of the constraint store (actually, stream variables). Moreover, the presence of non-constant rates allows to describe reactions with arbitrary chemical kinetics. More specifically, to each reaction like (2), we associate the following sCCP agent:

$$f\text{-reaction}(\mathbf{R}, \mathbf{X}, \mathbf{P}, \mathbf{k}) :-$$
$$\text{tell}_{f(\mathbf{R}, \mathbf{X}; \mathbf{k})} \left(\bigwedge_{i=i}^{n} (R_i - 1) \wedge \bigwedge_{j=i}^{m} (P_j + 1) \right).$$
$$f\text{-reaction}(\mathbf{R}, \mathbf{X}, \mathbf{P}, \mathbf{k})$$

This agent is a simple recursive loop, modifying the value of reactants' and products' variables at a speed given by the kinetic law. Note that the boundary conditions for the rate function f imply that no stream variable will ever become negative, as all reactions that may produce this effect have rate zero[3]. In Table 4 we give a list of some of the most common kinetics: mass action, Michaelis-Menten and Hill kinetics.

In order to describe genetic regulatory networks, instead, we use a modeling style mixing the *reaction-centric* point of view with the more classical *molecular-centric* one. Essentially, genes are described by sCCP agents, while proteins are associated to stream variables, like for biochemical reactions. An example of a genetic network can be found in Section 3.2. More information and examples on modeling biological systems in sCCP can be found in [15].

2.3   Restricted sCCP

The mapping between sCCP and ODE's is not defined for the whole sCCP language, but rather for a restricted version of it, which is, however, sufficient to describe biochemical reaction and genetic networks.

This restricted version of sCCP will be denoted in the following by RE-STRICTED($sCCP$), and is formally specified by the following definition:

---

[2] In case of mass action kinetics, this condition means that the rate for $R + R \to P$ must be $kR(R-1)$ and not kR^2. This is, however, consistent with the definition of the mass action principle in the stochastic setting.

[3] Boundary conditions for f may be relaxed by checking explicitly with ask instructions that variables stay within their domain. For instance, for the reaction $R + R \to P$, we can precede `tell` by `ask`$(R > 1)$. This allows us to use the more common kR^2 as rate function.

Table 4. List of some of the most common types of biochemical reaction, taken from [50]. The first three are first and second order mass-action-like reactions. The second arrow corresponds to a reaction with Michaelis-Menten kinetics. The last arrow replaces Michaelis-Menten kinetics with Hill's one (see [20]).

$$R \to_k P_1 + \ldots + P_m \qquad f_{ma}(R; k) = kR$$

$$R_1 + R_2 \to_k P_1 + \ldots + P_m \qquad f_{ma}(R_1, R_2; k) = kR_1 R_2$$

$$R + R \to_k P_1 + \ldots + P_m \qquad f_{ma}(R; k) = kR(R-1)$$

$$S \mapsto^E_{k,v} P \qquad f_{MM}(S, E; k, v) = \frac{vES}{k+S}$$

$$S \mapsto^E_{K,V_0,h} P \qquad f_{Hill}(S, E; h, k, v) = \frac{vES^h}{k+S^h}$$

Definition 1. *A* RESTRICTED(*sCCP*) *program is a tuple* $(Prog, \mathbf{X}, init(\mathbf{X}))$ *satisfying:*

1. *Prog is an sCCP-program respecting the grammar defined in Table 5.*
2. *The variables used in the definition of agents are taken from a finite set* $\mathbf{X} = \{X_1, \ldots, X_n\}$ *of global stream-variables, each with the same domain* $\mathbb{D}$, *usually* $\mathbb{D} = \mathbb{N}$ *or, more generally,* $\mathbb{D} = \mathbb{Z}$.
3. *The only admissible updates for variables* $\{X_1, \ldots, X_n\}$ *are constraints of the form* $X_i = X_i + k$ *or* $X_i = X_i - k$, *with* $k \in \mathbb{D}$ *constant.*
4. *Constraints that can be checked by ask instructions are finite conjunctions of linear equalities and inequalities.*
5. *The initial configuration of the store is specified by the formula* $init(\mathbf{X})$, *consisting in the following conjunction of constraints:* $(X_1 = x_1^0) \wedge \ldots \wedge (X_n = x_n^0)$, *with the constants* $x_i^0 \in \mathbb{D}$ *referred to as the* initial values *of the sCCP-program.*

This definition can be justified looking at the sCCP-agent associated to a biochemical reaction and also at the sCCP-model of genes considered in [15].

In fact, in these cases all employed variables are numerical variables of the stream-type[4], while all updates in the store add or subtract them a predefined constant quantity. Guards, instead, usually check if some molecules are present in the system ($X > 0$), though we consider here the more general case of linear equalities and inequalities. The use of global variables only, instead, can be justified noting that the existential operator $\exists_x$ is never used (neither in the

---

[4] We do not need further types of variables, as we just need to count the number of different molecules in the system.

Table 5. Syntax of the restricted version of sCCP

$$
\begin{aligned}
Prog &= Def.N & Def &= \varepsilon \mid Def.Def \mid p : -A \\
\pi &= \text{tell}_\lambda(c) \mid \text{ask}_\lambda(c) & M &= \pi.G \mid M + M \\
G &= \text{tell}_\infty(c).G \mid p \mid M & A &= \mathbf{0} \mid M \\
& & N &= A \mid A \parallel N
\end{aligned}
$$

reaction agent nor in gene models of [15]), as the scope of molecular interactions is system-wide. The suppression of the operator $\exists_x$, as a side consequence, guarantees that we can avoid to pass parameters to procedure calls: in fact, each procedure can be defined as operating on a specific subset of global variables. However, parameter passing is used in Section 2.2 to define parametrically the reaction agent. Therefore, we agree that each instance of a reaction agent, say $f\text{-reaction}(\mathbf{R}, \mathbf{X}, \mathbf{P}, \mathbf{k})$, is replaced with the corresponding ground form $f\text{-reaction}_{(\mathbf{R}, \mathbf{X}, \mathbf{P}, \mathbf{k})}$. The same trick will be used for other agents. We demand further comments on the restrictions in Section 3.4.

In order to fix the notation in the rest of the paper, we give the following definition:

Definition 2. *A* RESTRICTED($sCCP$) *agent A not containing any occurrence of the parallel operator $\parallel$ is called a* sequential component *or a* sequential agent. *A* RESTRICTED($sCCP$) *agent N is called an* sCCP-network *if it is the parallel composition of sequential agents.*

Inspecting the grammar of Table 5, we can observe that the initial configuration of a RESTRICTED($sCCP$) program is indeed an sCCP-network. The following property is straightforward:

Lemma 1. *The number of sequential components forming an $sCCP$-program $(Prog, \mathbf{X}, init(\mathbf{X}))$ remains constant at run-time and equals the number of sequential agents in the $sCCP$-network of the initial configuration.*

Proof. As sequential components do not contain any parallel operator, no new agents can be forked at run-time.[5]

In the rest of the paper, for notational convenience, we usually identify an sCCP-program with the corresponding sCCP-network.

Moreover, forbidding the definition of local variables implies the following property:

Lemma 2. *The number of variables involved in the evolution of an $sCCP$-network*[6] *is a subset of $\{X_1, \ldots, X_n\}$, hence finite.*

---

[5] We are counting also deadlocked agents.

[6] A variable is involved in the evolution of the network if one of the following things happen: it is updated in a tell instruction, it is part of a guard checked in an ask instruction, or it is used in the definition of a rate function.

The restrictions of RESTRICTED($sCCP$) are in the spirit of those introduced in [31]: we are forbidding an infinite unfolding of agents and we are considering global interactions only, forcing the speed of each action to depend on the whole state of the system. Indeed, also in [16] we find similar restrictions, though the comparison with sCCP is subtler. First of all, the version of π-calculus presented in [16] does not allow the use of the restriction operator, meaning that interactions have a global scope. However, agents in the π-calculus of [16] are not sequential, as each process is associated to a single molecule and the production of new molecules is essentially achieved by forking processes at run-time. This is not necessary in sCCP, as sCCP-agents model reactions, while molecules are identified by variables of the system. What is finite in [16], however, is the number of syntactically different agents that can be present in a system.

On the Restrictions of the Language. RESTRICTED($sCCP$) limits the full language in three main aspects: the allowance of sequential agents only, the suppression of local variables and the simplifications on the constraint store, cf. Definition 1. We will, however, comment here only on the first one, as the other two are forced by the translation to ODEs, hence their discussion will be postponed to Section 3.4.

As observed, in RESTRICTED($sCCP$) we constrain all the agents to be sequential, i.e. no occurrence of the parallel operator is allowed. Essentially, sequential agents are automata cooperating together, a property that will be exploited in the next section to represent them graphically in a simple way. Indeed, this restriction is only apparent: we can always convert a non-sequential agent into a network of sequential ones using additional (stream) variables of the constraint store. The idea is simply that of identifying all the syntactically different terms that are stochastic choices, associating a new variable to each of them. These variables are used to count the number of copies of each term that are in parallel. Each agents is modified consequently: all agents will only call recursively themselves, while the variations induced in the number of terms by transitions are dealt with by updating the new state variables. Finally, rates are corrected by multiplying them by the multiplicity variable associated to the agent executing the corresponding transition. This is justified by the fact that in Markovian models, the global rate of a set of actions is computed by adding all basic rates together—ultimately, a consequence of the properties of the exponential distribution [39]. For instance, consider the agents x and y, defined by $x :\text{-} \text{tell}_1(true).(y \parallel y)$, $y :\text{-} \text{tell}_1(true).x + \text{tell}_1(true).\mathbf{0}$. They can be made sequential by introducing two variables, X and Y, counting the number of copies of x and y respectively and by replacing x by $x' :\text{-} \text{ask}_X(X > 0).\text{tell}_\infty(X' = X - 1 \wedge Y' = Y + 2).x'$ and y by $y' :\text{-} \text{ask}_Y(Y > 0).\text{tell}_\infty(X' = X + 1 \wedge Y' = Y - 1).y' + \text{ask}_Y(Y > 0).\text{tell}_\infty(Y' = Y - 1).y'$.

3 From sCCP to ODE

In this section we define a translation method associating a set of ordinary differential equations to an sCCP program. This translation applies precisely to

RESTRICTED($sCCP$), as defined above. The procedure is organized in several simple steps, illustrated in the following paragraphs. Essentially, we first associate a finite graph to each sequential component of an sCCP network and then, analyzing the graph, we define an interaction matrix similar to the one defined in [31] or to action matrices of (stochastic) Petri nets (see, for instance, [27]). Writing ODE's from this matrix is then almost straightforward.

After defining this translation, in Section 3.1 we investigate how it relates to biochemical kinetics and we show that the ODE's associated to an sCCP-model of a set of biochemical reactions are the ones generally considered in standard biochemical praxis [17,20]. Some considerations on dynamical properties are then put forward in Section 3.2, while in Section 3.3 the focus is moved on the concept of behavioral equivalence. Finally, in Section 3.4 we reconsider the restrictions applied to sCCP in the light of the described transformation procedure.

Step 1: Reduced Transition Systems

The first step consists in associating a labeled graph, called *reduced transition system* [8], to each sequential agent composing the network. As a working example, we consider the following simple sCCP agent:

$$RW_X :-$$
$$\quad ask_1(X > 0).tell_\infty(X' = X - 1).RW_X$$
$$+ \ tell_1(X' = X + 2).RW_X$$
$$+ \ ask_{f(X)}(true).(\quad ask_1(X > 1).tell_\infty(X' = X - 2).RW_X$$
$$\qquad\qquad\qquad + \ tell_1(X' = X + 1).RW_X$$

$$f(X) = \tfrac{1}{X^2 + 1}$$

This agent performs a sort of random walk in one variable, increasing or decreasing its value by 1 or 2 units, depending on its inner state.

Inspecting Table 5, where the syntax of RESTRICTED($sCCP$) is summarized, we observe that each branch of a stochastic choice starts with a stochastic timed instruction, i.e. an $ask_\lambda(c)$ or a $tell_\lambda(c)$, followed by zero or more $tell_\infty(c)$, followed by a procedure call or by another stochastic choice. The first operation that we need to perform, in order to simplify the structure of agents, is that of collapsing each timed instruction with all the instantaneous `tell` instructions following it and replacing everything with one "action" of the form

$$action(c, d, \lambda),$$

where c is a guard that must be entailed by the store for the branch to be entered, d is the constraint that will be posted to the store, and λ is the stochastic rate of the branch, i.e. a function $\lambda : \mathcal{C} \to \mathbb{R}^+ \cup \{\infty\}$. The presence of ∞ among possible values of λ is needed to simplify the treatment of instantaneous tells.

To achieve this goal, we formally proceed by defining a conversion function, named COMPACT, by structural induction on terms. The result of this function

Table 6. Syntax of COMPACT($sCCP$)

$$\begin{array}{c} Prog = \hat{Def}.\hat{A} \\ \hat{Def} = \varepsilon \mid \hat{Def}.\hat{Def} \mid [\![p]\!] : -\hat{A} \\ \hat{\pi} = action(g, c, \lambda) \qquad \hat{M} = \hat{\pi};\hat{G} \mid \hat{M} \oplus \hat{M} \\ \hat{G} = [\![p]\!] \mid \hat{M} \qquad \hat{A} = [\![\mathbf{0}]\!] \mid \hat{M} \\ \hat{N} = \hat{A} \mid \hat{A} \parallel \hat{N} \end{array}$$

is that of transforming an agent written in RESTRICTED($sCCP$) into an agent of a simpler language, called COMPACT($sCCP$), where `ask` and `tell` are replaced by the instruction `action`. In order to distinguish between the two languages, we denote stochastic summation in COMPACT($sCCP$) by "$\oplus$", sequential composition by ";", and we surround procedure calls and nil agent occurrences by double square brackets "$[\![\cdot]\!]$". The syntax of COMPACT($sCCP$) is formally defined in Table 6; its constraint store, instead, follows the same prescriptions of Definition 1.

In defining the function COMPACT, we use a concatenation operator $\bowtie$ to merge instantaneous tells with the preceding stochastic action. Formally, COMPACT is defined as follows:

Definition 3. *The function* COMPACT: RESTRICTED*($sCCP$)* $\rightarrow$ COMPACT *($sCCP$) is defined by structural induction through the following rules:*

1. COMPACT($\mathbf{0}$) = $[\![\mathbf{0}]\!]$;
2. COMPACT(p) = $[\![p]\!]$.
3. COMPACT($ask_\lambda(c).G$) = $action(c, true, \lambda)$ $\bowtie$COMPACT(G).
4. COMPACT($tell_\lambda(c).G$) = $action(true, c, \lambda)$ $\bowtie$COMPACT(G).
5. COMPACT($tell_\infty(c).G$) = $action(true, c, \infty)$ $\bowtie$COMPACT(G).
6. COMPACT($M + M$) =COMPACT(M)$\oplus$COMPACT(M).

where $\infty : \mathcal{C} \rightarrow \mathbb{R}^+ \cup \{\infty\}$ *is defined by* $\infty(c) = \infty$, *for all* $c \in \mathcal{C}$.

We now define the concatenation operator $\bowtie$:

Definition 4. *The operator* $\bowtie$ *is defined by:*

1. $action(g, c, \lambda) \bowtie [\![p]\!] = action(g, c, \lambda);[\![p]\!]$.
2. $action(g, c, \lambda) \bowtie \hat{M} = action(g, c, \lambda);\hat{M}$.
3. $action(g_1, c_1, \lambda_1) \bowtie action(g_2, c_2, \lambda_2) = action(g_1 \wedge g_2, c_1 \wedge c_2, \min(\lambda_1, \lambda_2))$.

where $\min(\lambda_1, \lambda_2):\mathcal{C} \rightarrow \mathbb{R}^+\cup\{\infty\}$ *is defined by* $\min(\lambda_1, \lambda_2)(c)=\min\{\lambda_1(c), \lambda_2(c)\}$.

Going back to the agent RW$_X$ previously defined; if we apply the function COMPACT to it, we obtain the following agent:

COMPACT(RW$_X$) :-
$\qquad action(X > 0, true, 1) \bowtie action(true, X' = X - 1, \infty) \bowtie [\![RW_X]\!]$
$\qquad \oplus \quad action(true, X' = X + 2, 1) \bowtie [\![RW_X]\!]$

$\oplus$ action$(true, true, f(X))$ $\bowtie$
(action$(X > 1, true, 1)$ $\bowtie$ action$(true, X' = X - 2, \infty)$ $\bowtie$ $[\![\mathrm{RW}_X]\!]$
$\oplus$ action$(true, X' = X + 1, 1)$ $\bowtie$ $[\![\mathrm{RW}_X]\!]$)

After the removal of $\bowtie$ operator according to the rules in Definition (4), the agent COMPACT(RW_X) becomes a COMPACT(sCCP) agent:

COMPACT$(\mathrm{RW}_X) = [\![\mathrm{RW}_X]\!]$:-
action$(X > 0, X' = X - 1, 1)$; $[\![\mathrm{RW}_X]\!]$
$\oplus$ action$(true, X' = X + 2, 1)$; $[\![\mathrm{RW}_X]\!]$
$\oplus$ action$(true, true, f(X))$;
(action$(X > 1, X' = X - 2, 1)$; $[\![\mathrm{RW}_X]\!]$
$\oplus$ action$(true, X' = X + 1, 1)$; $[\![\mathrm{RW}_X]\!]$)

The above example shows clearly that the function COMPACT simply collapses all the actions performed on the store after one execution of the stochastic transition, as defined in [7,8]. It is a simple exercise to define a stochastic transition relation for COMPACT(sCCP) similar to the one for RESTRICTED(sCCP) and to successively prove the strong equivalence [30] between agents A and COMPACT(A).[7] Hence, from a semantic point of view, the application of function COMPACT is safe, as stated in the following

Lemma 3. *For each sequential agent A of* RESTRICTED(sCCP)*, A and $\hat{A} =$* COMPACT(A) *are strongly equivalent.*

Let $\hat{A} =$COMPACT(A) be an agent of COMPACT(sCCP). We want to associate a graph to such agent, containing all possible actions that $\hat{A}$ may execute. Nodes in such graph will correspond to different internal states of $\hat{A}$, i.e. to different stochastic branching points. Edges, on the other hand, will be associated to actions: each edge will correspond to one action(g, c, λ) instruction and will be labeled consequently by the triple (g, c, λ).

To define such graph, we proceed in two simple steps:

1. First we define an equivalence relation $\equiv_c$ over the set of COMPACT(sCCP) agents, granting associativity and commutativity to $\oplus$ and reducing procedure calls to automatic "macro-like" substitutions (a reasonable move as we do not pass any parameter). We will then work on the set $\mathcal{A}$ of COMPACT(sCCP) agents modulo $\equiv_c$, called the set of *states*; notably, all agents in $\mathcal{A}$ are stochastic summations.
2. Then, we define a structural operational semantics [42] on COMPACT(sCCP), whose labeled transition system (LTS) will be exactly the target graph.

Definition 5. *The equivalence relation $\equiv_c$ between* COMPACT(sCCP) *agents is defined as the minimal relation closed with respect to the following three rules:*

---

[7] Strong equivalence [30] is a form of bisimulation preserving probabilities: two agents are strongly equivalent if their exit rates are the same and transitions of one agent can be matched by transitions of the other having the same probability.

1. $\hat{M}_1 \oplus \hat{M}_2 \equiv_c \hat{M}_2 \oplus \hat{M}_1$;
2. $\hat{M}_1 \oplus (\hat{M}_2 \oplus \hat{M}_3) \equiv_c (\hat{M}_1 \oplus \hat{M}_2) \oplus \hat{M}_3$;
3. $[\![p]\!] \equiv_c \hat{A}$ if $[\![p]\!] : -\hat{A}$ belongs to the declarations $\hat{D}$.

The space of COMPACT($sCCP$) agents modulo $\equiv_c$ is denoted by $\mathcal{A}$, and is referred to as the space of states.

Definition 6. The transition relation $\rightsquigarrow \subseteq \mathcal{A} \times (\mathcal{C} \times \mathcal{C} \times \mathbb{R}^{\mathcal{C}}) \times \mathcal{A}$ is defined in the SOS style as the minimal relation closed with respect to the following rule:

$$action(g, c, \lambda); \hat{G} \oplus \hat{M} \overset{(g,c,\lambda)}{\rightsquigarrow} \hat{G}.$$

The transition relation $\rightsquigarrow$ encodes the possible actions that a COMPACT($sCCP$) agent can undertake. Notice that procedure calls are automatically solved as we are working modulo $\equiv_c$. The relation $\rightsquigarrow$ induces a labeled graph, its labeled transition system (LTS), whose nodes are agents in $\mathcal{A}$ and whose edges are labeled by triples (g, c, λ), where $g \in \mathcal{C}$ is a guard, $c \in \mathcal{C}$ is the update of the store, and λ the functional rate of the edge.

Definition 7. Let $\hat{A}$ be an agent of COMPACT($sCCP$); the portion of the labeled transition system reachable from the state $\hat{A}$ is denoted by $LTS(\hat{A})$.

Theorem 1. For any agent $\hat{A}$ of COMPACT($sCCP$) (modulo $\equiv_c$), $LTS(\hat{A})$ is finite.

Proof. The agents reachable from $\hat{A}$ are subagents of $\hat{A}$ or subagents of $\hat{A}'$, where $[\![p]\!] : -\hat{A}'$ is a procedure called by an agent reachable from $\hat{A}$. The number of subagents of $\hat{A}$ (modulo $\equiv_c$) corresponds to the number of summations present in $\hat{A}$, and it is finite for any definable agent. The proposition follows because there is only a finite number of agents defined in the declarations $\hat{D}$.

We are finally ready to define the reduced transition system for an agent A of RESTRICTED($sCCP$).

Definition 8. Let A be an agent of RESTRICTED $(sCCP)$. Its reduced transition system $RTS(A)$ is a finite labeled multigraph $(S(A), T(A), \ell_A)$ defined by

$$RTS(A) = LTS(\text{COMPACT}(A)).$$

Given $RTS(A) = (S(A), T(A), \ell_A)$, $S(A) \subseteq \mathcal{A}$ is the set of RTS-states reachable from agent COMPACT(A), finite for Theorem 1, $T(A)$ is the set of RTS-edges or RTS-transitions and $\ell_A : T(A) \rightarrow \mathcal{C} \times \mathcal{C} \times \mathbb{R}^{\mathcal{C}}$ is the label function assigning to each RTS-edge the triple (g, c, λ), $g, c \in \mathcal{C}$, $\lambda : \mathcal{C} \rightarrow \mathbb{R}^+$.
   In order to effectively compute the RTS of an agent $\hat{A}$ of COMPACT($sCCP$), we can proceed as follows:

1. Given an COMPACT($sCCP$) program, write the syntactic tree of the agent $\hat{A}$ and of all the agents $\hat{A}'$ such that $[\![p]\!]:-\hat{A}'$ is in $\hat{Def}$.

2. Nodes corresponding to action(g, c, λ) instructions will have one single incoming edge and one single outgoing edge. Remove them, connecting the entering and exiting edges and labeling them by the triple c, g, λ).
3. Leaves of the obtained labeled tree correspond either to nil agents or to procedure calls. The latter are replaced according to the following rule: if the syntactic tree of the procedure $[p]$ has not been added in the current tree, replace the leaf labeled with $[p]$ with the corresponding syntactic tree; otherwise, remove the leaf and redirect the incoming edge to the root of the copy of the syntactic tree of $[p]$. Iterate the application of the rule until no more leaves corresponding to procedure calls are available.

The previous procedure always terminates, as the number of different procedures $[p]$ in $\hat{Def}$ is finite, hence the algorithm needs to process only a finite number of leaves.

Going back to our running example, $RTS(\text{RW}_X) = LTS(\text{COMPACT}(\text{RW}_X))$ is shown in the figure below. Note that it has one RTS-edge for every action that can be performed by $\text{COMPACT}(\text{RW}_X)$, and just two RTS-states, corresponding to the two summations present in $\text{COMPACT}(\text{RW}_X)$. Three intermediate steps in the construction of the RTS can be visualized in Figure 1.

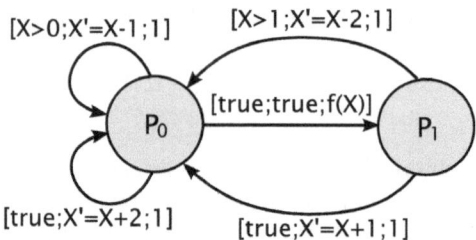

Step 2: The Interaction Matrix

Our next step consists in encoding all the information about the dynamics in a single *interaction matrix* and in a *rate vector*. Consider the initial sCCP-network $N = A_1 \parallel \ldots \parallel A_h$ of a RESTRICTED($sCCP$) program $(Prog, \mathbf{X}, init(\mathbf{X}))$, with sequential components $A_1, \ldots, A_h$. First of all, we construct the reduced transition system for all the components, i.e. $RTS(A_1) = (S(A_1), T(A_1), \ell_{A_1}), \ldots,$ $RTS(A_h) = (S(A_h), T(A_h), \ell_{A_h})$. Then we construct the set of RTS-states and RTS-transitions of the network (we agree that states and transitions belonging to different components $A_1, \ldots, A_h$ are distinct[8]), putting:

$$S(N) = S(A_1) \cup \ldots \cup S(A_h) \tag{3}$$

and

$$T(N) = T(A_1) \cup \ldots \cup T(A_h).^9 \tag{4}$$

---

[8] If the same component is present in multiple copies, we distinguish among them by suitable labels.

[9] The labeling function acting on $T(N)$ will be denoted consistently by ℓ_N.

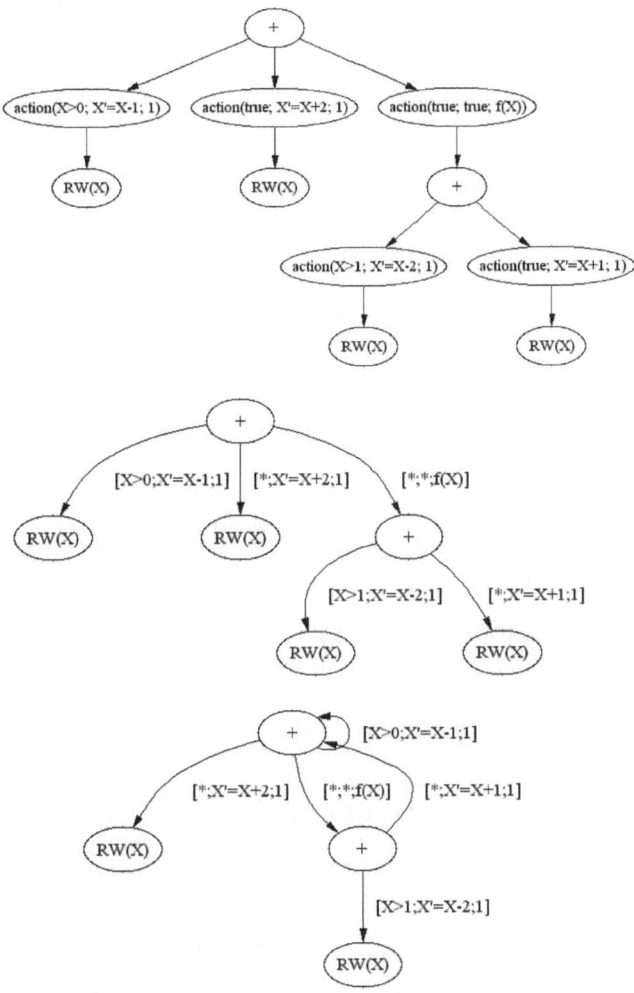

Fig. 1. Three intermediate steps of the construction of RTS for the agent $RW(X)$, according to the procedure sketched in the main body of the paper. The top one is the outcome of point 1, the middle one is the labeled tree obtained after step 2, while the bottom one is the result of applying twice the rule of step 3.

Suppose now that there are m RTS-states in $S(N)$ and k RTS-transitions in $T(N)$. We conveniently fix an ordering of these two sets, say $S(N)=\{\sigma_1,\ldots,\sigma_m\}$ and $T(N) = \{t_1,\ldots,t_k\}$.

The variables $\mathbf{Y}$ of the differential equations are of two different kinds, $\mathbf{Y} = \mathbf{X} \cup \mathbf{P}$. The first type corresponds to the global stream variables of the store, i.e $\mathbf{X} = \{X_1,\ldots,X_n\}$ (see Definition 1). In addition, we associate a variable of the second type $P_{\sigma_i} = P_i$ to each RTS-state of $S(N) = \{\sigma_1,\ldots,\sigma_m\}$, so $\mathbf{P} = \{P_1,\ldots,P_m\}$. For the manipulations to follow, we assume the existence of a lexicographic ordering among all variables, so that vectors and matrices depending on this ordering are defined uniquely and manipulated consistently. Moreover, variables will be also used to index of these objects.

Consider now an RTS-transition $t_j \in T(N)$, connecting RTS-states σ_{j_1} and σ_{j_2}, and suppose $\ell_N(t_j) = (g_j, c_j, \lambda_j)$. We introduce the following notation:

- $\mathrm{rate}_j^N(\mathbf{X}) = \lambda_j(\mathbf{X})$ is the rate function of t_j;
- $\mathrm{guard}_j^N(\mathbf{X})$ is the indicator function of g_j (by Definition 1, g_j is a conjunction of linear equalities and inequalities), i.e.

$$\mathrm{guard}_j^N(\mathbf{X}) = \begin{cases} 1 & \text{if } g_j \text{ is true for } \mathbf{X}, \\ 0 & \text{otherwise.} \end{cases} \tag{5}$$

We are now able to define the *rate vector*:

Definition 9. *The* rate vector r^N *for transitions* $T(N) = \{t_1, \ldots, t_k\}$ *is a k-dimensional vector of functions, whose components r_j^N are defined by*

$$r_j^N(\mathbf{Y}) = \mathrm{rate}_j^N(\mathbf{X}) \cdot \mathrm{guard}_j^N(\mathbf{X}) \cdot P_{j_1}, \tag{6}$$

where P_{j_1} is the variable associated to the source state σ_{j_1} of transition t_j.

Consider again a transition $t_j \in T(N)$, going from σ_{j_1} to σ_{j_2} and with label $\ell_N(t_j) = (g_j, c_j, \lambda_j)$, and consider the updates c_j, a conjunction of constraints of the form $X_i = X_i \pm k$, according to Definition 1. We can assume that each variable X_i appears in at most one conjunct of c_j.[10] We are now ready to define the *interaction matrix*.

Definition 10. *The* interaction matrix $I_{\mathbf{Y}}^N$ *for an sCPP-network N with respect to variables* $\mathbf{Y}$ *is an integer-valued matrix with $n+m$ rows (one for each variable of $\mathbf{Y}$) and k columns (one for each RTS-transition $T(N)$), defined by:*

1. *If $\sigma_{j_1} \neq \sigma_{j_2}$, then $I_{\mathbf{Y}}^N[P_{j_1}, t_j] = -1$ and $I_{\mathbf{Y}}^N[P_{j_2}, t_j] = 1$.*
2. *If $X_h = X_h \pm k$ is a conjunct of c_j, then $I_{\mathbf{Y}}^N[X_h, t_j] = \pm k$.*
3. *All entries not set by points 1,2 above are equal to zero.*

For the agent RW_X, the interaction matrix $I_{\mathbf{Y}}^{RW_X}$ for the variables $\mathbf{Y} = \{X, P_0, P_1\}$ is:

$$I_{\mathbf{Y}}^{RW_X} = \begin{pmatrix} -1 & +2 & 0 & -2 & +1 \\ 0 & 0 & -1 & +1 & +1 \\ 0 & 0 & +1 & -1 & -1 \end{pmatrix} \begin{matrix} (X) \\ (P_0) \\ (P_1) \end{matrix} \tag{7}$$

Similarly, the rate vector r^{RW_X} is

$$r^{RW_X} = \left(P_0 \langle X > 0 \rangle, \, P_0, \, f(X)P_0, \, P_1 \langle X > 1 \rangle, \, P_1, \right)^T, \tag{8}$$

where $\langle \cdot \rangle$ denotes the logical value of a formula (i.e., 1 if the formula is true, 0 otherwise).

---

[10] If, for instance, both $X_i = X_i + k_1$ and $X_i = X_i + k_2$ are in c_j. Then we can replace these two constraints with $X_i = X_i + (k_1 + k_2)$.

Step 3: Writing ODE's

Once we have the interaction matrix, writing the set of ODE's is very simple: we just have to multiply matrix $I_{\mathbf{Y}}^N$ by the (column) rate vector r^N, in order to obtain the vector $ode_{\mathbf{Y}}^N$:

$$ode_{\mathbf{Y}}^N = I_{\mathbf{Y}}^N \cdot r^N. \tag{9}$$

Each row of the $ode_{\mathbf{Y}}^N$ vector gives the differential equation for the corresponding variable. Specifically, the equation for variable Y_i is

$$\dot{Y}_i = ode_{\mathbf{Y}}^N[Y_i] = \sum_{j=1}^{k} I_{\mathbf{Y}}^N[Y_i, j] \cdot r_j^N(\mathbf{Y})$$

$$= \sum_{j=1}^{k} \left(I_{\mathbf{Y}}^N[Y_i, j] \cdot \text{guard}_j(\mathbf{X}) \cdot \text{rate}_j(\mathbf{X}) \cdot P_{j_1} \right)$$

For instance, the set of ODE's associated to the agent RW_X is

$$\begin{cases} \dot{X} = P_0(2 - \langle X > 0 \rangle) + P_1(1 - 2\langle X > 1 \rangle) \\ \dot{P}_0 = -\frac{1}{X^2+1}P_0 + P_1(1 + \langle X > 1 \rangle) \\ \dot{P}_1 = \frac{1}{X^2+1}P_0 + P_1(1 + \langle X > 1 \rangle) \end{cases}$$

In order to solve a set of ODE's, we need to fix the *initial conditions*. The variables $\mathbf{Y} = \mathbf{X} \cup \mathbf{P}$ of $ode_{\mathbf{Y}}^N$ are of two distinct types: $\mathbf{P}$, denoting states of the reduced transition systems of the components, and $\mathbf{X}$, representing stream variables of the store. The initial conditions for $\mathbf{P}$ are easily determined: we set to one all the variables corresponding to the initial states of RTS of each component, and to 0 all the others. Regarding $\mathbf{X}$, instead, initial conditions are given in the formula $init(\mathbf{X})$ of Definition 1, specifying the values assigned to stream variables before starting the execution of the sCCP program.

Elimination of Redundant State Variables

Consider an sCCP component A whose reduced transition system $RTS(A)$ has just one RTS-state. Then, the $ode_{\mathbf{Y}}^N$ vector of an sCCP-network N having A as one of its components will contain a variable corresponding to this RTS-state, say P_i, with equation $\dot{P}_i = 0$ and initial value $P_i(0) = 1$. Clearly, such variable is redundant, and we can safely remove it by setting $P_i \equiv 1$ in all equations containing it and by eliminating its equation from the $ode_{\mathbf{Y}}^N$ vector. From now on, we assume that this simplification has always been carried out. As an example, consider the following agent

$$\begin{aligned} A &:\text{-}\ \text{tell}_{f_1(X)}(X = X + 1).A \\ &+\ \text{tell}_{f_2(X)}(X = X - 1).A \end{aligned}$$

Its RTS contains just one state, corresponding to the only summation present in it, with associated variable P. As the other variable of the agent is X, the vector $ode^A_{\{X,P\}}$ contains two equations, namely

$$\begin{pmatrix} \dot{X} \\ \dot{P} \end{pmatrix} = \begin{pmatrix} f_1(X)P - f_2(X)P \\ 0 \end{pmatrix}.$$

The simplification introduced above just prescribes to remove the equation for P, setting its value to 1 in the other equations; therefore we obtain

$$ode^A_{\{X\}} = (f_1(X) - f_2(X)).$$

Notice that the set of variables $\mathbf{Y}$ is updated consistently, i.e. removing the canceled variables from it.

We summarize the whole method just presented defining the following operator.

Definition 11. *Let N be the sCCP-network of an* RESTRICTED(*sCCP*) *program. With*

$$ODE(N)$$

we denote the vector $ode^N_{\mathbf{Y}}$ associated to N by the translation procedure previously defined, after applying the removal of state variables coming from network components with just one RTS-state.

Compositionality of the Transformation Operator

In order to clearly state formal properties of the transformation, we need a version of the $ODE(N)$ indicating explicitly the variables $\mathbf{X}$ for which the differential equations are given. In the following, the variables for the equations $ODE(N)$ are indicated by $VAR(ODE(N))$.

Definition 12. *Let N be the sCCP-network of an* RESTRICTED(*sCCP*) *program and let $\mathbf{Y} = VAR(ODE(N))$ and $ODE(N) = ode^N_{\mathbf{Y}}$. The ordinary differential equations of N with respect to the set of variables $\mathbf{X}$, denoted by $ODE(N, \mathbf{X})$, is defined as*

$$ODE(N, \mathbf{X})[X_i] = \begin{cases} ode^N_{\mathbf{Y}}[Y_j] & \text{if } X_i = Y_j \in \mathbf{Y}, \\ 0 & \text{otherwise.} \end{cases}$$

The operations performed on $ODE(N)$ by $ODE(N, \mathbf{X})$ simply consist in the elimination of the equations of $ODE(N)$ for the variables not in $\mathbf{X}$, and in the addition of equations $\dot{X} = 0$ for all variables X in $\mathbf{X}$ but not in $\mathbf{Y}$. We can also associate a new interaction matrix $I^N_{\mathbf{X}}$ to $ODE(N, \mathbf{X})$, whose rows are derived according to Definition 12. As the set of RTS-transitions $T(N)$ is unaltered by $ODE(N, \mathbf{X})$, the equation $ODE(N, \mathbf{X}) = I^N_{\mathbf{X}} \cdot r^N$ continues to hold.

We can now prove the following theorem, stating compositionality of ODE operator.

Theorem 2. *Let N_1, N_2 be two sCCP-networks, and let $N = N_1 \parallel N_2$ be their parallel composition. If $\mathbf{Y_1} = VAR(ODE(N_1))$, $\mathbf{Y_2} = VAR(ODE(N_2))$, and $\mathbf{Y} = \mathbf{Y_1} \cup \mathbf{Y_2}$, then*[11]

$$ODE(N_1 \parallel N_2, \mathbf{Y}) = ODE(N_1, \mathbf{Y}) + ODE(N_2, \mathbf{Y}).$$

Proof. The components in $N_1 \parallel N_2$ are the components of N_1 plus the components of N_2. Therefore, the set of RTS-transitions (i.e. edges in the RTS of the components) $T(N_1 \parallel N_2)$ of $N_1 \parallel N_2$ is equal to $T(N_1) \cup T(N_2)$. As each column of $I_{\mathbf{Y}}^{N_1 \parallel N_2}$ is either a transition of $T(N_1)$ or a transition of $T(N_2)$, it clearly holds $I_{\mathbf{Y}}^{N_1 \parallel N_2}[Y_i, t_j] = I_{\mathbf{Y}}^{N_h}[Y_i, t_j]$ if $t_j \in T(N_h)$, $h = 1, 2$. The following chain of equalities then follows easily from the definitions:

$$
\begin{aligned}
ODE(N_1 \parallel N_2, \mathbf{Y})[Y_i] &= \sum_{t_j \in T(N_1 \parallel N_2)} I_{\mathbf{Y}}^{N_1 \parallel N_2}[Y_i, t_j] r_j^{N_1 \parallel N_2} \\
&= \sum_{t_j \in T(N_1)} I_{\mathbf{Y}}^{N_1}[Y_i, t_j] r_j^{N_1} + \sum_{t_j \in T(N_2)} I_{\mathbf{Y}}^{N_2}[Y_i, t_j] r_j^{N_2} \\
&= ODE(N_1, \mathbf{Y}) + ODE(N_2, \mathbf{Y}).
\end{aligned}
$$

3.1 Preservation of Rate Semantics

In Section 2.2 we discussed how biochemical reaction with general kinetic laws can be modeled in sCCP. Given a list of reactions, the standard praxis in computational chemistry is that of building a corresponding differential (set of ODE's) or stochastic (CTMC) model. The definition of such models is *canonical*, and it is fully specified by the reaction list, see [53] for further details.

sCCP models of biochemical reactions are generated by associating an agent, call it *biochemical agent*, to each reaction in the list. These agents are rather simple: they can execute in one single way, namely an infinite loop consisting of activation steps, where the agents compete stochastically for execution, with rate given by the kinetic law specified in the reaction arrow, and update steps, in which the store is modified according to the prescriptions of the reaction. This bijective mapping between reactions and sCCP biochemical agents soon implies that the stochastic model for sCCP is identical to the continuous-time Markov chain generated in classical stochastic simulations with Gillespie algorithm [26], for instance like those obtainable with a program like Dizzy [46,19].[12] A different question is wether the ODE's that are associated to an sCCP model of biochemical reactions coincide with the canonical ones. In the rest of the section, we show that this is indeed the case.

Following the approach by Cardelli in [17], we can then say that the translation from sCCP to ODE's *preserves the rate semantics*. The sense of this sentence is better visualized in Figure 2, graphically depicting the correspondence between

---

[11] Here "+" denotes the usual sum of vectors.

[12] Stochastic simulations with Michaelis-Menten and Hill rate functions have been considered, for instance, in [47].

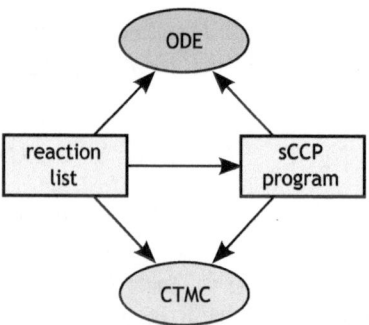

Fig. 2. Diagram of relations intervening between stochastic and ODE-based semantics of chemical reactions and sCCP agents

stochastic and differential models of biochemical reactions and of the derived sCCP agents. Preservation of rate semantics essentially means that the arrows in the diagram commute.

As a matter of fact, in [17] the author deals only with mass action kinetics, due to the intrinsic properties of π-calculus (all definable rates are mass-action like). In our setting, instead, functional rates and the constraint store allow us to deal with arbitrary chemical kinetics, including also Michaelis-Menten and Hill ones (cf. [20]). In the following, we formally prove the equivalence of ODE's obtained from sCCP agents with the corresponding classical ones.

In general, the stochastic and the deterministic rate of a reaction are not the same, because ODE's variables measure concentration, while sCCP variables count the number of molecules. Therefore, in passing from one model to the other, we need to convert numbers to concentrations, dividing by the volume V times the Avogadro number N_A ($\gamma = V N_A$ will be referred as *system size*). Rates need also to be scaled consistently, see [53] for further details. In the rest of this section, however, we get rid of scaling problems simply by assuming $\gamma = 1$. In any case, system size can be reintroduced without difficulties, by change the scale of rates and variables after the derivation of ODE's.

We now put forward some notation, in order to specify how to formally derive a set of ODE's given a set of reactions $\mathcal{R} = \{\rho_1, \ldots, \rho_k\}$, where each ρ_i denotes a single reaction. Each reaction ρ has some attributes: a multiset of reactants (species can have a specific multiplicity), denoted by $REACT(\rho)$, a multiset of products, $PROD(\rho)$, and a real-valued rate function, $RATE(\rho)$, depending on the variables associated to the molecules involved in the reaction, $VAR(\rho)$. This last function, VAR, can be easily extended to sets of reactions by letting $VAR(\{\rho_1, \ldots, \rho_k\}) = VAR(\rho_1) \cup \ldots \cup VAR(\rho_k)$.

Now let $\mathcal{R}$ be a set of reactions and $\mathbf{X} = VAR(\mathcal{R})$. The (canonical) differential equations associated to $\mathcal{R}$ w.r.t. variables $\mathbf{X}$, denoted by $ODE(\mathcal{R}, \mathbf{X})$[13] are defined for each variable X_i as $\dot{X}_i = ODE(\mathcal{R}, \mathbf{X})[X_i]$, where:[14]

---

[13] We overload here the symbol introduced in Definitions 11 and 12; however, the two cases can be easily distinguished looking at their first argument.

[14] We conveniently identify each variable with the molecule it represents.

$$ODE(\mathcal{R}, \mathbf{X})[X_i] = \sum_{\substack{\rho \in \mathcal{R}: \\ X_i \in PROD(\rho)}} RATE(\rho) - \sum_{\substack{\rho \in \mathcal{R}: \\ X_i \in REACT(\rho)}} RATE(\rho). \quad (10)$$

If X_i is not involved in any reaction of $\mathcal{R}$, then $ODE(\mathcal{R}, \mathbf{X})[X_i] = 0$. We observe that the correct stoichiometry is automatically dealt with by the fact that we are using multisets to list reactants and products. Restricting this construction to a fixed variable ordering allows to state the following straightforward compositionality lemma.

Lemma 4. *Let* $\mathcal{R} = \mathcal{R}_1 \cup \mathcal{R}_2$ *be a partition of* $\mathcal{R}$ *and let* $\mathbf{X}_1 = VAR(\mathcal{R}_1)$, $\mathbf{X}_2 = VAR(\mathcal{R}_2)$, *and* $\mathbf{X} = \mathbf{X}_1 \cup \mathbf{X}_2$. *Then*

$$ODE(\mathcal{R}_1 \cup \mathcal{R}_2, \mathbf{X}) = ODE(\mathcal{R}_1, \mathbf{X}) + ODE(\mathcal{R}_2, \mathbf{X}).$$

We turn now to formally define the encoding of reactions into sCCP agents. For each reaction ρ, its sCCP agent $SCCP(\rho)$ is constructed according to Section 2.2. Operator $SCCP$ is extended compositionally to sets of reactions $\mathcal{R} = \{\rho_1, \ldots, \rho_k\}$ by letting $SCCP(\mathcal{R}) = SCCP(\rho_1) \parallel \ldots \parallel SCCP(\rho_k)$.

We are finally ready to state the theorem of preservation of rate semantics:

Theorem 3 (Preservation of rate semantics). *Let* $\mathcal{R}$ *be a set of biochemical reactions, with* $\mathbf{X} = VAR(\mathcal{R})$. *Then*

$$ODE(\mathcal{R}, \mathbf{X}) = ODE(SCCP(\mathcal{R}), \mathbf{X}) \quad (11)$$

Proof. We prove the theorem by induction on the size k of the set of reactions $\mathcal{R}$. For the base case $k = 1$, consider a reaction ρ

$$R_1 + \ldots + R_n \rightarrow_{f(\mathbf{R}, \mathbf{X}; k)} P_1 + \ldots + P_m$$

and its associated sCCP agent $SCCP(\rho)$

$$SCCP(\rho) :\text{-- tell}_{f(\mathbf{R}, \mathbf{X}; k)} \left(\bigwedge_{i=i}^{n} (R_i - 1) \wedge \bigwedge_{j=i}^{m} (P_j + 1) \right) . SCCP(\rho).$$

Clearly, the reduced transition system of such agent is

$$\left. P \right\rangle \quad [\text{true}; \bigwedge_{i,j} (R'_i = R_i \text{-} 1, P'_j = P_j \text{+} 1); f(R, X; k)\,]$$

Let $\mathbf{Y} = VAR(\rho)$, then the interaction matrix $I_{\mathbf{Y}}^{SCCP(\rho)}$ has $|\mathbf{Y}|$ rows and 1 column, with entries corresponding to the stoichiometry of the reaction:

$$I_{\mathbf{Y}}^{SCCP(\rho)}[Y_i] = \sum_{Y_i \in PROD(\rho)} 1 - \sum_{Y_i \in REACT(\rho)} 1.$$

The rate vector $r^{SCCP(\rho)}$, instead, is the scalar $f(\mathbf{R}, \mathbf{X}; \mathbf{k})$. Hence, the equation for variable Y_i is

$$\dot{Y}_i = \sum_{Y_i \in PROD(\rho)} f(\mathbf{R}, \mathbf{X}; \mathbf{k}) - \sum_{Y_i \in REACT(\rho)} f(\mathbf{R}, \mathbf{X}; \mathbf{k}),$$

which is equal to equation (10).

The inductive case follows easily from compositionality properties of ODE operators. Suppose the theorem holds for lists up to $k-1$ reactions, and let $\mathcal{R} = \mathcal{R}_0 \cup \{\rho\}$ be a set of k chemical reactions (hence $|\mathcal{R}_0| = k-1$). Then

$$
\begin{aligned}
ODE(SCCP(\mathcal{R}), \mathbf{X}) &= ODE(SCCP(\mathcal{R}_0 \cup \{\rho\}), \mathbf{X}) \\
&= ODE(SCCP(\mathcal{R}_0) \parallel SCCP(\rho), \mathbf{X}) \\
&= ODE(SCCP(\mathcal{R}_0), \mathbf{X}) + ODE(SCCP(\rho), \mathbf{X}) \\
&= ODE(\mathcal{R}_0, \mathbf{X}) + ODE(\rho, \mathbf{X}) \\
&= ODE(\mathcal{R}, \mathbf{X}),
\end{aligned}
$$

where the second equality follows from the definition of $SCCP$, the third follows from Theorem 2, the fourth is implied by the induction hypothesis on $SCCP(\mathcal{R}_0)$ and by the base case proof on $SCCP(\rho)$, while the last is a consequence of Lemma 4.

3.2   Preservation of Dynamic Behavior

In Theorem 3 we proved that the ODE map, when applied to sCCP-models of biochemical networks, satisfies a condition of coherence: it preserves the kinetic principles used in the construction of the model (i.e., the rate semantics).

A different question is whether an sCCP-network N (evolving stochastically according to the prescriptions of its semantic) shows a dynamic behavior equivalent to the one exhibited by the equations $ODE(N)$. This problem is the sCCP-counterpart of the famous mathematical issue concerning the relation between stochastic and differential models [35,36], studied deeply also in the context of biochemical reactions [26,23]. It is well-known that stochastic and differential models of biochemical reactions are behaviorally equivalent only in some cases.

These results are significant also for sCCP. Theorem 3, in fact, states that, when biochemical reactions are concerned, the stochastic process underlying the sCCP-models and the associated ODE's are exactly the classical ones. Therefore, in the mapping from sCCP to ODE's we have the same phenomenology as in the classical case. However, the logical structure of sCCP-agents makes the problem of behavioral preservation subtler.

In the following, we discuss this problem with different examples, especially of situations in which an sCCP-network and the corresponding ODE's show a different behavior. In particular, we are interested in sketching a brief, and plausibly incomplete, classification of the causes of behavioral divergence.

An important issue is the concept of behavioral equivalence itself, which is difficult to formalize, as already discussed in the introduction. We will return on this problem in the next section.

Oregonator. The Oregonator is a chemical systems showing an oscillatory behavior, devised by Field and Noyes [40] as a simplified version of the Belousov-Zhabotinsky oscillator[15]. Essentially, Oregonator is composed of three chemical substances, call them A, B, C, subject to the following reactions:

$$
\begin{aligned}
B &\to_{k_1} A \\
A + B &\to_{k_2} \emptyset \\
A &\to_{k_3} 2A + C \\
2A &\to_{k_4} \emptyset \\
C &\to_{k_5} B
\end{aligned}
\tag{12}
$$

Actually, other chemical substances are involved, but they are kept constant in the experiment. The differential equations associated to (12) are known to possess a *stable limit cycle* for a wide range of parameter's values [28], containing an *unstable equilibrium*. The limit cycling behavior is clearly visible in Figure 3(a), where the numerical solution of Oregonator's ODE's is shown.

In Figure 3(b), instead, we plot a stochastic simulation of the sCCP model associated to (12) according to prescriptions of Section 2.2. In this case, the stochastic model shows the same pattern as the differential one. Theorem 3 guarantees that the graph in Figure 3 depicts the numerical solution of ODE's associated to the sCCP program by the transformation previously defined. In this case, the behavior is preserved.

We remark two things regarding Oregonator. First, the size of each molecular species is of the order of thousands, hence the relative variation induced by one reaction in the stochastic model is small. Under this condition, stochastic and deterministic models of biochemical reactions usually coincide [26]. Another property of the Oregonator that can be important for behavioral preservation is that the limit cycle is an *attractor* in the *phase space*: nearby trajectories asymptotically converge to it (see [51]). This means that a relatively small perturbation is not willing to change the overall dynamics: stochastic fluctuations have a negligible effect. Things are different if we start from the unstable equilibrium of the system. The numerical solution of ODE's shows a constant evolution (Figure 4(a)), while the stochastic simulation (Figure 4(b)) essentially evolves as the limit cycle of Figure 3. In fact, stochastic fluctuations, in this case, make the sCCP system move away from the instable equilibrium into the basin of attraction of the limit cycle. This shows another well known fact: *stochastic and differential models usually differ near instabilities* [26].

Lotka-Volterra system. The Lotka-Volterra system is a famous simple model of population dynamics, see for example [26] and references therein. There are

---

[15] This chemical system is called "Oregonator" because its inventors where working at the University of Oregon.

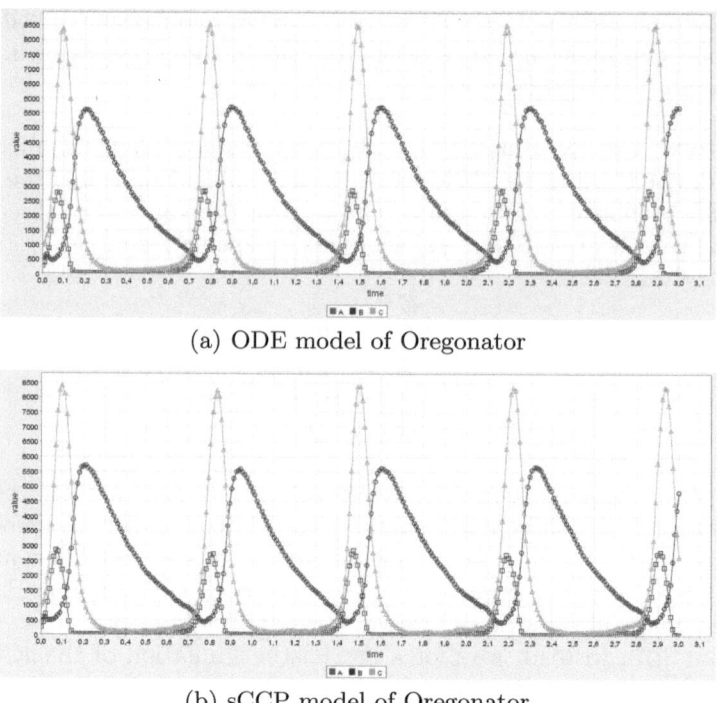

(a) ODE model of Oregonator

(b) sCCP model of Oregonator

Fig. 3. 3(a): Numerical simulation of the differential equation model of the Oregonator, with parameters determined according to the method presented in [26]. Specifically, let $A_s = 500$, $B_s = 1000$ and $C_s = 2000$ be an equilibrium of the system of equations, and let $R_1 = 2000$, $R_2 = 50000$. Then parameters are equal to $k_1 = R_1/B_s = 2$, $k_2 = R_2/(A_s B_s) = 0.1$, $k_3 = (R_1 + R_2)/A_s = 104$, $k_4 = ((2R_1)/(A_s^2))/2 = 4e^{-7}$, and $k5 = (R_1 + R_2)/C_s = 26$. The starting point is $A_0 = A_s/2$, $B_0 = B_s/2$, $C_0 = C_s/2$. The system soon approaches an attractive limit cycle. **3(b)**: Stochastic simulation with Gillespie's method of the sCCP network associated to reactions (12). Parameters and initial conditions are those specified above. The effect of stochastic fluctuations is negligible, and the plot essentially coincide with its deterministic counterpart.

two species: preys and predators. Preys eat some natural resource, supposed unbounded, and reproduce at a rate depending only on their number. Predators, instead, can reproduce only if they eat preys, otherwise they die. To keep the model simple, we admit predation as the only source of prey's death. The previous hypotheses can be summarized in the following set of reactions, where E refer to preys and C to predators:

$$
\begin{aligned}
E &\to_{k_b} 2E \\
C &\to_{k_d} \emptyset \\
E + C &\to_{k_p} 2C
\end{aligned}
\tag{13}
$$

If we consider the standard mass action ODE's (they coincide with the equations derived from the sCCP model due to Theorem 3), a typical solution shows

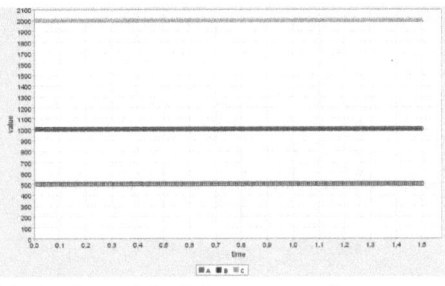 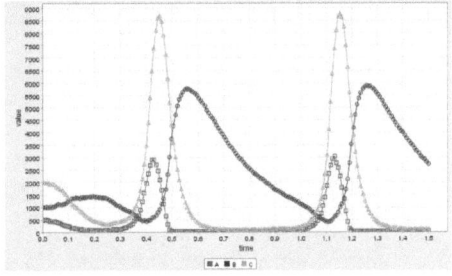

(a) ODE model of Oregonator from unsta- (b) sCCP model of Oregonator from un-
ble equilibrium   stable equilibrium

Fig. 4. 4(a): Numerical simulations of ODE's derived from reactions (12), with parameters given in caption of Figure 3, starting from an unstable equilibrium of the system. **4(b)**: Stochastic simulation of sCCP model associated to reactions (12), with the same parameters and initial conditions than the differential counterpart. As we can see, stochastic fluctuations drive the system away from the unstable equilibrium, so that its surrounding limit cycle is approached.

oscillations in which high values of preys and predators alternate. An example of such a solution is given in Figure 5(a). Inspecting equations, it can be shown that the point $E_s = k_d/k_p$, $C_s = k_b/k_p$ is an equilibrium of a rather special kind: it is stable (trajectories starting nearby it stay close) but not asymptotically stable (trajectories starting nearby do not converge to it as time approaches infinity). This behavior is easily understood looking at the phase space (Figure 5(b)), in which we can see that trajectories form closed orbits around the equilibrium, whose amplitude increases with distance from equilibrium. More details can be found, for instance, in [51].

What kind of behavior can we expect from the stochastic evolution of the sCCP model for (13)? Stochastic fluctuations will make the system jump from one trajectory to nearby ones, without any force pulling it towards the equilibrium. Therefore, fluctuations can, in the long run, make the system wander in the phase plane, eventually reaching a borderline trajectory (corresponding to E or C axis in the phase plane). Whenever this happens, then both preys and predators go extinct (C-axis trajectory), or just predators do, while preys go to infinity (E-axis trajectory). This intuition is confirmed in Figure 6, where we compare the ODE solution starting from equilibrium (dotted lines), and a trace of the sCCP model, starting from the same initial configuration. As we can see, the stochastic system starts oscillating until both species go extinct.

This is another well known case in which stochastic and differential dynamics differ, again induced by properties of the phase space [26].

A negatively auto-regulated system. The effect of stochastic fluctuations is mostly remarkable in biological phenomena where gene expression is involved. This is because the transcription of a gene is usually a slower process than protein-protein interaction, and often the number of mRNA strands for a given

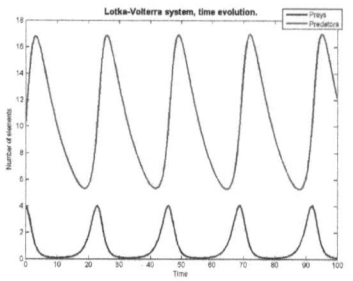
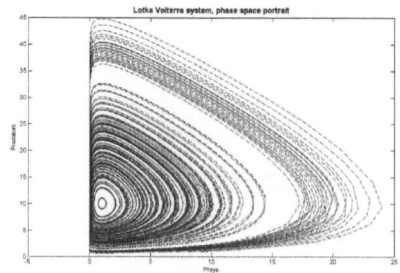

(a) ODE solution of Lotka-Volterra model   (b) Phase space of Lotka-Volterra model

Fig. 5. **5(a)**: Numerical solution of ODE's associated to reactions (13), with parameters $k_b = 1$, $k_p = 0.1$, $k_d = 0.1$ and initial conditions $E_0 = 4$ and $C_0 = 10$. **5(b)**: phase portrait of the Lotka Volterra system, for the same value of parameters as above. As we can see, all the solutions show an oscillating behaviour. The system has an (instable) equilibrium for $E = 1$, $C = 10$, at the center of the circles.

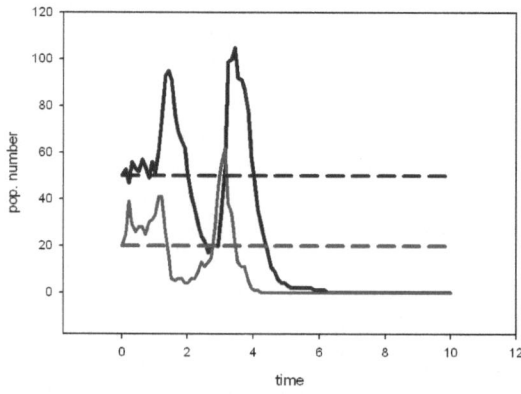

Fig. 6. Effect of stochastic fluctuations for the Lotka-Volterra system. The dotted lines are an equilibrium solution for the ODE model (parameters are as in caption of Figure 5). A stochastic trace of the sCCP model is drawn with solid lines: both species fluctuate around the equilibrium values until they both get extinct.

gene present in the cell is very small, of the order of some units. As the production of one single mRNA is a rare event (compared to other cellular events), stochastic variability in its happening can induce behaviors difficult to capture if mRNA is approximated with its concentration. Stochasticity in gene expression is indeed a phenomenon that has received a lot of attention, see for instance [38,6].

We present here a simple, artificial example taken from [53] and depicted in Figure 7. The biological network shown represents a simple autoregulatory mechanism in gene expression of a procaryotic cell. Gene g produces, via mRNA r, a

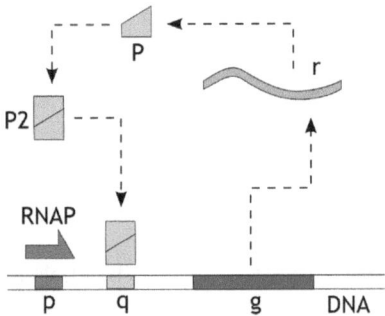

Fig. 7. Diagram of a simple self-regulated gene network. Gene g produces mRNA r and, from it, protein P. P can dimerise and its dimer can bind to a promoter region of gene g, downstream of RNA polymerase binding site. The P_2-binding blocks polymerase activity, thus inhibiting gene expression.

protein P that, as a dimer, can bind to a promoter region of gene g, preventing RNA-polymerase activity and thus inhibiting its own production.

Following the approach of [5], genes can be modeled as logical gates having a fixed output (the produced mRNA or protein), and several inputs, corresponding to different proteins of the system, exerting a positive or negative regulatory function. A gene gate with one inhibitory input is called in [5] *neg gate*, and can be modeled in RESTRICTED(sCCP) simply as:

$$\text{neg\_gate}_{P,I} :\text{-}$$
$$\text{tell}_{k_p}(P = P + 1).\text{neg\_gate}_{P,I}$$
$$+ \quad \text{ask}_{k_b \cdot I}(I \geq 1).\text{ask}_{k_u}(true).\text{neg\_gate}_{P,I},$$

where k_p is the basic production rate, k_b is the binding rate of the repressor to the promoter region of the gene and k_u is its unbinding rate.

In order to model the system of Figure 7 we can combine one neg gate with some reactions. This is an example of the modeling style mixing the reaction-centric and the molecular-centric point of view, see Section 2.2. The model is the following:

$$
\begin{aligned}
&\text{neg\_gate}_{r,P_2}\\
&r \rightarrow_{k_t} r + P\\
&P \rightarrow_{k_{dim_1}} P_2\\
&P_2 \rightarrow_{k_{dim_2}} P\\
&r \rightarrow_{k_{d_1}} \emptyset\\
&P \rightarrow_{k_{d_2}} \emptyset
\end{aligned}
\tag{14}
$$

In Figure 8 we compare a stochastic simulation of the sCCP model of reactions (14) with the numerical solution of the associated ODE's. As we can readily see, the two plots are completely different. In particular, in the stochastic simulation, P_2 is produced in short bursts; normally it is slowly degraded. The bursts correspond to mRNA production events, shown in Figure 8(a) as blue peaks. The

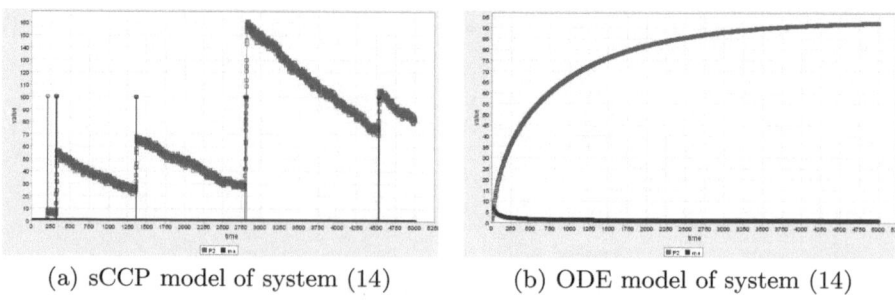

(a) sCCP model of system (14)   (b) ODE model of system (14)

Fig. 8. 8(a): Simulation of the sCCP model of (14). The red line corresponds to P_2, while the blue line shows the evolution of r, multiplied for a factor 100 (for visualization purposes). Note that the increases in P_2 expression immediately follow mRNA production events. Parameters of the models are the following: $k_p = 0.01$, $k_b = 1$, $k_u = 10$, $k_t = 10$, $k_{dim_1} = 1$, $k_{dim_2} = 1$, $k_{d_1} = 0.1$, and $k_{d_2} = 0.01$. All molecules are set initially equal to 0. **8(b)**: Numerical simulation of ODE's associated to the sCCP model of (14), for the same parameters just given. The evolution of P_2 is tamer than in the stochastic counterpart, as it converges quickly to an asymptotic value.

ODE's system, however, presents a much simpler pattern of evolution, in which the quantity of P_2 converges to an asymptotic value. This divergence is caused by the fact that, approximating continuously the number of RNA molecules, we lose the discrete information that seems to characterize its dynamics, i.e. the fact that mRNA can be present in one unit of completely absent from the system.

Staten otherwise, continuously approximating molecular species present in low quantities may lead to errors inducing a completely divergent observable behavior.

Repressilator. The *Repressilator* [21] is an artificial biochemical clock composed of three genes expressing three different proteins, **tetR**, **λcI**, **LacI**, exerting a regulatory function on each other's gene expression. In particular, protein **tetR** represses the expression of protein **λcI**, protein **λcI** represses the gene producing protein **LacI**, and, finally, protein **LacI** is a repressor for protein **tetR**. The expected behavior is an oscillation of the concentrations of the three proteins. A simple stochastic model of Repressilator can be found in [5], where the authors describe it with three neg gates (see the previous paragraph) cyclically connected, in such a way that the product of one gate inhibits the successive gene gate in the cycle. In addition, they introduce degradation mechanisms for the three repressors. More formally, the model is the following

$$
\begin{aligned}
&\text{neg\_gate}_{A,C} \\
&\text{neg\_gate}_{B,A} \\
&\text{neg\_gate}_{C,B} \\
&A \to_{k_d} \emptyset \\
&B \to_{k_d} \emptyset \\
&C \to_{k_d} \emptyset
\end{aligned}
\tag{15}
$$

In Figure 9(a) we show a trace of the stochastic model generated by a simulator of sCCP based on Gillespie algorithm. The oscillatory behavior is manifest.

If we apply the translation procedure discussed in Section 3 to this particular model, we obtain the ODE's shown in Table 7, while their numerical integration is shown in Figure 9(b). As we can readily see there is no oscillation at all, but rather the three proteins converge to an asymptotic value, after an initial adjustment.

Inspecting the ODE's, we note the presence of six variables (Y_A, Y_B, Y_C and Z_A, Z_B, Z_C) in addition to those representing the quantity of repressors in the system (A, B, C). Such variables correspond to states of genes gates, and they are used to model the change of configuration of the gates, from active to repressed and vice versa.

This scenario seems rather unjustified here: there is no argument to support the introduction of these variables, especially because we are continuously approximating boolean quantities.

An interesting point regarding Repressilator is the relation between the solution of the ODE's and the average trace of the stochastic system (i.e. $\mathbb{E}[\mathbf{X}(t)]$, returning the average value of system variables as a function of time). In fact, we may expect that the behavior preserved by the differential equations is the average dynamics of the stochastic system, rather than that shown by one of its traces. Interestingly, also the average value of the Repressilator model does not oscillate, as can be seen from Figure 10. This can be explained by noticing that the oscillations' period in the stochastic model is not constant, but it varies considerably. Hence, for every instant (when the Markov chain is at the stationary regime), we will observe one of the proteins at its peak value approximatively only in one third of the traces. Hence its average value will tend to stabilize at one third of the peak value, as confirmed by Figure 10. In fact, when we average Repressilator, we measure the fraction of traces in which a certain gene is

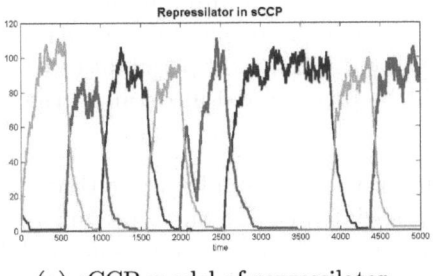

(a) sCCP model of repressilator

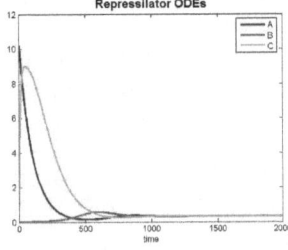

(b) sCCP model of repressilator

Fig. 9. 9(a): Stochastic time trace for the Repressilator system of described by reactions 15. Parameters are $k_p = 1$, $k_d = 0.01$, $k_b = 1$, $k_u = 0.01$. **9(b):** Solution of the differential equations of Table 7, automatically derived from sCCP program associated to reactions 15. Parameters are the same as in stochastic simulation. The stochastic simulation lasts longer than the ODE one in order to better underline its oscillatory behavior.

Table 7. ODE's derived for the Repressilator, generated by the method of Section 3.1

$$\dot{A} = k_p Y_A - k_d A \qquad \dot{Y}_1 = k_u Z_A - k_b Y_A C \qquad \dot{Z}_1 = k_b Y_A C - k_u Z_A$$
$$\dot{B} = k_p Y_B - k_d B \qquad \dot{Y}_2 = k_u Z_B - k_b Y_B A \qquad \dot{Z}_2 = k_b Y_B A - k_u Z_B$$
$$\dot{C} = k_p Y_C - k_d C \qquad \dot{Y}_3 = k_u Z_C - k_b Y_C B \qquad \dot{Z}_3 = k_b Y_C B - k_u Z_C$$

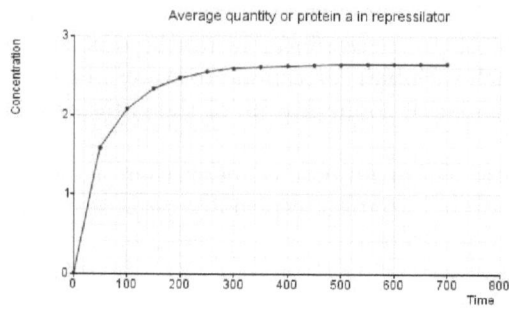

Fig. 10. Average value of the sCCP model for Repressilator, computed using model checker PRISM [37]. See [8] for further details.

active and the fraction of traces in which it is inactive, for every time instant. In this way, however, we lose any information regarding the sequence of gene gate's state changing. The different behavior existing between a trace of a stochastic system and its average trace suggests that the switching dynamics of genes can be the driving force behind oscillations. This implies that another source of non-equivalence between sCCP models and the associated ODE's can appear due to the representation of RTS-states with continuous RTS-state variables.

Indeed, this example suggested us to preserve part of the discrete dynamics, mapping the sCCP Repressilator into an hybrid automaton. The work put forward in [?] shows that this move is enough to maintain oscillations. The translation to hybrid automata opens an entire range of possibilities to combine discreteness and continuity. These will be investigated in detail in the planned second part of this paper.

Sources of non-equivalence. In the previous examples we outlined different cases in which an sCCP model and its associated ODE's fail to be equivalent from a dynamical viewpoint. We remark that most of these examples are well known, as they have been studied in detail in theoretical and applicative contexts, like biochemical reactions [26,53] and our main interest here is in their connection with the sCCP translation machinery. For sake of clarity, we summarize the different sources of non-equivalence.

1. In some cases, non-equivalence is a direct consequence of *properties of the phase space*. For instance, instable trajectories are destroyed by small fluctuations, like the equilibrium trajectory of the Oregonator. Also stable but

not asymptotically stable trajectories can be troublesome, as stochastic fluctuations are not counterbalanced by any attracting force, and so they can bring the stochastic system far away from the initial trajectory. This is the case of the Lotka-Volterra system.

2. Another well-known problem is related to the *approximation by continuous quantities of integer variables having small (absolute) values*. In this case, in fact, the effect of a single stochastic fluctuation has a relative magnitude that is relevant, so the dynamics can change quite dramatically. A typical example appearing in Biology is related to the transcription of genes, as shown in the simple example of a self-regulated gene.

3. A final source of non-equivalence is, instead, characteristic of the translation procedure defined for sCCP. In fact, in this case we *represent each RTS-state of a component of the system with a continuous variable*, which can take values in the real interval $[0, 1]$. RTS-states represent, in some sense, logical structures that control the activity of the system, while a change of state is an event triggered by some condition of the system. Moreover, in each sCCP trace, each component can be in only one state, hence RTS-state variables are boolean quantities. Continuous approximation, in this case, can have dramatic consequences, as the example of Repressilator seems to suggest.

3.3 Behavioral Equivalence

Comparing the dynamical evolution of a deterministic and a stochastic system is a delicate issue, because stochastic processes have a noisy evolution, hence we need to remove noise from their traces, before attempting any comparison with time traces evolving deterministically. In the previous discussion, in fact, we appealed to the concept of "behavioral equivalence" always in a vague sense, essentially leaving to the reader the task of visually comparing plots and recognizing similarities and differences. Clearly, a mathematical definition is needed in order to prove theorems and automatize comparisons.

We first consider the comparison of traces generated by ODE's with the average trace of the stochastic system, taken as the representative of its whole ensemble of traces. In practice, for each time instant t we need to compute the average value $\mathbb{E}(X(t))$ of each stream variable X w.r.t. the probability distribution on states of the system at time t. This probability can be obtained as the solution of the *Chapman-Kolmogorov forward equation* [39], a system of differential equations of the size of the state space. This equation, known in biochemical literature as the *chemical master equation* [25], can rarely be solved analytically, and it is also very difficult to integrate numerically [26]. A more efficient approach to compute an estimate of the average consists in generating several (thousands of) stochastic traces and in computing pointwise their sample mean. Alternatively, the average value of one or more variables can be computed for a small sample of time points $\{t_1, \ldots, t_k\}$ using numerical techniques, as those implemented in the model checker PRISM [37].

Whatever the method chosen, the computation (even approximate) of the average trace of a stochastic system is a difficult matter. Whenever such trace is known, we can compare it with the trace of the ODE's, generated using standard numerical techniques [43], using quantitative measures (essentially computing a distance between the two curves). Indeed, in [9] it is shown that the ODE associated to a sCCP program is a *first order approximation* of the true equation for the average.

However, the average trace of a stochastic system is not necessarily a good representative of its evolution. A paradigmatic example is the Repressilator, whose average trace (sampled with PRISM, see caption of Figure 10) converges to an asymptotic value, while all its stochastic traces show persistent oscillations. Hence, *even when averaging a stochastic system, we may lose the characterizing qualitative features of its dynamics.*

The example of Repressilator suggests that the notion of behavioral equivalence is probably better captured in a qualitative setting. Qualitative comparison requires a formal definition of the *features* of dynamical evolution, like oscillations, convergence to a stable value, and so on. A possibility we suggest in this direction is to describe these features as logical formulae of a suitable logical language $\mathcal{L}$, for instance temporal logic, as done in Simpathica [3]. The concepts below are just sketched; this subject is currently under investigation and we will deal with it in detail in future works. Let Φ denote the set of formulae describing all dynamical features of interest. Associating a Kripke structure K_1 to the trace of an ODE and another structure K_2 to a stochastic trace, then we may declare these traces equivalent whenever their Kripke structures satisfy the same subset of formulae of Φ (possibly restricting the attention to formulae of degree $\leq n$).

Below we give three examples of temporal logic formulas expressing infinite oscillations:

$$G(Z = z_m \rightarrow F(Z = z_M)) \ \wedge \ G(Z = z_M \rightarrow F(Z = z_m)) \ \wedge$$
$$G(z_m \leqslant Z \leqslant z_M) \ \wedge \ F(Z = z_m);$$

$$G(Z = z_m \rightarrow X(Z > z_m \ U \ Z = z_M)) \wedge G(Z = z_M \rightarrow X(Z < z_M \ U \ Z = z_m)) \ \wedge$$
$$G(z_m \leqslant Z \leqslant z_M) \ \wedge \ F(Z = z_m);$$

$$\left(\neg G\left(\frac{dZ}{dt} > 0\right)\right) \ \wedge \ \left(\neg G\left(\frac{dZ}{dt} = 0\right)\right) \ \wedge \ \left(\neg G\left(\frac{dZ}{dt} < 0\right)\right).$$

In the above formulas X stands for *next*, G stands for *always (globally)*, F stands for *sometimes (in the future)*, U stands for *until*, z_m and z_M are minimum and maximum values, and the thirds formula uses propositional formulas taking values according to the sign of the first derivative.[16]

This idea seems promising, as it gives a considerable freedom in the definition of formulae Φ, hence allowing to privilege some aspects of dynamical evolution

---

[16] In order to use meaningfully the notion of "next" for ODE's we need to consider a discretization of the time and of the state space, such as that performed in [3].

more than others. However, the real problem is in the definition of a reasonable Kripke structure for a stochastic trace (and for sets of traces). In fact, Kripke structures for ODE's can be constructed starting from one or more traces, as done in Simpathica [3], in the following way: the bounded (product) domain of all variables is divided in small, compact regions; a state of the Kripke automaton consists of one of such regions; edges connect two states if a trajectory crosses the corresponding regions consecutively. This construction, however, is not reasonable for stochastic traces, as noise would force the addition of many edges that may introduce spurious behaviors. Of course, it is possible to model check directly on CTMC formulae written in CSL [4,37]. However, the complexity of this latter approach makes the definition of non-deterministic Kripke structures interesting also for stochastic traces. We are currently investigating this direction, considering the introduction of a bounded form of memory to tame noise.

3.4   More on the Restrictions of the Language

RESTRICTED($sCCP$) restricts the full language in several aspects, see Section 2.3. Actually, these restrictions have been introduced in order to define in a reasonably simple way the mapping to ODE's. We discuss them in detail in the following.

First, all agents must be sequential, i.e. not containing any occurrence of the parallel operator. As already remarked at the end of Section 2.3, this does not constitute a real limitation, as each non-sequential agent can be transformed into a network of sequential ones. Here we note that the same trick of Section 2.3 can be used to transform each sCCP-network into an equivalent network where each sequential agent has an RTS with one single state; indeed, this is done implicitly by the transformation to ODE's itself. However, writing programs in this form is less natural.

Another syntactic restriction regards the definition of local variables. Actually, variables in ODE's have a global scope. Of course, any local variable can be made global by suitably renaming it. There is a problem, however, concerning the fact that at run-time we may generate an unbounded number of local variables. This implies that their use may lead to a set of ODE's with an infinite number of variables (although each equation will depend only on a finite number of them). The uprising of an infinite number of variables requires more complex mathematical techniques, and it prevents the use of standard numerical solvers.

Finally, the third class of restrictions regards the constraint store. The restriction to numeric variables is obviously necessary, as we are mapping to ODE's. The restriction on the admissible constraints for the updating of variables, on the other hand, is related to the fact that each update in a sCCP program needs to be considered as a flux acting on some variables. Indeed, even a simple update like $X' = 0$ is difficult to render within ODE's framework, as it is inherently discrete. A possible way out is to mix the continuous ingredient of ODE's with discreteness, mapping sCCP programs to hybrid automata [29]. Within this formalism, updates like $X' = 0$ are perfectly admissible: they are *resets* associated to discrete transitions.

4   From ODE's to sCCP

In this section we define a transformation $SCCP$, associating an sCCP network to a generic set of ordinary differential equations, analyzing both its mathematical properties and the relation with the the map ODE defined in the previous section.

Before entering into the mathematical details, we need to make a preliminary remark. Essentially, the main obstacle we have to face in defining the map $SCCP$ is the fact that ODE's are an *aggregate* description of a system. To be more precise, if a system can be described by a set of fluxes acting on the different entities into play (i.e. on the system variables), then the ODE's hide part of the logical structure of such fluxes by combining them into the equations. To clarify the concept, consider the following two sCCP agents:

$$A \coloncolon (\text{tell}_1 (X = X + 1) + \text{tell}_1 (Y = Y + 1)).A$$

$$B \coloncolon (\text{tell}_1 (X = X + 1 \wedge Y = Y + 1)).B$$

When we apply the ODE operator to the networks $N_1 = A$ and $N_2 = B$, we obtain, in both cases, the following equations:

$$\begin{pmatrix} \dot{X} \\ \dot{Y} \end{pmatrix} = \begin{pmatrix} 1 \\ 1 \end{pmatrix}$$

Therefore, two different sCCP agents can be mapped into the same set of ODE's. Note that A and B are "semantically" different, as they induce two different CTMC. The chain associated to A has edges connecting a state (i, j) to $(i+1, j)$ and $(i, j + 1)$ (hence the exit rate from (i, j) is 2), while the chain of B has transitions only from (i, j) to $(i + 1, j + 1)$ (with exit rate 1).

This information pertains the logical structure of the system, which is manifest in the sCCP program, but irremediably lost in the associated ODE's.

An even worse situation happens for the following agent, implementing a one-dimensional random walk [39]:

$$C \coloncolon (\text{tell}_1 (X = X + 1) + \text{tell}_1 (X = X - 1)).C$$

The equation associated to C by ODE is $\dot{X} = 0$, as the production and degradation rate cancel out when summed together. This equation predicts a constant evolution for X, thus failing to capture its erratic behavior. Note, however, that the average value of X is constant also in the stochastic model for C.

Therefore, the structural information lost in passing from sCCP agents to ODE's makes impossible to recover the original sCCP network; stated otherwise, the map $ODE(\cdot)$ is not injective. Indeed, the lack of injectivity of $ODE(\cdot)$ means

that an sCCP program is more informative than a set of ODE's: it defines not
only the fluxes, but also their logical relation.

The previous discussion suggests that the map $SCCP$ must be defined with
care. Given an ODE set $\dot{\mathbf{x}} = \mathbf{f}(\mathbf{x})$, there are many sCCP networks that can
be associated to it, i.e. at least all those belonging to the set $ODE^{-1}(\mathbf{f}(\mathbf{x}))$.
In order to choose one among them, additional discriminating information is
required, essentially related to the structure of the fluxes, hence to the logic of
the system modeled.

As suggested by the previous discussion, we will therefore define not a single
$SCCP$ map, but rather a class of maps, parametric w.r.t. the additional information required to sort out the logical structure of fluxes. We will then show that,
independently of this additional information, the transformation scheme satisfies
properties guaranteeing a form of coherence w.r.t. the ODE mapping and also
a form of behavioral equivalence. Finally, we will provide two instantiations of
such scheme, assuming specific conditions on the system modeled.

4.1   The Translation to sCCP

In the conversion from ODE's to sCCP, we approximate continuous quantities by
discrete variables. Therefore, this mapping will depend on an additional parameter, the *step* δ, specifying the *granularity* of the approximation of continuous
variables. The magnitude of δ has a strong impact on the preservation of dynamical behavior; this point will be the content of Section 4.3.

Consider a system of first order ODE's with n variables $\mathbf{x} = (x_1, \ldots, x_n)$:

$$\dot{\mathbf{x}} = \mathbf{f}(\mathbf{x}).$$

We will now define the notion of *set of covering functions*, which captures the idea
of external knowledge required to solve the ambiguity about the logical structure
inherent in the ODE's. Essentially, a set of covering functions corresponds to a
plausible choice of a set of fluxes, generating the given ODE's.

Definition 13. *A set of covering functions $\mathcal{G}$ for the ODE $\dot{\mathbf{x}} = \mathbf{f}(\mathbf{x})$ is a set of
pairs $\{(g_i, \mathbf{h_i}) \mid i = 1, \ldots, k\}$, such that each g_i is a function $g_i : \mathbb{R}^n \to \mathbb{R}$, each
$\mathbf{h_i}$ is a vector of $\mathbb{Z}^n$, and, for each $\mathbf{x} \in \mathbb{R}^n$,*

$$\sum_{i=1}^{k} \mathbf{h_i} g_i(\mathbf{x}) = f(\mathbf{x}).$$

Example 1. Consider the following simple system of ODE's with two variables:

$$\begin{cases} \dot{x} = a - by \\ \dot{y} = c + dx \end{cases} \tag{16}$$

One possible covering set is the following:

$$\mathcal{G} = \{(a, (1, 0)), (by, (-1, 0)), (c, (0, 1)), (dx, (0, 1))\},$$

which corresponds to the choice of disentangling all addends of the equations. Another possibility, instead, is the following: $\mathcal{G}' = \{(-a + by, (-1, 1)), (a - by, (0, 1,)), (c + dx, (0, 1,))\}$, as easily verified.

In the sCCP program associated to the ODE's $\dot{\mathbf{x}} = \mathbf{f}(\mathbf{x})$, we approximate the continuous variables $\mathbf{x}$ with discrete stream variables $\mathbf{X}$. Definition 1, however, requires variables $\mathbf{X}$ to have integer values. In order to set the size of the basic increment to an arbitrary step δ, we can change variables, setting $\mathbf{x} = \delta\mathbf{X}$ and expressing $\mathbf{f}$ with respect to $\mathbf{X}$ (in this way, a unit increment of X_i corresponds to an increment of δ of x_i). The equation for $\mathbf{X}$ thus becomes

$$\dot{\mathbf{X}} = \frac{1}{\delta}\mathbf{f}(\delta\mathbf{X}) = \mathbf{F}(\mathbf{X}; \delta).$$

If we are given a set of covering functions $\mathcal{G}$ for $\dot{\mathbf{x}} = \mathbf{f}(\mathbf{x})$, we can apply the same variable's substitution to each g_i, obtaining new covering functions $G_i(\mathbf{X}; \delta) = \frac{1}{\delta}g_i(\delta\mathbf{X})$ such that

$$\dot{\mathbf{X}} = \mathbf{F}(\mathbf{X}; \delta) = \frac{1}{\delta}\mathbf{f}(\delta\mathbf{X}) = \frac{1}{\delta}\sum_{i=1}^{k} g_i(\delta\mathbf{X}) = \sum_{i=1}^{k} G_i(\mathbf{X}; \delta).$$

The translation to sCCP simply proceeds associating an agent to each element of the set of covering functions $\mathcal{G}$:

Definition 14. *Let* $\dot{\mathbf{x}} = \mathbf{f}(\mathbf{x})$ *be a set of ODE's, and* $\mathcal{G}$ *be a set of covering functions for it. Let* $g_i \in \mathcal{G}$ *and* $\delta \in \mathbb{R}^+$. *The agent* $\mathrm{man}_{G_i,\delta}$ *is defined as*[17]

$$\mathrm{man}_{G_i,\delta} :- \mathrm{ask}_{|G_i(\mathbf{X};\delta)|} (G_i(\mathbf{X}; \delta) > 0). \ \mathrm{tell}_\infty (\mathbf{X}' = \mathbf{X} + \mathbf{h_i}).\mathrm{man}_{G_i,\delta}$$
$$+ \ \mathrm{ask}_{|G_i(\mathbf{X};\delta)|} (G_i(\mathbf{X}; \delta) < 0). \ \mathrm{tell}_\infty (\mathbf{X}' = \mathbf{X} - \mathbf{h_i}).\mathrm{man}_{G_i,\delta}$$

The agent $\mathrm{man}_{G_i,\delta}$ is a summation with two branches: both have rate equal to the modulus of function G_i, but one is active when $G_i > 0$, and it increments the value of $\mathbf{X}$ according to the vector $\mathbf{h_i}$, while the other is active when $G_i < 0$, decrementing $\mathbf{X}$ by $\mathbf{h_i}$.

In order to construct the sCCP network associated to a set of ODE $\dot{\mathbf{x}} = \mathbf{f}(\mathbf{x})$, we simply need to define an agent $\mathrm{man}_{G_i,\delta}$ for each function G_i of the covering set $\mathcal{G}$, putting these agents in parallel. We can render this procedure in the following $SCCP$ operator:

Definition 15. *Let* $\dot{\mathbf{x}} = \mathbf{f}(\mathbf{x})$ *be a set of ODE's,* $\mathcal{G}$ *be a set of covering functions for* $\mathbf{f}$, *and* $\delta \in \mathbb{R}$, $\delta > 0$. *The sCCP-network associated to* $\mathbf{f}(\mathbf{x})$, *with respect to the set of covering functions* $\mathcal{G}$ *and the increment's step* δ, *indicated by* $SCCP(\mathbf{f}(\mathbf{x}), \mathcal{G}, \delta)$, *is*

$$SCCP(\mathbf{f}(\mathbf{x}), \mathcal{G}, \delta) = \mathrm{man}_{G_1,\delta} \| \dots \| \mathrm{man}_{G_k,\delta}, \qquad (17)$$

with $\mathbf{x} = \delta\mathbf{X}$.

---

[17] The name "man" stands for manager.

The initial conditions of the sCCP program, given by $init(\mathbf{X})$, are $\mathbf{X}(0) = \frac{1}{\delta}\mathbf{x_0}$, where $\mathbf{x_0}$ are the initial conditions of the ODE's.

Functional rates of sCCP are central in the definition of this translation: *each function of the covering set becomes a rate in a branch of an sCCP summation.* This is made possible only due to the freedom in the definition of rates, because differential equations and covering functions considered here are general.

The possibility of having general rates in sCCP is intimately connected with the presence of the constraint store, which contains information external to the agents. This means that part of the description of interactions can be moved from the logical structure of agents to the functional form of rates. Common stochastic process algebras like stochastic π-calculus [44] or PEPA [30], on the other hand, have simple numerical rates and they rely just on the structure of agents (and on additivity of the exponential distribution [39]) to compute the global rate. This restricts severely the class of functional rates that they can model. Indeed, in a recent work [18] Hillston introduces general rates in PEPA essentially through the addition of information external to the model, an approach similar in spirit to sCCP.

4.2   Invertibility

We turn now to study the relation between the two translations defined, i.e. ODE and $SCCP$. Specifically, we will show that $(ODE \circ SCCP)$ returns the original differential equations, independently from the covering set $\mathcal{G}$. The other direction, instead, cannot hold, as pointed out at the beginning of this section. In fact, we have seen that several sCCP agents can be mapped by ODE to the same equations, hence the map ODE cannot be inverted.

Theorem 4. *Let $\dot{\mathbf{x}} = \mathbf{f}(\mathbf{x})$ be a set of differential equations, with $\mathbf{x}=(x_1,\ldots,x_n)$ and $\mathbf{X} = \frac{1}{\delta}\mathbf{x}$, and let $\mathcal{G}$ be a set of covering functions for $\mathbf{f}$. Then*

$$ODE(SCCP\,(\mathbf{f}(\mathbf{x}), \mathcal{G}, \delta)\,, \mathbf{X}) = \mathbf{f}(\mathbf{x}).$$

Proof. From Definition 15 we know that $SCCP\,(\mathbf{f}(\mathbf{x}), \delta) = \mathrm{man}_{G_1,\delta} \parallel \cdots \parallel \mathrm{man}_{G_n,\delta}$, and by Theorem 2,

$$ODE\,(\mathrm{man}_{G_1,\delta} \parallel\cdots\parallel \mathrm{man}_{G_n,\delta}, \mathbf{X})=ODE(\mathrm{man}_{G_1,\delta}, \mathbf{X})+\ldots+ODE(\mathrm{man}_{G_n,\delta}, \mathbf{X}).$$

Now, the agent $\mathrm{man}_{G_i,\delta}$ can modify several variables X_i, according to the vector $\mathbf{h_i}$ coupled with the function G_i. The RTS of $\mathrm{man}_{G_i,\delta}$ is easily seen to have the following form

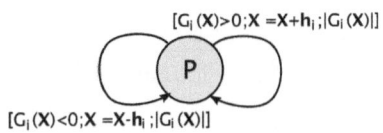

Therefore, the interaction matrix associated to $\mathrm{man}_{G_i,\delta}$ has just two columns, corresponding to the vectors $\mathbf{h_i}$ and $-\mathbf{h_i}$, and $ODE(\mathrm{man}_{G_i,\delta}, \mathbf{X})$ is equal to

$$\mathbf{h_i}|G_i(\mathbf{X};\delta)|\langle G_i(\mathbf{X};\delta) > 0\rangle - \mathbf{h_i}|G_i(\mathbf{X};\delta)|\langle G_i(\mathbf{X};\delta) < 0\rangle.$$

In the previous equation, $\langle\cdot\rangle$ denotes, as in Section 3, the logical value of a formula. The previous equation can be simplified by noting that

$$|G_i(\mathbf{X};\delta)|\langle G_i(\mathbf{X};\delta) > 0\rangle - |G_i(\mathbf{X};\delta)|\langle G_i(\mathbf{X};\delta) < 0\rangle = G_i(\mathbf{X};\delta),$$

hence

$$ODE(\mathrm{man}_{G_i,\delta}, \mathbf{X}) = \mathbf{h_i}G_i(\mathbf{X};\delta).$$

By applying Theorem 2 we then obtain

$$ODE(SCCP(\mathbf{f}(\mathbf{x}),\mathcal{G},\delta),\mathbf{X}) = \sum_{i=1}^{k} \mathbf{h_i}G_i(\mathbf{X};\delta) = F(\mathbf{X}),$$

which is equal to $\dot{\mathbf{x}} = \mathbf{f}(\mathbf{x})$ when changing the variables back to $\mathbf{x}$.

4.3   Behavioral Equivalence

We start this section by presenting an example showing how the translation from ODE's to sCCP works. In particular we will be concerned with the behavior exhibited by both systems and with the dependence on the step size δ, governing the size of the basic increment or decrement of variables. Intuitively, δ controls the "precision" of the sCCP agents w.r.t. the original ODE's. Hence, varying the size of δ, we can *calibrate the effect of the stochastic fluctuations*, reducing or increasing it. This is evident in the following example, where we compare solutions of ODE's and the simulation of the corresponding sCCP processes.

Let's consider the following system of equations, representing another model of the Repressilator (see Section 3.2), a synthetic genetic network having an oscillatory behavior (see [21,3]):

$$\begin{aligned}
\dot{x}_1 &= \alpha_1 x_3^{-1} - \beta_1 x_1^{0.5}, & \alpha_1 &= 0.2, & \beta_1 &= 0.01 \\
\dot{x}_2 &= \alpha_2 x_1^{-1} - \beta_2 x_2^{0.5}, & \alpha_2 &= 0.2, & \beta_2 &= 0.01 \\
\dot{x}_3 &= \alpha_3 x_2^{-1} - \beta_3 x_3^{0.5}, & \alpha_3 &= 0.2, & \beta_3 &= 0.01.
\end{aligned} \qquad (18)$$

We fix the following set $\mathcal{G}$ of covering functions: $g_1 = \alpha_1 x_3^{-1}$, $\mathbf{h_1} = (1,0,0)$, $g_2 = \alpha_2 x_1^{-1}$, $\mathbf{h_2} = (0,1,0)$, $g_3 = \alpha_3 x_2^{-1}$, $\mathbf{h_3} = (0,0,1)$, $g_4 = \beta_1 x_1^{0.5}$, $\mathbf{h_4} = (-1,0,0)$, $g_5 = \beta_2 x_2^{0.5}$, $\mathbf{h_5} = (0,-1,0)$, $g_6 = \beta_3 x_3^{0.5}$, $\mathbf{h_6} = (0,0,-1)$.

The corresponding sCCP process, after changing variables according to $X_i = \frac{x_i}{\delta}$, is:

$$\mathrm{man}_{G_1,\delta} \parallel \mathrm{man}_{G_2,\delta} \parallel \mathrm{man}_{G_3,\delta} \parallel \mathrm{man}_{G_4,\delta} \parallel \mathrm{man}_{G_5,\delta} \parallel \mathrm{man}_{G_6,\delta}, \qquad (19)$$

where, for instance, the agent $\mathrm{man}_{G_1,\delta}$ is

$$\mathrm{man}_{G_1,\delta} :- \mathrm{ask}_{\left|\frac{\alpha_1}{\delta}(\delta X_3)^{-1}\right|}\left(\frac{\alpha_1}{\delta}(\delta X_3)^{-1} > 0\right).\mathrm{tell}_\infty(X_1' = X_1 + 1).\mathrm{man}_{G_1,\delta}$$
$$+ \mathrm{ask}_{\left|\frac{\alpha_1}{\delta}(\delta X_3)^{-1}\right|}\left(\frac{\alpha_1}{\delta}(\delta X_3)^{-1} < 0\right).\mathrm{tell}_\infty(X_1' = X_1 - 1).\mathrm{man}_{G_1,\delta}$$

In Figure 11, we study the dependence on δ of the sCCP network obtained from equations (19). From the plots, we note that the smaller the δ, the closer the stochastic trace is to the solution of ODE's. However, increasing δ, the effect of stochastic perturbations gets stronger and stronger, making the system change dynamics radically.

Reducing the value of δ seems to be essentially the same as working with a sufficiently high number of molecules in standard biochemical networks, see [26,24] and the discussion in Section 3.2. It is thus reasonable to expect that, by taking δ smaller and smaller, the deterministic and the stochastic dynamics will eventually coincide. In fact, reducing δ we are diminishing the magnitude of stochastic fluctuations, hence their perturbation effects.

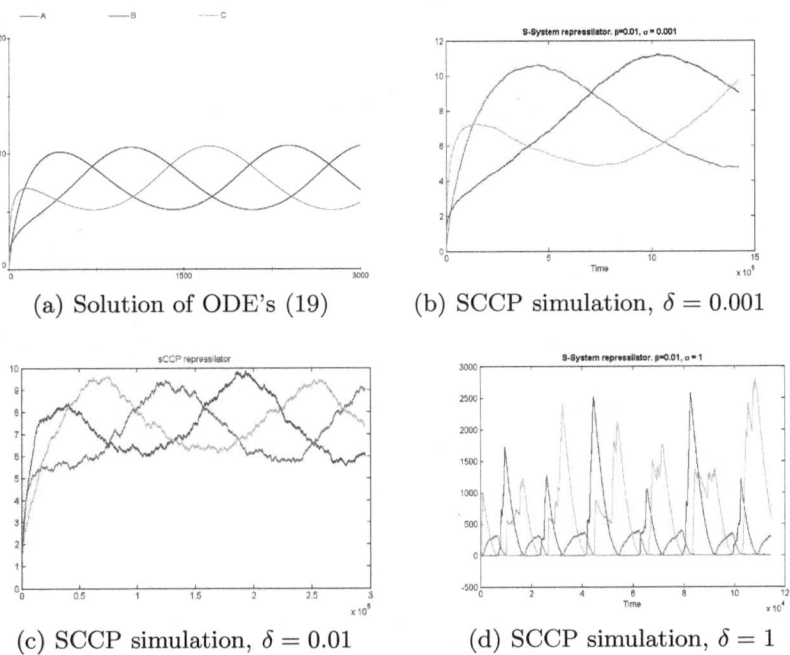

(a) Solution of ODE's (19)   (b) SCCP simulation, $\delta = 0.001$

(c) SCCP simulation, $\delta = 0.01$ (d) SCCP simulation, $\delta = 1$

Fig. 11. Different simulations of sCCP agent obtained from S-Systems equations of repressilator (19), as basic step δ varies. Specifically, in Figure 11(a) we show the solution of ODE's (19), while in Figures 11(b), 11(c), 11(d) we present three simulations of the sCCP agent corresponding to ODE's (19), for $\delta = 0.001, 0.01, 1$ respectively. In the last diagram, the behavior of S-System's equations is destroyed. Note that in Figure 11(c) the time axis is stretched by a factor of 100, while in Figure 11(b) the time axis is stretched by a factor of 1000, consistently with the rescaling of variables by $\frac{1}{\delta}$ performed in the translation from ODE to sCCP.

This conjecture is indeed true: in the rest of the section we prove that, under mild conditions on the ODE's and of the functions of $\mathcal{G}$, the trajectories of the stochastic simulation converge to the solution of the ODE's, independently of the choice of the covering set $\mathcal{G}$. In fact, the set of stochastic traces whose distance from the solution of the ODE's is greater than a fixed arbitrary constant has zero probability in the limit $\delta \to 0$.

Kurtz theorem. In 1970 Thomas Kurtz proved a theorem giving conditions for a family of density dependent Continuous Time Markov Chains to converge to a solution of a system of ODE's [35,36]. In fact, under mild assumptions on the smoothness of functions into play, the trajectories of the CTMC remain, in the limit, close to the solution of a particular set of ODE's with probability one. Our mapping $SCCP$ easily fits into Kurtz's framework, with the step δ playing the role of the density. We start by recalling the Kurtz's theorem.

Let V be a positive parameter, playing the role of the "size" of the system, and $\mathbf{X}_V(t)$ be a family of CTMC with state space $\mathbb{Z}^n$, depending on the parameter V. Suppose that there exist a continuous positive real function $\varphi : \mathbb{R}^n \times \mathbb{Z}^n \to \mathbb{R}$, such that the infinitesimal generator matrix [39] $Q = (q_{\mathbf{X},\mathbf{Y}})$ for $\mathbf{X}_V(t)$ is given by

$$q_{\mathbf{X},\mathbf{X}+\mathbf{h}} = V\varphi(\frac{1}{V}\mathbf{X}, \mathbf{h}), \quad \mathbf{h} \neq \mathbf{0}.$$

In addition, let $\Phi(\mathbf{x}) = \sum_{\mathbf{h} \in \mathbb{Z}^n} \mathbf{h}\varphi(\mathbf{x}, \mathbf{h})$.

Theorem 5 (Kurtz [35]). *Fix a bounded time interval $[0, T]$. Suppose there exists an open set $E \subseteq \mathbb{R}^n$ and a constant $M_E \in \mathbb{R}^+$ such that*

1. $|\Phi(\mathbf{x}) - \Phi(\mathbf{y})| < M_E|\mathbf{x} - \mathbf{y}|$, $\forall \mathbf{x}, \mathbf{y} \in E$ *(i.e. Φ satisfies the Lipschitz condition);*
2. $\sup_{\mathbf{x} \in E} \sum_{\mathbf{h} \in \mathbb{Z}^n} |\mathbf{h}|\varphi(\mathbf{x}, \mathbf{h}) < \infty$;
3. $\lim_{d \to \infty} \sup_{\mathbf{x} \in E} \sum_{|\mathbf{h}| > d} |\mathbf{h}|\varphi(\mathbf{x}, \mathbf{h}) = 0$.

Then, for every trajectory $\mathbf{x}(t)$ that is a solution of $\dot{\mathbf{x}} = \Phi(\mathbf{x})$ satisfying $\mathbf{x}(0) = \mathbf{x_0}$ and $\mathbf{x}(t) \in E$, $t \in [0, T]$, if

$$\lim_{V \to \infty} \frac{1}{V}\mathbf{X}_V(0) = \mathbf{x_0},$$

then for every $\varepsilon > 0$,

$$\lim_{V \to \infty} \mathbb{P}\left\{\sup_{t \leq T} \left|\frac{1}{V}\mathbf{X}_V(t) - \mathbf{x}(t)\right| > \varepsilon\right\} = 0.$$

This theorem states that the trajectories of $\mathbf{X}_V(t)$ converge, in a bounded time interval, to the solution of $\dot{\mathbf{x}} = \Phi(\mathbf{x})$, when $V \to \infty$. The function Φ is essentially the sum of all fluxes of the system.

Convergence for *SCCP*

Our framework can be easily adapted to fit this theorem. Consider a system of ODE's $\dot{\mathbf{x}} = \mathbf{f}(\mathbf{x})$ and a set of covering functions $\mathcal{G}$. Denote by $\mathbf{X}_\delta(t)$ the CTMC associated to the sCCP-network $SCCP(\mathbf{f}(\mathbf{x}), \mathcal{G}, \delta)$.

Theorem 6. *Let $\dot{\mathbf{x}} = \mathbf{f}(\mathbf{x})$ be a system of ODE's, with $\mathbf{x} \in \mathbb{R}^n$, and $[0, T]$ a bounded time interval. Let $\mathcal{G} = \{(g_i, \mathbf{h_i}) \mid i = 1 \ldots, \mu\}$ be a set of covering functions for $\mathbf{f}$.*

If there exists an open set $E \subseteq \mathbb{R}^n$ such that $\mathbf{f}$ satisfies the Lipschitz condition in E and $\sup_{\mathbf{x} \in E} |g_i(\mathbf{x})| < \infty$, for each $i = 1 \ldots, \mu$, then for every $\varepsilon > 0$

$$\lim_{\delta \to 0} \mathbb{P} \left\{ \sup_{t \leq T} |\delta \mathbf{X}_\delta(t) - \mathbf{x}(t)| > \varepsilon \right\} = 0,$$

where $\mathbf{x}(t)$ is the solution of $\dot{\mathbf{x}} = \mathbf{f}(\mathbf{x})$ with initial condition $\mathbf{x}(0) = \mathbf{x_0}$ and $\delta \mathbf{X}_\delta(0) = \mathbf{x_0}$.

Proof. In order to prove the theorem, we simply need to show that we satisfy all the hypothesis of the Kurtz's theorem. First of all, in this setting the density V is equal to $\frac{1}{\delta}$, so that $\frac{1}{\delta} \to \infty$ when $\delta \to 0$.

Consider now the function $g_i(\mathbf{x})$, and define as customary $g_i^+(\mathbf{x}) = g_i(\mathbf{x})\langle g_i(\mathbf{x}) \geq 0\rangle$ and $g_i^-(\mathbf{x}) = g_i(\mathbf{x})\langle g_i(\mathbf{x}) \leq 0\rangle$, so that $g_i(\mathbf{x}) = g_i^+(\mathbf{x}) - g_i^-(\mathbf{x})$ and $|g_i(\mathbf{x})| = g_i^+(\mathbf{x}) + g_i^-(\mathbf{x})$, where $\langle \cdot \rangle$ denotes the logical value as before.

Consider now the infinitesimal generator matrix $Q^\delta = (q_{\mathbf{X},\mathbf{Y}}^\delta)$ of the CTMC $\mathbf{X}_\delta(t)$. It is straightforward to prove that, for each $\mathbf{h} \in \mathbb{Z}^n$,

$$q_{\mathbf{X},\mathbf{X}+\mathbf{h}}^\delta = \frac{1}{\delta} \sum_{i \mid \mathbf{h_i}=\mathbf{h}} g_i^+(\delta \mathbf{X}) + \frac{1}{\delta} \sum_{i \mid \mathbf{h_i}=-\mathbf{h}} g_i^-(\delta \mathbf{X}),$$

where the sum must be intended equal to zero if the index set is empty. Clearly, these are density dependent rates, with density $\frac{1}{\delta}$. Note that there is a finite number of vectors for which $q_{\mathbf{X},\mathbf{X}+\mathbf{h}}^\delta$ is different from zero, as the set $\{\mathbf{h_1}, \ldots, \mathbf{h_\mu}\}$ is finite.

Therefore, the function φ of the Kurtz theorem is simply defined as

$$\varphi(\mathbf{x}, \mathbf{h}) = \sum_{i \mid \mathbf{h_i}=\mathbf{h}} g_i^+(\mathbf{x}) + \sum_{i \mid \mathbf{h_i}=-\mathbf{h}} g_i^-(\mathbf{x}).$$

Then, the function $\Phi(\mathbf{x})$ is

$$\Phi(\mathbf{x}) = \sum_{\mathbf{h}} \mathbf{h}\varphi(\mathbf{x}, \mathbf{h}) = \sum_{j=1}^{\mu} \mathbf{h_j}(g_j^+(\mathbf{x}) - g_j^-(\mathbf{x})) = \sum_{j=1}^{\mu} \mathbf{h_j}g_j(\mathbf{x}) = \mathbf{f}(\mathbf{x}).$$

It only remains to prove that conditions 1–3 of Theorem 5 are satisfied. Condition 1 is obvious because $\Phi = \mathbf{f}$ is Lipschitz by hypothesis, while condition 3 hold because $|\mathbf{h}| > M$ implies $\varphi(\mathbf{x}, \mathbf{h}) = 0$, where $M = \max_{1 \leq i \leq \mu} |\mathbf{h_i}|$. Finally, condition 2 follows because

$$\sum_{\mathbf{h}} |\mathbf{h}| \varphi(\mathbf{x}, \mathbf{h}) \leq M \sum_{j=1}^{\mu} (g_j^+(\mathbf{x}) + g_j^-(\mathbf{x})) = M \sum_{j=1}^{\mu} |g_j(\mathbf{x})|,$$

hence

$$\sup_{\mathbf{x} \in E} \sum_{\mathbf{h}} |\mathbf{h}| \varphi(\mathbf{x}, \mathbf{h}) \leq \sup_{\mathbf{x} \in E} M \sum_{j=1}^{\mu} |g_j(\mathbf{x})| \leq \sum_{j=1}^{\mu} \sup_{\mathbf{x} \in E} M |g_j(\mathbf{x})| \leq \infty,$$

due to the condition on g_i functions.

Comments and examples on Theorem 6. Theorem 6 states that sCCP networks are able to simulate ODE's with an arbitrary precision. The cost of an exact stochastic simulation of the sCCP-network of Definition 15, however, is proportional to $\frac{1}{\delta}$, hence accurate stochastic simulations of ODEs are computationally impractical. On the other hand, there is no apparent reason to generate stochastic trajectories indistinguishable from the solution of the ODE's, as the latter can be generally obtained with much less computational effort.

In a work related to ours [22], Hillston et al. used the same Kurtz theorem to prove an analogous result for the equations that can be obtained from a PEPA program. Theorem 6 can be seen as a generalization of their result. Moreover, in [22] the authors suggest that a stochastic approximation of ODE's can be used together with analysis techniques typical of CTMC, like steady state analysis. This is a promising direction, but extreme care must be used.

Kurtz theorem, in fact, guarantees convergence only in a fixed and bounded time interval $[0, T]$, hence it does say nothing about asymptotic convergence of stochastic trajectories to ODE's.

Intuitively, the step δ may not be the only responsible for asymptotic convergence; an important role should also be played by initial conditions through topological properties of the phase space. If the ODE-trajectory we are considering is stable, i.e. resistant to small perturbations, then we can expect it to be reproduced in sCCP along the whole time axis, given a step δ small enough. On the other hand, if the trajectory is unstable, then even small perturbations can drive the dynamics far away from it; stochasticity, in this case, will unavoidably produce a trace dramatically different from the one of ODE's. Of course, by taking the interval $[0, T]$ of the theorem big enough (hence δ small enough), we can postpone arbitrarily far away in time the moment in which a stochastic and an unstable deterministic trajectories will diverge.

As an example of instability, let's consider a simple linear system of differential equations:

$$\begin{pmatrix} \dot{X} \\ \dot{Y} \end{pmatrix} = \begin{pmatrix} X + Y \\ 4X + Y \end{pmatrix} \tag{20}$$

The theory of dynamical systems [51] tells us that the point $(0,0)$ is a *saddle node*, i.e. an unstable equilibrium whose phase space resembles the one depicted in Figure 12(a). The two straight lines are the directions spanned by the eigenvectors of the matrix of coefficients $\begin{pmatrix} 1 & 1 \\ 4 & 1 \end{pmatrix}$, and are called *stable* and *unstable manifolds*. Motion in the stable manifold converges to the equilibrium $(0,0)$, while the unstable manifold and all other trajectories diverge to infinity. However, small perturbations applied to the stable manifold can bring the system on a divergent hyperbolic trajectory, so we expect that ODE's and the associated sCCP agent, when starting from the stable manifold (say from point $(1,-2)$), will eventually jump on a divergent trajectory. Moreover, we expect that the smaller δ the later this event will happen. This intuition is confirmed in Figures 12(b), 12(c), 12(d).

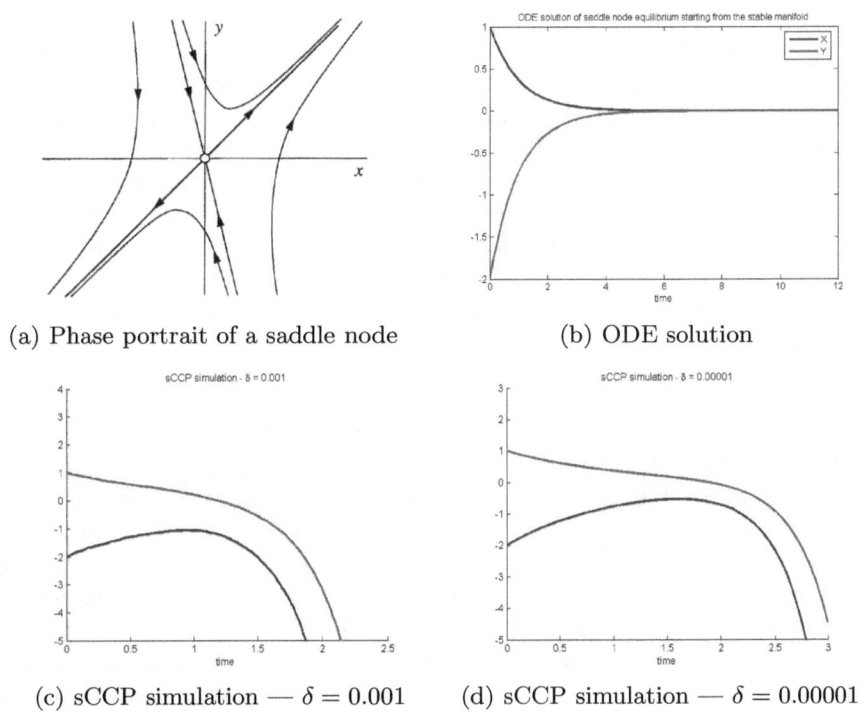

(a) Phase portrait of a saddle node   (b) ODE solution

(c) sCCP simulation — $\delta = 0.001$ (d) sCCP simulation — $\delta = 0.00001$

Fig. 12. **12(a)**: Phase space of the linear system (20). The origin is a saddle node; the stable manifold is displayed with arrows pointing towards the origin, while the unstable manifold has arrows diverging from it. **12(b)**: Solution of the ODE's (20), starting from $(1,-2)$, a point belonging to the stable manifold. **12(c),12(d)**: Simulation of the sCCP agent associated to the linear system (20), w.r.t. the set of convering functions $\{(X,(1,0)),(Y,(1,0)),(4X,(0,1)),(Y,(0,1))\}$ with initial conditions $(1,-2)$. The step δ is equal respectively to 0.001 and 0.00001. The time in which these trajectories diverge from the solution of the ODE's increases as δ becomes smaller.

This example shows that convergence issues need to be investigate further. In particular, conditions taking into account the topology of the phase space of the ODE's are required in order to guarantee also asymptotic convergence. Another interesting direction to investigate is to exploit other results by Kurtz [36] in order to state error bounds in the approximation.

Examples of Sets of Covering Functions. All the results of this Section have been given parametrically w.r.t. a set of covering functions $\mathcal{G}$. When we motivated the introduction of such concept, we stated that a choice of a specific $\mathcal{G}$ corresponds to a specific logical structure of the fluxes generating the ODEs. We discuss now two possible choices of $\mathcal{G}$, one natural in absence of information, and the other tailored for ODEs coming from sets of biochemical reactions for which it is possible to reconstruct the reactions from the ODEs.

Example 2. The simplest choice of a covering set is the one in which all the addends of ODEs are treated as independent flux sources. To be more specific, consider a set of ODE $\mathbf{f}(\mathbf{x}) = (f_1(\mathbf{x}), \ldots, f_k(\mathbf{x}))$, with $f_i(\mathbf{x}) = \sum_{j=1}^{k_i} f_{ij}(\mathbf{x})$, where f_{ij} are the single addends of the ODE. The idea is to treat independently each such f_{ij}, so that $\mathcal{G}_D = \{(f_{ij}, \mathbf{e_i}) \mid 1 \leq i \leq k, 1 \leq j \leq k_i\}$. Such covering set $\mathcal{G}_D$ will be called in the following the *disentangled covering set*. This was the choice adopted, for instance, when discussing the Repressilator example in Section 4.3. This choice is reasonable in absence of any further information on the system modeled by the ODE's, and it does not preserve structural properties of the system, like mass conservation. For instance, consider the system defined by the single reaction $A \rightarrow_{k \cdot A} B$, which preserves the total mass $A + B$. With the disentangled covering set, however, we would reconstruct the logical structure of the following system of biochemical reactions: $A \rightarrow_{k \cdot A}$ and $\rightarrow_{k \cdot A} B$, which does not preserve the total mass $A + B$.

Example 3. Assume now we have a system generated by a set of mass action reactions such that the left hand side (i.e., the list of reactants) of each such reaction is unique. This has the consequence that each reaction is uniquely identified by the algebraic structure as a monomial of its rate function. For instance, the only reaction with A and B as reagents, $A + B \rightarrow_k ?$, is uniquely identified by the signature as a monomial of its rate function kAB, i.e. by the monomial AB with coefficient 1. This property has the immediate consequence that, whenever we find two addends of the ODE with the same signature as a monomial, we are guaranteed that they are two instances of the same flux. That is to say, if the condition is satisfied, we know how to reconstruct the set of reactions that originated the ODE's.

Formally, given $\mathbf{f}(\mathbf{x}) = (f_1(\mathbf{x}), \ldots, f_k(\mathbf{x}))$, with $f_i(\mathbf{x}) = \sum_{j=1}^{k_i} f_{ij}(\mathbf{x})$, we can construct the covering set $\mathcal{G}_R$ as follows: list all the different support monomials $\{p_1, \ldots, p_s\}$ in the set $\{f_{ij} \mid 1 \leq i \leq k, 1 \leq j \leq k_i\}$, and, for each l, define the terms $\alpha_{l,j} \in \mathbb{Z}$ and $\beta_l \in \mathbb{R}^+$ such that:

1. if p_l occurs in the equation for variable i, then its occurrence is equal to $\alpha_{l,i} \beta_l p_l$;

2. if p_l does not occur in the equation for variable i, then $\alpha_{l,i} = 0$;
3. all $\alpha_{l,j} \neq 0$ are prime among them[18];

Then, letting $\alpha_l = (\alpha_{l,1}, \ldots, \alpha_{l,k})$, the covering set $\mathcal{G}_R$ can be defined as $\mathcal{G}_R = \{(\beta_l p_l, \alpha_l) \mid l = 1, \ldots, s\}$.

For instance, consider the ODEs

$$\begin{cases} \dot{X} = -2kX^2Y \\ \dot{Y} = -kX^2Y \\ \dot{Z} = 3kX^2Y \end{cases}$$

The associated covering set $\mathcal{G}_R$ is simply $\mathcal{G}_R = \{(kX^2Y, (-2, -1, 3))\}$, corresponding to the single reaction $2X + Y \rightarrow_k 3Z$.

5   Final Discussion

In this paper we presented a method to associate ordinary differential equations to sCCP programs (written with a restricted syntax), and also a method that generates an sCCP-network from a set of ODE's. The translation from sCCP to ODE's is based on the construction of a graph, called RTS, whose edges represent all possible actions performable by sCCP-agents. Properties of RE-STRICTED($sCCP$) guarantee that the graph is always finite. From an RTS, we can construct an interaction matrix containing the modifications that each action makes to each variable. Writing the corresponding ODE's is simply a matter of combining the interaction matrix with the rate of each action. The inverse translation, from ODE's to sCCP, exploits the functional form that rates have in sCCP. In this way, we can associate sCCP-agents to general ODE's. An important feature of this method is that it is parametric w.r.t. the basic increment of variables, meaning that we can reduce the effect of stochastic fluctuations in the sCCP-model. Actually, we proved in Theorem 6 that, in the limit of an infinitesimal increment, the trajectories of the ODE's and of the corresponding sCCP-system coincide.

In Section 3.1, we showed that the translation from sCCP to ODE's, when applied to models of biochemical reactions, preserves the rate semantics in the sense of [17]. This condition, however, is not sufficient to guarantee that the translation maintains also the dynamical behavior of the sCCP-model. In fact, in Section 3.2, we provided several examples where an sCCP-network and the associated ODE's manifest a different behavior. This divergence can be caused by many factors, all qualitatively different.

Preserving dynamical behavior, however, is not just a mathematical game, but is is a central property that a translation from sCCP to ODE should have in order to be used as an analysis technique for stochastic process algebras. In this light, also the mapping from ODE to sCCP can be seen as a tool to investigate behavioral preservation.

---

[18] This condition guarantees that α_{lj} are uniquely defined.

In Section 4, when we introduced the notion of set of covering functions, we noted that in the passage to ODE's we unavoidably lose something of the logic of the sCCP model. This also suggest that the preservation of behavior may be reasonable only from a qualitative point of view. Indeed, this weaker approach fits better with the management of stochastic noise, see the discussion at the end of Section 3.2. The loss of precision in passing to ODE's is, however, counterbalances by the computational gain: simulating stochastic processes is undoubtedly much more expensive than numerically solving ODE's [24].

There are several open problems related to the question of behavioral equivalence. We list hereafter some of the most important ones, according to us.

– We need to identify the class of sCCP models (and their regions of parameter space/initial conditions) for which the mapping ODE preserves dynamics. Intuitively, according to discussion of Section 3.2, this may happen if all variables have big absolute values and if the phase space of the ODE's has asymptotically stable trajectories with ample basins of attraction.
– The repressilator and the simple self-inhibited genetic network of Section 3.2 suggest that the discrete ingredient cannot be continuously approximated so easily. In particular, associating continuous variables to RTS-states seems rather arbitrary. A possible solution can be that of transforming an sCCP network into a hybrid system, in which continuous and discrete dynamics coexist. In this way, we may be able to preserve part of the discrete structure of an sCCP-network, possibly just that fundamental for maintaining the behavior. We are investigating this direction, mapping sCCP-programs to hybrid automata [29,2]. The first results are encouraging, see [13,14]
– The notion of behavioral equivalence needs to be specified formally. At the end of Section 3.2, we suggested an approach based on a suitable temporal logic, in which equivalence would mean equi-satisfiability of the same set of formulae.

As a final remark, we would like to consider this work under the perspective of the study of systemic properties. In fact, when we model a biological system, we are concerned mainly with the understanding of its systemic properties, especially what they are and how they emerge from basic interactions. In this direction, a modeler needs a formal language to specify biological systems, possibly provided with different semantics, related to one another and stratified in several layers of increasing approximation and abstraction. For example, sCCP has a natural CTMC-based semantics, but an ODE-based one can be assigned to it via the ODE operator. A possible layer in the middle consists in a semantic based, for instance, on hybrid automata. Finally, we need also a language to specify system's properties, automatically verifying them on the different semantics, or better, on the simpler semantic where answers are correct (i.e., on the simpler semantic showing the same dynamical behavior of the most general one). All these features must clearly be part of the same operative framework (and of the same software tool), hence all the open questions presented above can be seen as steps in this direction.

References

1. Converging sciences. Trento (2004), `http://www.unitn.it/events/consci/`
2. Alur, R., Belta, C., Ivancic, F., Kumar, V., Mintz, M., Pappas, G., Rubin, H., Schug, J.: Hybrid modeling and simulation of biomolecular networks. In: Di Benedetto, M.D., Sangiovanni-Vincentelli, A.L. (eds.) HSCC 2001. LNCS, vol. 2034, pp. 19–32. Springer, Heidelberg (2001)
3. Antoniotti, M., Policriti, A., Ugel, N., Mishra, B.: Model building and model checking for biochemical processes. Cell Biochemistry and Biophysics 38(3), 271–286 (2003)
4. Aziz, A., Singhal, V., Balarin, F., Brayton, R., Sangiovanni-Vincentelli, A.: Verifying continuous time markov chains. In: Alur, R., Henzinger, T.A. (eds.) CAV 1996. LNCS, vol. 1102. Springer, Heidelberg (1996)
5. Blossey, R., Cardelli, L., Phillips, A.: A compositional approach to the stochastic dynamics of gene networks. T. Comp. Sys. Biology, 99–122 (2006)
6. Blossey, R., Cardelli, L., Phillips, A.: Compositionality, stochasticity and cooperativity in dynamic models of gene regulation. HFPS Journal (2007) (in print)
7. Bortolussi, L.: Stochastic concurrent constraint programming. In: Proceedings of 4th International Workshop on Quantitative Aspects of Programming Languages (QAPL 2006). ENTCS, vol. 164, pp. 65–80 (2006)
8. Bortolussi, L.: Constraint-based approaches to stochastic dynamics of biological systems. PhD thesis, PhD in Computer Science, University of Udine (2007), `http://www.dmi.units.it/~bortolu/files/reps/Bortolussi-PhDThesis.pdf`
9. Bortolussi, L.: A master equation approach to differential approximations of stochastic concurrent constraint programming. In: Proceedings of QAPL 2008. ENTCS (2008) (to appear)
10. Bortolussi, L., Fonda, S., Policriti, A.: Constraint-based simulation of biological systems described by molecular interaction maps. In: Proceedings of IEEE conference on Bioinformatics and Biomedicine, BIBM 2007 (2007)
11. Bortolussi, L., Policriti, A.: Relating stochastic process algebras and differential equations for biological modeling. In: Proceedings of PASTA 2006 (2006)
12. Bortolussi, L., Policriti, A.: Stochastic concurrent constraint programming and differential equations. In: Proceedings of Fifth Workshop on Quantitative Aspects of Programming Languages, QAPL 2007. ENTCS, vol. 16713 (2007)
13. Bortolussi, L., Policriti, A.: Hybrid approximation of stochastic concurrent constraint programming. In: Proceedings of IFAC 2008 (2008)
14. Bortolussi, L., Policriti, A.: The importance of being (a little bit) discrete. In: Proceedings of FBTC 2008. ENTCS (2008) (to appear)
15. Bortolussi, L., Policriti, A.: Modeling biological systems in concurrent constraint programming. Constraints 13(1) (2008)
16. Cardelli, L.: From processes to odes by chemistry (2006), `http://lucacardelli.name/`
17. Cardelli, L.: On process rate semantics. Theoretical Computer Science 391(3), 190–215 (2008)
18. Ciocchetta, F., Hillston, J.: Bio-PEPA: an extension of the process algebra PEPA for biochemical networks. In: Proceeding of FBTC 2007. Workshop of CONCUR 2007 (2007)
19. Seattle CompBio Group, Institute for Systems Biology. Dizzy home page
20. Cornish-Bowden, A.: Fundamentals of Chemical Kinetics, 3rd edn. Portland Press (2004)

21. Elowitz, M.B., Leibler, S.: A synthetic oscillatory network of transcriptional regulators. Nature 403, 335–338 (2000)
22. Geisweiller, N., Hillston, J., Stenico, M.: Relating continuous and discrete pepa models of signalling pathways. Theoretical Computer Science (2008) (in print)
23. Gillespie, D.: The chemical langevin equation. Journal of Chemical Physics 113(1), 297–306 (2000)
24. Gillespie, D., Petzold, L.: Numerical Simulation for Biochemical Kinetics. In: System Modelling in Cellular Biology. MIT Press, Cambridge (2006)
25. Gillespie, D.T.: A general method for numerically simulating the stochastic time evolution of coupled chemical reactions. J. of Computational Physics 22 (1976)
26. Gillespie, D.T.: Exact stochastic simulation of coupled chemical reactions. J. of Physical Chemistry 81(25) (1977)
27. Haas, P.J.: Stochastic Petri Nets. Springer, Heidelberg (2002)
28. Hastings, S.P., Murray, J.D.: The existence of oscillatory solutions in the field-noyes model for the belousov-zhabotinskii reaction. SIAM Journal on Applied Mathematics 28(3), 678–688 (1975)
29. Henzinger, T.A.: The theory of hybrid automata. In: LICS 1996: Proceedings of the 11th Annual IEEE Symposium on Logic in Computer Science (1996)
30. Hillston, J.: A Compositional Approach to Performance Modelling. Cambridge University Press, Cambridge (1996)
31. Hillston, J.: Fluid flow approximation of PEPA models. In: Proceedings of the Second International Conference on the Quantitative Evaluation of Systems, QEST 2005 (2005)
32. Kitano, H.: Foundations of Systems Biology. MIT Press, Cambridge (2001)
33. Kitano, H.: Computational systems biology. Nature 420, 206–210 (2002)
34. Kohn, K.W., Aladjem, M.I., Weinstein, J.N., Pommier, Y.: Molecular interaction maps of bioregulatory networks: A general rubric for systems biology. Molecular Biology of the Cell 17(1), 1–13 (2006)
35. Kurtz, T.G.: Solutions of ordinary differential equations as limits of pure jump markov processes. Journal of Applied Probability 7, 49–58 (1970)
36. Kurtz, T.G.: Limit theorems for sequences of jump markov processes approximating ordinary differential processes. Journal of Applied Probability 8, 244–356 (1971)
37. Kwiatkowska, M., Norman, G., Parker, D.: Probabilistic symbolic model checking with prism: A hybrid approach. International Journal on Software Tools for Technology Transfer 6(2), 128–142 (2004)
38. Mcadams, H.H., Arkin, A.: Stochastic mechanisms in gene expression. PNAS 94, 814–819 (1997)
39. Norris, J.R.: Markov Chains. Cambridge University Press, Cambridge (1997)
40. Noyes, R.M., Field, R.J.: Oscillatory chemical reactions. Annual Review of Physical Chemistry 25, 95–119 (1974)
41. Nurse, P.: Understanding cells. Nature 24 (2003)
42. Plotkin, G.D.: A structural approach to operational semantics. J. Log. Algebr. Program., 60-61, 17–139 (2004)
43. Press, W.H., Teukolsky, S.A., Vetterling, W.T., Flannery, B.P.: Numerical Recipes in C++: The Art of Scientific Computing. Cambridge University Press, Cambridge (2002)
44. Priami, C.: Stochastic π-calculus. The Computer Journal 38(6), 578–589 (1995)
45. Priami, C., Regev, A., Shapiro, E.Y., Silverman, W.: Application of a stochastic name-passing calculus to representation and simulation of molecular processes. Inf. Process. Lett. 80(1), 25–31 (2001)

46. Ramsey, S., Orrell, D., Bolouri, H.: Dizzy: stochastic simulation of large-scale genetic regulatory networks. Journal of Bioinformatics and Computational Biology 3(2), 415–436 (2005)
47. Rao, C.V., Arkin, A.P.: Stochastic chemical kinetics and the quasi-steady state assumption: Application to the gillespie algorithm. Journal of Chemical Physics 118(11), 4999–5010 (2003)
48. Regev, A., Shapiro, E.: Cellular abstractions: Cells as computation. Nature 419 (2002)
49. Saraswat, V.A.: Concurrent Constraint Programming. MIT press, Cambridge (1993)
50. Shapiro, B.E., Levchenko, A., Meyerowitz, E.M., Wold, B.J., Mjolsness, E.D.: Cellerator: extending a computer algebra system to include biochemical arrows for signal transduction simulations. Bioinformatics 19(5), 677–678 (2003)
51. Strogatz, S.H.: Non-Linear Dynamics and Chaos, with Applications to Physics, Biology, Chemistry and Engeneering. Perseus books, Cambridge (1994)
52. Vilar, J.M.G., Yuan Kueh, H., Barkai, N., Leibler, S.: Mechanisms of noise resistance in genetic oscillators. PNAS 99(9), 5991 (2002)
53. Wilkinson, D.J.: Stochastic Modelling for Systems Biology. Chapman & Hall, Boca Raton (2006)

Computing Equilibrium Points of Genetic Regulatory Networks

Graziano Chesi

Department of Electrical and Electronic Engineering
University of Hong Kong
chesi@eee.hku.hk
http://www.eee.hku.hk/~chesi

Abstract. Computing equilibrium points of genetic regulatory networks is a problem of primary importance for numerous investigations in these systems. This paper addresses this problem for differential equation models, with the regulation function expressed in a general form which includes both SUM form and PROD form for saturation functions of any type. Specifically, a recursive algorithm is proposed, which provides at each recursion a region guaranteed to contain all equilibrium points. This region progressively shrinks, and asymptotically converges to the sought set of equilibrium points. Moreover, the proposed algorithm can also allow one to delimit and find limit cycles. Some numerical examples are reported to illustrate and validate the proposed algorithm, including examples where standard mathematical tools fail to compute the sought equilibrium points.

Keywords: Genetic regulatory network, Differential equation, Saturation, Equilibrium point, Limit cycle.

1 Introduction

Genetic regulatory networks explain the interactions between genes and proteins to form complex systems that perform complicated biological functions, see for instance [1,2,3,4,5,6,7,8]. Basically, there are two types of genetic regulatory network models, i.e., the Boolean model (or discrete model) and the differential equation model (or continuous model). In Boolean models, the activity of each gene is expressed in one of two states, ON or OFF, and the state of a gene is expressed by a Boolean function of the states of other related genes. In the differential equation models, the variables describe the concentrations of gene products, such as mRNAs and proteins, as continuous values of the gene regulation systems. See for example [9,10,11,12,13] and references therein for a wider categorization of genetic regulatory networks models.

This paper focuses on genetic regulatory networks described through differential equation models. In these models the dynamics of each concentration is expressed by a function of all concentrations of the system. This function typically consists of two parts: a linear part which defines the natural decay rate of

C. Priami et al. (Eds.): Trans. on Comput. Syst. Biol. XI, LNBI 5750, pp. 268–282, 2009.

the concentration itself, and a nonlinear part which defines the influence by all the other concentrations. The nonlinear part can be either described via sum of saturation functions (in this case the system is said to be in SUM form) or via product of saturation functions (in this case the system is said to be in PROD form). See for instance [14,15,16,17].

A fundamental problem in these networks consists of determining the equilibrium points, i.e. the amounts of concentrations for which the regulation process results complete. This is a necessary step for several investigations, such as steady-state, stability, disturbance rejection, etc. Unfortunately, to determine equilibrium points of genetic regulatory networks is a difficult problem because these systems contain saturation functions, and hence the calculation of the equilibrium points amounts to solving a system of nonlinear equations. Indeed, there do not exist techniques able to guarantee to find all solutions of such a system, except in the case of polynomial equations, which however can be addressed only for small degrees and small number of variables, see for instance [18,19,20,21] and references therein.

In this paper we address the problem of computing equilibrium points of genetic regulatory networks described through differential equation models. We consider a general model which includes both SUM form and PROD form for saturation functions of any type. The contribution consists of a recursive algorithm which holds the following properties. First, at each recursion the algorithm provides a region containing all equilibrium points, i.e. no equilibrium is lost. Second, this region progressively shrinks, i.e. the conservatism does not increase. Third, this region asymptotically converges to the set of equilibrium points, i.e. all equilibrium points are found. The proposed algorithm is illustrated and validated through some numerical examples with synthetic and real genetic regulatory networks. In these examples it is also shown that standard mathematical tools for solving systems of nonlinear equations may fail to compute the sought equilibrium points. Moreover, in these examples it is also explained that the proposed algorithm can be useful to delimit and find limit cycles.

The paper is organized as follows. Section 2 introduces some preliminaries on genetic regulatory networks. Section 3 describes the proposed results. Section 4 presents some numerical examples. Finally, Section 5 reports some concluding remarks.

2   Preliminaries

First of all, let us introduce the notation used throughout the paper:

- $\mathbb{R}_+$: positive real number set, i.e. $\{x \in \mathbb{R} : x \geq 0\}$;
- 0_n: null vector of size $n \times 1$;
- I_n: identity matrix of size $n \times n$;
- e_i: i-th column of I_n;
- $\mathrm{diag}(x_1, \ldots, x_n)$: diagonal matrix with x_i at the (i, i) entry;
- X^T: transpose of vector/matrix X;
- TF: transcription factor.

The genetic regulatory networks considered in this paper are described by the differential equation model

$$
\begin{cases}
\dot{m}_i(t) = -a_i m_i(t) + b_i(p_1(t), \ldots, p_n(t)) \\
\dot{p}_i(t) = -c_i p_i(t) + d_i m_i(t) \\
\quad i = 1, \ldots, n
\end{cases}
\tag{1}
$$

where $m_i(t), p_i(t) \in \mathbb{R}_+$ are the concentrations of mRNA and protein of the i-th gene, $a_i, c_i \in \mathbb{R}_+$ are the degradation rates, $d_i \in \mathbb{R}_+$ expresses the effect of $m_i(t)$ on $p_i(t)$, and $b_i : \mathbb{R}_+^n \to \mathbb{R}_+$ is the regulatory function of the i-th gene. This function is typically nonlinear, and either always increases or always decreases with respect to any component of $p(t)$ whenever its other components are fixed, i.e.

$$
(-1)^{k_i} b_i(p_1, \ldots, p_{i-1}, x_2, p_{i+1}, \ldots, p_n) \geq (-1)^{k_i} b_i(p_1, \ldots, p_{i-1}, x_1, p_{i+1}, \ldots, p_n)
$$
$$
\forall x_1, x_2 : \ x_1 \leq x_2 \ \forall p_1(t), \ldots, p_n(t) \in \mathbb{R}_+ \ \forall i = 1, \ldots, n
\tag{2}
$$

for some $k_1, \ldots, k_n \in \{0, 1\}$.

In genetic regulatory networks with SUM form, the function $b_i(p_1(t), \ldots, p_n(t))$ is expressed as the sum of functions of a single variable, i.e.

$$
b_i(p_1(t), \ldots, p_n(t)) = \sum_{j=1}^{n} \alpha_{i,j} b_{i,j}(p_j(t))
\tag{3}
$$

where $\alpha_{i,j} \in \mathbb{R}_+$ is the contribution of TF j to the transcriptional rate for gene i, and $b_{i,j} : \mathbb{R}_+ \to \mathbb{R}_+$ is a monotonic function, i.e. $b_{i,j}(p_j(t))$ either always increases or always decreases with respect to $p_j(t)$.

In genetic regulatory networks with PROD form, the function $b_i(p_1(t), \ldots, p_n(t))$ is expressed as the product of the functions $b_{i,j}(p_j(t))$, i.e.

$$
b_i(p_1(t), \ldots, p_n(t)) = \alpha_i \prod_{j=1}^{n} b_{i,j}(p_j(t))
\tag{4}
$$

where $\alpha_i \in \mathbb{R}_+$ represents the transcriptional rate for gene i.

Each function $b_{i,j}(p_j(t))$ in (3) and (4) is typically expressed as

$$
b_{i,j}(p_j(t)) = \begin{cases}
f(p_j(t)) & \text{if TF } j \text{ is an activator of gene } i \\
1 - f(p_j(t)) & \text{if TF } j \text{ is a repressor of gene } i \\
\gamma & \text{otherwise}
\end{cases}
\tag{5}
$$

where $\gamma \in \mathbb{R}$ is a constant depending on the model which expresses the independence of gene i on TF j ($\gamma = 0$ for SUM form, $\gamma = 1$ for PROD form), and the function $f(p_j(t))$ is a saturation function. For saturation function we mean a function satisfying the following properties:

$$
\begin{cases}
f : \mathbb{R}_+ \to [0, 1] \\
f(0) = 0 \\
\lim_{x \to \infty} f(x) = 1 \\
f(x_2) \geq f(x_1) \ \forall x_1, x_2 : \ x_1 \leq x_2
\end{cases}
\tag{6}
$$

Hence, a saturation function is an increasing function between 0 and 1 defined for positive value of the variable. For instance, in the case of regulatory functions with Hill form, the function $f(p_j(t))$ is given by

$$f(p_j(t)) = \frac{p_j(t)^H}{\beta^H + p_j(t)^H} \tag{7}$$

where $\beta \in \mathbb{R}_+$ and H is an integer known as Hill coefficient.

In order to describe the results of this paper in a more compact form, we introduce a matrix version of the model (1) according to

$$\begin{cases} \dot{m}(t) = Am(t) + b(p(t)) \\ \dot{p}(t) = Cp(t) + Dm(t) \end{cases} \tag{8}$$

where

$$\begin{aligned} m(t) &= (m_1(t), \ldots, m_n(t))' \\ p(t) &= (p_1(t), \ldots, p_n(t))' \end{aligned} \tag{9}$$

are the vectors containing the concentrations of mRNA and protein, and

$$\begin{aligned} A &= \mathrm{diag}(-a_1, \ldots, -a_n) \\ C &= \mathrm{diag}(-c_1, \ldots, -c_n) \\ D &= \mathrm{diag}(d_1, \ldots, d_n) \end{aligned} \tag{10}$$

are diagonal matrices containing the decay rates (matrices A and C) and the effect of $m(t)$ on $p(t)$ (matrix D). The function $b : \mathbb{R}_+^n \to \mathbb{R}_+^n$ is a nonlinear function representing the regulation of the process, whose i-th component $b_i(p(t))$ satisfies the monotonicity condition (2).

We observe that the model (8) under the assumption (2), which is an equivalent matrix version of the model (1), includes:

1. genetic regulatory networks with SUM form, by choosing the i-th component of $b(p(t))$ as in (3);
2. genetic regulatory networks with PROD form, by choosing the i-th component of $b(p(t))$ as in (4);
3. genetic regulatory networks that are neither in SUM form nor in PROD form, provided that (2) holds. For instance, the choice for $n = 3$ given by

$$b(p(t)) = \begin{pmatrix} b_{1,1}(p_1(t)) + b_{1,2}(p_2(t))^3 \\ e^{b_{2,1}(p_1(t))} b_{2,3}(p_3(t)) \\ b_{3,2}(p_2(t))^3 + \sqrt{b_{3,3}(p_3(t))} \end{pmatrix} \tag{11}$$

defines a genetic regulatory networks which is neither in SUM form nor in PROD form, but which is included in the model (8) under the assumption (2).

The problem addressed in this paper consists of determining the equilibrium points of (8), i.e. the solutions of the system of nonlinear equations

$$\begin{cases} Am + b(p) = 0_n \\ Cp + Dm = 0_n \\ m, p \in \mathbb{R}_+^n \end{cases} \tag{12}$$

Remark 1. Before proceeding let us observe that existing mathematical tools for solving systems of nonlinear equations generally do not guarantee to find all solutions of such systems. Indeed, systems of nonlinear equations can be solved via either analytical techniques or numerical techniques. Analytical techniques can be used in the case of polynomial or rational equations, and provides the sought solutions as roots of a one-variable polynomial. Unfortunately, the degree of this polynomial is prohibitive (except for very small systems) since in the worst case coincides with the maximum number of solutions of the system, which is given by the degree of the equations to the power of the number of variables, see for instance [18,22,19,21]. Numerical techniques, which are either based on the numerical minimization of a suitable function via for example Newton's iterations starting from an initial point, or on homotopy methods which adopt continuation strategies, do not suffer of the previous problems. Unfortunately, these techniques cannot guarantee to find all sought solutions, see for instance [23,20] and Section 4.

Remark 2. Another remark concerns the fact that genetic regulatory networks can be also modeled as stochastic systems, where the input is represented by a stochastic process such as white noise. For instance, such an input could affect (8) according to

$$\begin{cases} \dot{m}(t) = Am(t) + b(p(t)) + w(t) \\ \dot{p}(t) = Cp(t) + Dm(t) \end{cases} \tag{13}$$

where $w(t) \in \mathbb{R}^n$ is a stochastic process. In these systems there are no equilibrium points in the classic sense since the input is a non-constant function of the time and hence the equation

$$Am + b(p) + w(t) = 0_n \tag{14}$$

would not admit solutions where m and p do not depend on the time (which is the classic definition of equilibrium point). Instead, one can consider equilibrium points corresponding to particular constant values of the stochastic process, such as its mean value, that the algorithm proposed in this paper allows one to compute. Indeed, these equilibrium points are defined analogously to (12) as

$$\begin{cases} Am + b(p) + \bar{w} = 0_n \\ Cp + Dm = 0_n \\ m, p \in \mathbb{R}_+^n \end{cases} \tag{15}$$

where $\bar{w} \in \mathbb{R}^n$ is the the stochastic expectation of $w(t)$.

3   Equilibria Computation

In this section we describe the proposed algorithm. Specifically, in Theorems 1 and 2 we introduce two preliminary functions and we describe their properties. Then, in Theorem 3 we provide the main algorithm to be used to compute the sought equilibrium points.

Before proceeding, let us observe that the m-component of any solution of (12) is related to its p-component by the relationship $Cp + Dm = 0_n$ where C, D are nonsingular diagonal matrices with C negative definite. This means that (12) can be equivalently rewritten as

$$\begin{cases} -AD^{-1}Cp + b(p) = 0_n \\ m = -D^{-1}Cp \\ p \in \mathbb{R}^n_+ \end{cases} \tag{16}$$

Therefore, in the sequel we will focus on the computation of the vectors p fulfilling (16). We indicate the set of such vectors as

$$\mathcal{E} = \{p \in \mathbb{R}^n_+ : \ -AD^{-1}Cp + b(p) = 0_n\} \tag{17}$$

Theorem 1. *Let $\mathcal{H}$ be the rectangle defined by*

$$\mathcal{H} = \{p \in \mathbb{R}^n_+ : \ p_i \in [p_{i,-}, p_{i,+}]\} \tag{18}$$

for some $p_{1,-}, p_{1,+}, \ldots, p_{n,-}, p_{n,+} \in \mathbb{R}_+$, and let us define the map $\mathcal{A}(\mathcal{H})$ as

$$\mathcal{A}(\mathcal{H}) = \{p \in \mathbb{R}^n_+ : \ p_i \in [q_{i,-}, q_{i,+}]\} \tag{19}$$

where $q_{1,-}, q_{1,+}, \ldots, q_{n,-}, q_{n,+} \in \mathbb{R}_+$ are computed according to

$$q_{i,-} = \max \left\{ p_{i,-} \ , \ \min_{z \in \mathcal{Z}} e_i^T C^{-1} D A^{-1} z \right\} \tag{20}$$

$$q_{i,+} = \min \left\{ p_{i,+} \ , \ \max_{z \in \mathcal{Z}} e_i^T C^{-1} D A^{-1} z \right\} \tag{21}$$

where $\mathcal{Z}$ is the set given by

$$\mathcal{Z} = \{b(p) : \ p_i \in \{p_{i,-}, p_{i,+}\}, \ i = 1, \ldots, n\}. \tag{22}$$

Then, the following properties hold:

- Property P1: $\mathcal{A}(\mathcal{H}) \subseteq \mathcal{H}$;
- Property P2: $p^* \in \mathcal{H} \cap \mathcal{E} \Rightarrow p^* \in \mathcal{A}(\mathcal{H})$;
- Property P3: $\mathcal{H} \cap \mathcal{A}(\mathcal{H}) = \emptyset \Rightarrow \mathcal{H} \cap \mathcal{E} = \emptyset$.

Proof. First, the property P1 holds because from (20)–(21) one has

$$q_{i,-} \geq p_{i,-} \text{ and } q_{i,+} \leq p_{i,+} \quad \forall i = 1, \ldots, n. \tag{23}$$

Second, the property P2 holds due to the monotonicity property (2) of $b_i(p)$ with respect to each component of p and to the linearity of the function $e_i^T C^{-1} D A^{-1} z$ with respect to z. In fact, we have

$$p^* \in \mathcal{H} \Rightarrow b_i(p^*) \in [\min_{z \in \mathcal{Z}} z_i, \max_{z \in \mathcal{Z}} z_i]. \tag{24}$$

Moreover,

$$p^* \in \mathcal{E} \Rightarrow e_i^T C^{-1} D A^{-1} b(p^*) = p_i^*. \tag{25}$$

Hence, it follows

$$p^* \in \mathcal{H} \cap \mathcal{E} \Rightarrow q_{i,-} \leq p_i^* \text{ and } q_{i,+} \geq p_i^*. \tag{26}$$

Lastly, the property P3 holds because, if one suppose for contradiction that $\mathcal{H} \cap \mathcal{A}(\mathcal{H}) = \emptyset$ and $\mathcal{H}$ contains a vector p^* of $\mathcal{E}$, then it would follow from the property P2 that p^* belongs to $\mathcal{A}(\mathcal{H})$, hence contradicting the assumption that $\mathcal{H} \cap \mathcal{A}(\mathcal{H}) = \emptyset$. □

Let us observe that map $\mathcal{A}(\cdot)$ requires trivial computations, i.e. evaluation of a linear function in some given points. In fact, let us observe that the set $\mathcal{Z}$ is finite. From the map $\mathcal{A}(\cdot)$ we define the map $\mathcal{B}(\cdot)$ in the following theorem.

Theorem 2. *Let $\mathcal{H}$ be a rectangle in (18), and let us define the map $\mathcal{B}(\mathcal{H})$ as follows:*

- *(Step 1) set $\mathcal{H}^{(0)} = \mathcal{H}$ and $k = 0$ (k denotes the iteration number);*
- *(Step 2) if $\mathcal{H}^{(k)} \cap \mathcal{A}(\mathcal{H}^{(k)}) = \emptyset$, set $\mathcal{B}(\mathcal{H}) = \emptyset$ and exit;*
- *(Step 3) if $\mathcal{A}(\mathcal{H}^{(k)})$ is a point, set $\mathcal{B}(\mathcal{H}) = \mathcal{A}(\mathcal{H}^{(k)})$ and exit;*
- *(Step 4) if $\mathcal{H}^{(k)} = \mathcal{A}(\mathcal{H}^{(k)})$, set $\mathcal{B}(\mathcal{H}) = \mathcal{H}^{(k)}$ and exit;*
- *(Step 5) set $\mathcal{H}^{(k+1)} = \mathcal{A}(\mathcal{H}^{(k)})$, $k = k + 1$, and go to 2.*

Then, $\mathcal{B}(\mathcal{H})$ returns either a rectangle, a point, or the empty set. Moreover:

- *Property P4: $\mathcal{B}(\mathcal{H}) \subseteq \mathcal{H}$;*
- *Property P5: $p^* \in \mathcal{H} \cap \mathcal{E} \Rightarrow p^* \in \mathcal{B}(\mathcal{H})$.*

Proof. First of all, let us observe that the output of $\mathcal{B}(\mathcal{H})$ can be either the empty set (output of Step 2), a point (output of Step 3), or a rectangle (output of Step 4). Then, the property P4 follows from the fact that the output of $\mathcal{B}(\mathcal{H})$ is a sequence of applications of the map $\mathcal{A}(\cdot)$ for which the property P1 ensures that the output is a subset of the input. Lastly, the property P5 holds since $\mathcal{B}(\mathcal{H})$ returns either a sequence of applications of the map $\mathcal{A}(\cdot)$ for which the property P2 ensures that no vector of $\mathcal{H} \cap \mathcal{E}$ can be lost, or the empty set in the case $\mathcal{H}^{(k)} \cap \mathcal{A}(\mathcal{H}^{(k)}) = \emptyset$ which however guarantees the absence of vectors of $\mathcal{E}$ in $\mathcal{H}^{(k)}$ (and hence in $\mathcal{H}$) due to the properties P2 and P3. □

The map $\mathcal{B}(\cdot)$ transforms a given rectangle via a sequence of applications of the map $\mathcal{A}(\cdot)$, and returns a set which can be either a rectangle, a point, or the empty set. By exploiting the map $\mathcal{B}(\cdot)$ we derive the algorithm for the computation of the sought equilibrium points as follows.

Theorem 3. *(Algorithm for equilibrium points computation) Let $\mathcal{H}$ be a rectangle in (18) and let us define the map $\mathcal{C}(\mathcal{H})$ as follows:*

- *(Step 1) if $\mathcal{B}(\mathcal{H})$ is either the empty set or a point, then set $\mathcal{C}(\mathcal{H}) = \mathcal{B}(\mathcal{H})$ and exit;*
- *(Step 2) divide the rectangle $\mathcal{B}(\mathcal{H})$ in 2^k rectangles $\mathcal{H}_1, \dots, \mathcal{H}_{2^k}$ by taking the middle point on each side of $\mathcal{B}(\mathcal{H})$ with nonzero length;*
- *(Step 3) set $\mathcal{C}(\mathcal{H}) = \bigcup_{i=1,\dots,2^k} \mathcal{C}(\mathcal{H}_i)$ and exit.*

Then, the algorithm to be launched in $\mathcal{C}(\mathbb{R}_+^n)$, for which the following properties hold:

- *Property P6: the positive octant $\mathbb{R}_+^n$ is progressively shrunk without losing any point of $\mathcal{E}$;*
- *Property P7: the set provided by the algorithm asymptotically converges to the set $\mathcal{E}$.*

Proof. The property P6 holds because $\mathcal{B}(H)$ is guaranteed to include any vector in $\mathcal{H} \cap \mathcal{E}$ according to the property P5, moreover from the property P4 one has that the set returned by the algorithm cannot increase. Then, property P7 holds because no portion of $\mathbb{R}_+^n$ is lost in the division of each rectangle $\mathcal{B}(\mathcal{H})$. $\square$

Hence, the proposed algorithm for computing the equilibrium points of (8) is launched as $\mathcal{C}(\mathbb{R}_+^n)$, which means that the positive octant $\mathbb{R}_+^n$ is used as initial rectangle $\mathcal{H}$. This because $\mathbb{R}_+^n$ is clearly guaranteed to contain all solutions of (16). Then, the initial rectangle is passed to the map $\mathcal{B}(\cdot)$. If the output of this map is either the empty set or a point, then the algorithm stops as it is guaranteed that there are no equilibrium points inside the considered rectangle. Otherwise, the output is another rectangle, which is then divided in smaller ones. The rectangles obtained in this division are passed to the map $\mathcal{C}(\cdot)$ itself, hence realizing a recursive algorithm. As explained by the properties P6 and P7, the set provided by the algorithm is guaranteed to contain all points of $\mathcal{E}$ at each recursion, and to asymptotically converge to $\mathcal{E}$.

Remark 3. It is worth to remark that the proposed algorithm differs from existing techniques for computing the solutions of systems of nonlinear equations. A first difference is that the proposed algorithm does not rely on analytical techniques, which can be used only in special cases and typically for small systems. A second difference is that the proposed algorithm does not consider one possible initial point only contrary to some numerical techniques. Instead, the proposed algorithm consider the whole space of possible solutions, and progressively shrinks this space to the sought set of equilibrium points without losing any portion of it.

Remark 4. Lastly, it is interesting to observe that the proposed algorithm can also allow one to investigate limit cycles of (8), which are periodic solutions $m(t), p(t)$ of (8) satisfying the condition

$$\exists T \in \mathbb{R}: \quad \begin{cases} m(t) = m(t+T) \\ p(t) = p(t+T) \end{cases} \quad \forall t \geq 0 \tag{27}$$

where T represents the period. Indeed, at the first recursion of the proposed algorithm one obtains the rectangle $\mathcal{B}(\mathbb{R}^n_+)$ which is expected to contain existing limit cycles of (8) as they are periodic solutions of the system of differential equations. This suggests a strategy which can be useful to establish the existence of limit cycles in (8). In fact, once that $\mathcal{B}(\mathbb{R}^n_+)$ has been found at the first recursion of the algorithm, one can investigate the trajectories starting along its boundary (for instance, at the vertices) to reveal limit cycles. See for instance Example 3.

4   Illustrative Examples

In this section we present some examples where the proposed algorithm is used. We report only the p-component of each equilibrium point, being the m-component directly given by $D^{-1}Cp$ according to (16). The computational time for all examples is lesser than 5 seconds with an implementation of the proposed algorithm in Matlab 7 running under Windows XP on a personal computer with Pentium IV 2.2 GHz and 2 GB RAM.

4.1   Genetic Regulatory Network in PROD Form with Non-Hill Function

Let us start by considering the genetic regulatory network described in PROD form given by

$$\begin{cases} \dot{m}_1(t) = -0.17m_1(t) + 0.73f(p_2)(1 - f(p_3)) \\ \dot{m}_2(t) = -0.8m_2(t) + 0.95(1 - f(p_3)) \\ \dot{m}_3(t) = -0.52m_3(t) + 0.58(1 - f(p_1)) \\ \dot{p}_i(t) = -p_i(t) + m_i(t) \ \forall i = 1, 2, 3 \end{cases} \tag{28}$$

and the saturation function

$$f(p_i(t)) = 1 - e^{-p_i(t)^2}. \tag{29}$$

This genetic regulatory network is characterized by the fact that TF 1 is a regressor of gene 3, TF 2 is an activator of gene 1, and TF 3 is a regressor of genes 1 and 2.

Let us use the algorithm proposed in Theorem 3. At the first recursion of the algorithm we obtain that the positive octant $\mathbb{R}^+_3$ is shrunk to the rectangle shown in Figure 1a. At the second recursion, the rectangle previously found is divided in four equal rectangles, one of which is shown in Figure 1b, another one shrinks to the equilibrium point shown in Figure 1b, and the other two converges to the empty set. At the fourth recursion, another equilibrium point is found as shown in Figure 1c, and only one rectangle is left. Then, at the eight recursion the last equilibrium point is found and no rectangle is left as shown in Figure 1d. We

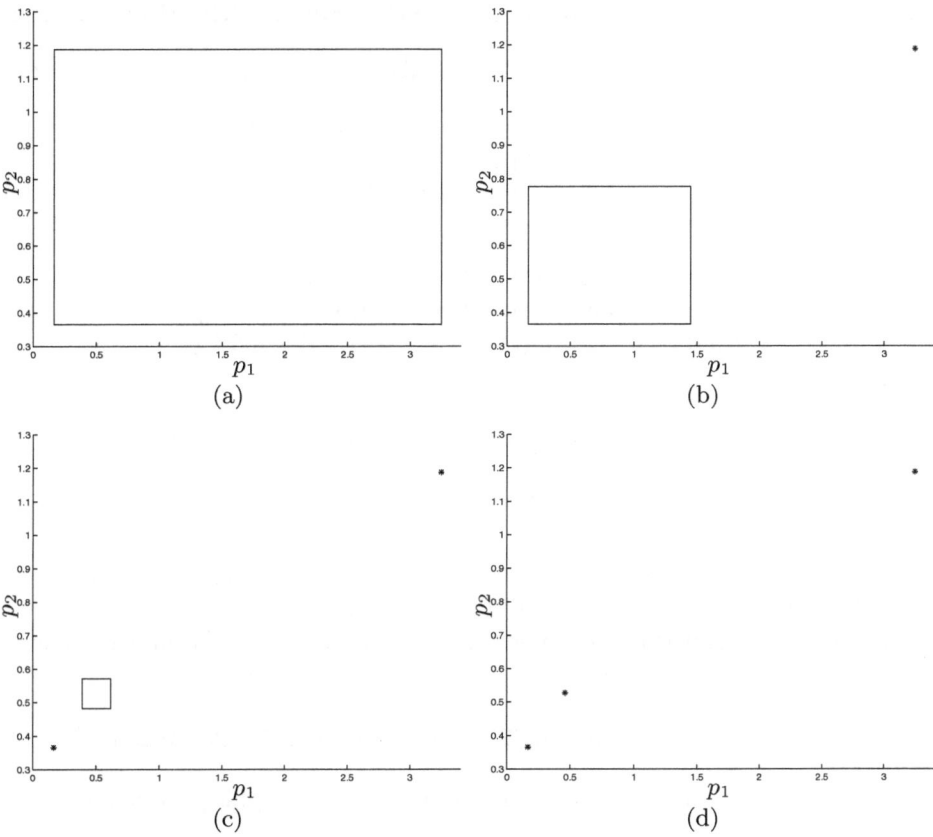

Fig. 1. Steps of the proposed algorithm for the example in Section 4.1 (shown in the plane p_1–p_2 for clarity of presentation): (a) first recursion, $\mathbb{R}^3_+$ is shrunk to a rectangle; (b) second recursion, an equilibrium point is found (denoted by the "$*$" mark); (c) fourth recursion, another equilibrium point is found. (d): ninth recursion, the last equilibrium point is found.

hence conclude that this system has three equilibrium points, in particular the set $\mathcal{E}$ in (17) is given by

$$\mathcal{E} = \left\{ (3.246, 1.189, 0.000)^T, (0.461, 0.527, 0.902)^T, (0.166, 0.366, 1.085)^T \right\}. \quad (30)$$

For comparison, we attempt to use standard mathematical tools, in particular via Matlab and Mathematica. We hence use the functions "solve" (Matlab function for both analytical and numerical techniques) and "findroot" (Mathematica function for numerical technique) which find only one solution. This happens because the equations in (12) are neither polynomial nor rational in this case, which means that no analytical technique exist for finding the solutions in this case. Existing tools therefore apply numerical techniques which allow to find a local solution starting from an initial point, but the other solutions are lost.

4.2   Genetic Regulatory Network in SUM Form with Hill Function

In this example we consider the genetic regulatory network in SUM form with

$$
\begin{cases}
\dot{m}_1(t) = -2.0m_1(t) + 0.9(1 - f(p_2)) + 0.5f(p_3) \\
\dot{m}_2(t) = -2.2m_2(t) + 0.9(1 - f(p_3)) + 0.5f(p_4) \\
\dot{m}_3(t) = -2.4m_3(t) + 0.9(1 - f(p_4)) + 0.5f(p_5) \\
\quad \vdots \\
\dot{m}_8(t) = -3.4m_8(t) + 0.9(1 - f(p_9)) + 0.5f(p_{10}) \\
\dot{m}_9(t) = -3.6m_9(t) + 0.9(1 - f(p_{10})) + 0.5f(p_1) \\
\dot{m}_{10}(t) = -3.8m_{10}(t) + 0.9(1 - f(p_1)) + 0.5f(p_2) \\
\dot{p}_i(t) = -p_i(t) + m_i(t) \quad \forall i = 1, \ldots, 10
\end{cases}
\tag{31}
$$

where the saturation function is chosen as the Hill function

$$
f(p_i(t)) = \frac{1}{1 + p_i(t)^6}.
\tag{32}
$$

This genetic regulatory network is characterized by the cyclic structure where gene i has TF $i + 1$ as regressor and TF $i + 2$ as activator.

By using the algorithm proposed in Theorem 3 we have that the positive octant $\mathbb{R}^{10}_+$ shrinks to the set

$$
\mathcal{E} = \left\{ (0.449, 0.408, 0.375, 0.346, 0.321, 0.300, 0.281, 0.267, 0.251, 0.236)^T \right\},
\tag{33}
$$

hence implying that there is one equilibrium point only in this genetic regulatory network.

Also in this case we attempt to use standard mathematical tools as done in the previous example. However, by using analytical techniques (which can be used since the equations in (12) are rational for this example) we do not obtain any solution. This happens because the degree of the one-variable polynomial that the analytical techniques allow one to find is prohibitive in this case since the equations in (12) have degree 12 (the degree of $b(p)$) and 10 variables (the p-components of the state), therefore there can be up to 12^{10} solutions. Also, we attempt to use numerical techniques, and find that they return the sought equilibrium point. Unfortunately, these techniques are not able to establish whether this solution is unique or not.

4.3   Repressilator Model in E. Coli

Here we consider the repressilator investigated in *Escherichia coli* [24]:

$$
\begin{cases}
\dot{m}_i(t) = -m_i(t) + \alpha^{rep}(1 - f(p_j(t))) \\
\dot{p}_i(t) = -\beta^{rep}(p_i(t) - m_i(t)) \\
i = lacl, tetR, cl; \quad j = cl, lacl, tetR
\end{cases}
\tag{34}
$$

where the saturation function is the Hill function

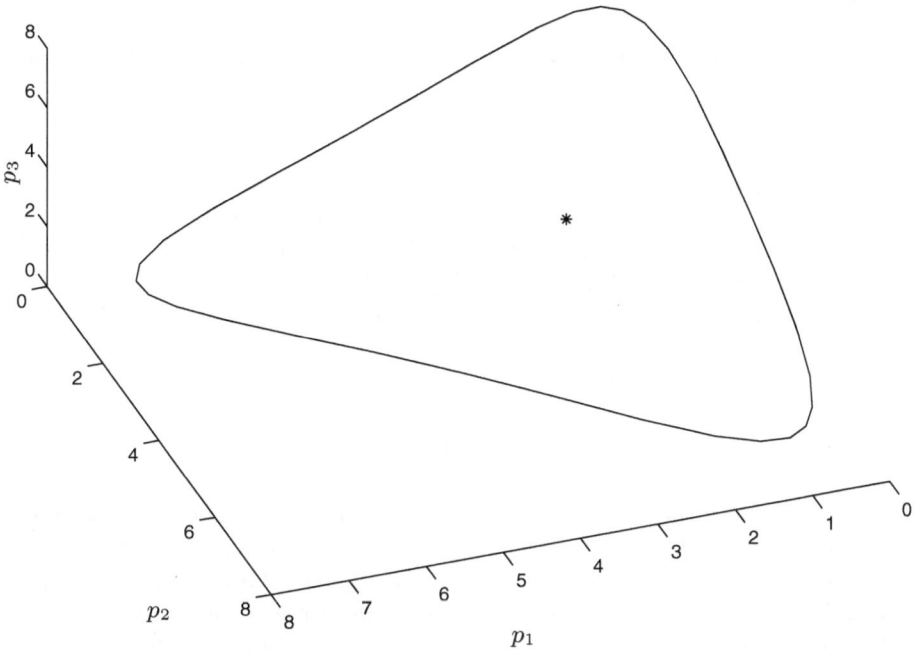

Fig. 2. Found equilibrium point and limit cycle in the example of Section 4.3

$$f(p_i(t)) = \frac{1}{1 + p_i(t)^2} \tag{35}$$

and $\alpha^{rep}, \beta^{rep} \in \mathbb{R}_+$ are positive constants.

Let us select the plausible values $\alpha^{rep} = 10$ and $\beta^{rep} = 1$. By using the algorithm proposed in Theorem 3 we find that there is a unique equilibrium point, in particular

$$\mathcal{E} = \left\{ (2, 2, 2)^T \right\}. \tag{36}$$

For this example it is interesting to observe that, in addition to the found equilibrium point, there exists a limit cycle that the proposed algorithm can help to find. Indeed, as explained in Remark 4, at the first recursion of the proposed algorithm one obtains the rectangle $\mathcal{B}(\mathbb{R}_+^3)$, which is equal to $[0.1010, 9.899]^3$. Then, the limit cycle is revealed by simply computing the trajectory of the system starting at the vertices of this rectangle. Figure 2 shows the projection on the plane p_1-p_2 of the found limit cycle.

4.4 Genetic Regulatory Network in SUM Form with Non-Hill Function

As last example, we consider the genetic regulatory network in SUM form described by

$$
\begin{cases}
\dot{m}_1(t) = -2m_1(t) + 0.5f(p_5) \\
\dot{m}_2(t) = -m_2(t) + 0.1(1 - f(p_2)) + 0.4(1 - f(p_4)) \\
\dot{m}_3(t) = -0.6m_3(t) + 0.2f(p_1) + 1.1(1 - f(p_4)) \\
\dot{m}_4(t) = -m_4(t) + 0.5(1 - f(p_3)) + 1.5f(p_4) \\
\dot{m}_5(t) = -2m_5(t) + 0.3f(p_2) + 0.3(1 - f(p_5)) \\
\dot{p}_i(t) = -p_i(t) + m_i(t) \quad \forall i = 1, \ldots, 5
\end{cases}
\tag{37}
$$

and the saturation function

$$
f(p_i(t)) = \frac{2}{\pi} \arctan(p_i(t)^2).
\tag{38}
$$

This genetic regulatory network is characterized by the fact that TF 1 is an activator of gene 3, TF 2 is an activator of gene 5 and a regressor of gene 2, TF 3 is a regressor of gene 4, TF 4 is a regressor of genes 2 and 3 and an activator of gene 4, and TF 5 is an activator of gene 1 and a regressor of gene 5.

By using the algorithm proposed in Theorem 3 as done in the previous examples we conclude that this system has three equilibrium points, in particular the set $\mathcal{E}$ in (17) is given by

$$
\mathcal{E} = \big\{ (0.0037, 0.1961, 0.4518, 1.566, 0.1515)^T, (0.0039, 0.3130, 1.003,
$$
$$
0.9278, 0.1570)^T, (0.0046, 0.4827, 1.821, 0.1035, 0.1691)^T \big\}.
\tag{39}
$$

However, by using standard mathematical tools, we obtain only one solution similarly to the example in Section 4.1.

5 Conclusion

We have proposed an algorithm which allows one to find the equilibrium points of genetic regulatory networks described by differential equation models and which include both SUM form and PROD form with saturation functions of any type. The proposed algorithm is guaranteed to find all sought equilibrium points, moreover as shown by some numerical examples the computation is reasonably fast also in cases where standard mathematical tools for solving systems of nonlinear equations may fail.

It is hence expected that the proposed algorithm represents a useful tool for researchers working in the area of genetic regulatory networks. In particular, the proposed algorithm can allow one to investigate issues such as stability, disturbance rejection, and robustness, for which the knowledge of the equilibrium points is required, see for instance [25,26,27,28].

Acknowledgement

The author would like to thank the Editor and the Reviewers for their time and useful comments.

References

1. D'haeseleer, P., Wen, X., Fuhrman, S., Somogyi, R.: Mining the gene expression matrix: Inferring gene relationships from large scale gene expression data. In: Paton, R.C., Holcombe, M. (eds.) Information Processing in Cells and Tissues. Plenum Publishing, New York (1998)
2. Davidson, E.H.: The Regulatory Genome: Gene Regulatory Networks In Development And Evolution. Academic Press, London (2006)
3. D'haeseleer, P.: Reconstructing Gene Networks from Large Scale Gene Expression Data. PhD thesis, University of New Mexico (2000)
4. D'haeseleer, P., Liang, S., Somogyi, R.: Genetic network inference: From co-expression clustering to reverse engineering. Bioinformatics 16(8), 707–726 (2000)
5. Li, C., Chen, L., Aihara, K.: A systems biology perspective on signal processing in genetic network motifs. IEEE Signal Processing Magazine 221(3), 136–142 (2007)
6. Yuh, C.H., Bolouri, H., Davidson, E.H.: Genomic cis-regulatory logic: Experimental and computational analysis of a sea urchin gene. Science 279, 1896–1902 (1998)
7. Tsai, H.K., Yang, J.M., Tsai, Y.F., Kao, C.Y.: An evolutionary approach for gene expression patterns. IEEE Transactions on Information Technology in Biomedicine 8(2), 69–78 (2004)
8. Maraziotis, I.A., Dragomir, A., Bezerianos, A.: Gene networks reconstruction and time-series prediction from microarray data using recurrent neural fuzzy networks. IET Systems and Biology 1(1), 41–50 (2007)
9. Smolen, P., Baxter, D.A., Byrne, J.H.: Mathematical modeling of gene networks. Neuron 26(3), 567–580 (2000)
10. Bower, J.M., Bolouri, H. (eds.): Computational Modeling of Genetic and Biochemical Networks. Computational Molecular Biology. MIT Press, Cambridge (2001)
11. Jong, H.D.: Modeling and simulation of genetic regulatory systems: A literature review. Journal of Computation Biology 9, 67–103 (2002)
12. D'haeseleer, P., Liang, S., Somogyi, R.: Gene expression data analysis and modeling. In: Proc. Pacific Symposium on Biocomputing, Hawaii, USA (1999)
13. Aracena, J., Lamine, S.B., Mermet, M.A., Cohen, O., Demongeot, J.: Mathematical modeling in genetic networks: Relationships between the genetic expression and both chromosomic breakage and positive circuits. IEEE Transactions on Systems, Man, and Cybernetics–Part b: Cybernetics 33(5), 825–834 (2003)
14. Bintu, L., Buchler, N.E., Garcia, H.G., Gerland, U., Hwa, T., Kondev, J., Phillips, R.: Transcriptional regulation by the numbers: models. Current Opinion in Genetics and Development 15(2), 116–124 (2005)
15. Li, C., Chen, L., Aihara, K.: Stability of genetic networks with sum regulatory logic: Lure system and lmi approach. IEEE Trans. on Circuits and Systems I 53(11), 2451–2458 (2006)
16. Li, C., Chen, L., Aihara, K.: Stochastic stability of genetic networks with disturbance attenuation. IEEE Transactions on Circuits and Systems II 54(10), 892–896 (2007)

17. Chesi, G., Hung, Y.S.: Stability analysis of uncertain genetic SUM regulatory networks. Automatica 44(9), 2298–2305 (2008)
18. Chesi, G., Garulli, A., Tesi, A., Vicino, A.: Characterizing the solution set of polynomial systems in terms of homogeneous forms: an LMI approach. Int. Journal of Robust and Nonlinear Control 13(13), 1239–1257 (2003)
19. Mora, T.: Solving Polynomial Equation Systems II. Cambridge University Press, Cambridge (2005)
20. Nocedal, J., Wright, S.: Numerical Optimization. Springer Series in Operations Research and Financial Engineering. Springer, Heidelberg (2006)
21. Chesi, G.: Optimal representation matrices for solving polynomial systems via LMI. Int. Journal of Pure and Applied Mathematics 45(3), 397–412 (2008)
22. Stetter, H.J.: Numerical Polynomial Algebra. SIAM, Philadelphia (2004)
23. Ortega, J.M., Rheinboldt, W.C.: Iterative Solution of Nonlinear Equations in Several Variables. SIAM, Philadelphia (1987)
24. Elowitz, M.B., Leibler, S.: A synthetic oscillatory network of transcriptional regulators. Nature 403, 335–338 (2000)
25. Khalil, H.K.: Nonlinear Systems, 3rd edn. Prentice Hall, Englewood Cliffs (2001)
26. Chesi, G., Garulli, A., Tesi, A., Vicino, A.: Homogeneous Lyapunov functions for systems with structured uncertainties. Automatica 39(6), 1027–1035 (2003)
27. Chesi, G., Garulli, A., Tesi, A., Vicino, A.: Solving quadratic distance problems: an LMI-based approach. IEEE Trans. on Automatic Control 48(2), 200–212 (2003)
28. Chesi, G., Garulli, A., Tesi, A., Vicino, A.: Homogeneous Polynomial Forms for Robustness Analysis of Uncertain Systems. Lecture Notes in Control and Information Sciences, vol. 390. Springer, London (2009)

Code, Context, and Epigenetic Catalysis in Gene Expression

Rodrick Wallace[1] and Deborah Wallace[2]

[1] The New York State Psychiatric Institute,
549 W. 123 St., Suite 16F, New York, NY, 10027. Tel.: (212) 865-4766
`wallace@pi.cpmc.columbia.edu`
[2] Consumers Union
`rdwall@ix.netcom.com`

Abstract. We examine a class of probability models describing how epigenetic context affects gene expression and organismal development, using the asymptotic limit theorems of information theory in a highly formal manner. Taking classic results on spontaneous symmetry breaking abducted from statistical physics in groupoid, rather than group, circumstances, the work suggests that epigenetic information sources act as analogs to a tunable catalyst, directing development into different characteristic pathways according to the structure of external signals. The results have significant implications for epigenetic epidemiology, in particular for understanding how environmental stressors, in a large sense, can induce a broad spectrum of developmental disorders in humans.

1 Introduction

1.1 Toward New Tools

Researchers have begun to explore a de-facto cognitive paradigm for gene expression in which contextual factors determine the behavior of what Cohen calls a 'reactive system', not at all a deterministic, or even stochastic, mechanical process (e.g., [18, 19, 74]). The different approaches, while highly formal, are nonetheless much in the spirit of the pioneering efforts of Maturana and Varela [53, 54] who foresaw the essential role that cognitive process must play in a vast realm of biological phenomena.

O'Nuallain [57] has recently placed gene expression firmly in the realm of complex linguistic behavior, for which context imposes meaning, claiming that the analogy between gene expression and language production is useful both as a fruitful research paradigm and also, given the relative lack of success of natural language processing (nlp) by computer, as a cautionary tale for molecular biology. First O'Nuallain argues that, at the orthographic or phonological level, depending on whether the language is written or spoken, we can map from phonetic elements to nucleotide sequence. His second claim is that Nature has designed highly ambiguous codes in both cases, and left disambiguation to the context.

C. Priami et al. (Eds.): Trans. on Comput. Syst. Biol. XI, LNBI 5750, pp. 283–334, 2009.

He notes that, given our concern with the Human Genome Project (HGP) and its implications for human health, only 2% of diseases can be traced back to a straightforward genetic cause. As a consequence the HGP will have to be redone for a variety of metabolic contexts in order to establish a sound technology of genetic engineering [58].

Here we investigate a broad class of probability models based on the asymptotic limit theorems of information theory that instantiate this perspective, finding a 'natural' means by which epigenetic context 'farms' gene expression in an inherently punctuated manner via a kind of tunable catalysis. The models will be used to study how normal developmental modes can be driven by external context into pathological trajectories often expressed, in humans, as comorbid psychiatric and physical disorders, expanding recent work [71]. It appears possible to convert such models to powerful tools for data analysis, much as those based on the Central Limit Theorem can be converted to parametric statistics. A more formal version of the underlying mathematics can be found in [34].

We will begin with a summary of the biological context, then examine the popular spinglass model of development taken from neural network studies that we will ultimately generalize using a cognitive paradigm. The expanded approach permits import of tools and methods from statistical physics via the homology between information source uncertainty and free energy density, and this leads directly to the idea of epigenetic catalysis.

It is worth keeping in mind throughout the formal mathematics that Feynman's basic measure of information is simply the free energy needed to erase it [31].

1.2 Epigenetic Epidemiology

What we attempt is itself embedded in a large and lively intellectual context. Jablonka and Lamb [41, 42] have long argued that information can be transmitted from one generation to the next in ways other than through the base sequence of DNA. It can be transmitted through cultural and behavioral means in higher animals, and by epigenetic means in cell lineages. All of these transmission systems allow the inheritance of environmentally induced variation. Such Epigenetic Inheritance Systems are the memory structures that enable somatic cells of different phenotypes but identical genotypes to transmit their phenotypes to their descendants, even when the stimuli that originally induced these phenotypes are no longer present.

In chromatin-marking systems information is carried from one cell generation to the next because it rides with DNA as binding proteins or additional chemical groups that are attached to DNA and influence its activity. When DNA is replicated, so are the chromatin marks. One type of mark is the methylation pattern a gene carries. The same DNA sequence can have several different methylation patterns, each reflecting a different functional state. These alternative patterns can be stably inherited through many cell divisions.

Epigenetic inheritance systems are very different from the genetic system. Many variations are directed and predictable outcomes of environmental changes.

Epigenetic variants are, in the view of [41, 42], often, although not necessarily, adaptive. The frequency with which variants arise and their rate of reversion varies widely and epigenetic variations induced by environmental changes may be produced coordinatedly at several loci.

Parenthetically, some authors, e.g., [39], disagree with the assumption of adaptiveness, inferring that input responsible for methylation effects simply produces a phenotypic variability then subject to selection. The matter remains open.

Jablonka and Lamb [42] conclude that epigenetic systems may therefore produce rapid, reversible, co-ordinated, heritable changes. However such systems can also underlie non-induced changes, changes that are induced but non-adaptive, and changes that are very stable.

What is needed, they feel, is a concept of epigenetic heritability comparable to the classical concept of heritability, and a model similar to those used for measuring the effects of cultural inheritance on human behavior in populations.

Following a furious decade of research and debate, this perspective received much empirical confirmation. Backdahl et al. [6], for example, write that epigenetic regulation of gene expression primarily works through modifying the secondary and tertiary structures of DNA (chromatin), making it more or less accessible to transcription. The sum and interaction of epigenetic modifications has been proposed to constitute an 'epigenetic code' which organizes the chromatin structure on different hierarchical levels [67]. Modifications of histones include acetylation, methylation, phosphorylation, ubiquitination, and sumoylation, but also other modifications have been observed. Some such modifications are quite stable and play an important part in epigenetic memory although DNA methylation is the only epigenetic modification that has maintenance machinery which preserves the marks through mitosis. This argues for DNA methylation to function as a form of epigenetic memory for the epigenome.

Codes and memory, of course, are inherent to any cognitive paradigm.

Jaenish and Bird [45] argue that cells of a multicellular organism are genetically homogeneous but structurally and functionally heterogeneous owing to the differential expression of genes. Many of these differences in gene expression arise during development and are subsequently retained through mitosis. External influences on epigenetic processes are seen in the effects of diet on long-term diseases such as cancer. Thus, epigenetic mechanisms seem to allow an organism to respond to the environment through changes in gene expression. Epigenetic modifications of the genome provide a mechanism that allows the stable propagation of gene activity states from one generation of cells to the next. Because epigenetic states are reversible they can be modified by environmental factors, which may contribute to the development of abnormal responses. What needs to be explained, from their perspective, is the variety of stimuli that can bring about epigenetic changes, ranging from developmental progression and aging to viral infection and diet.

Jaenish and Bird conclude that the future will see intense study of the chains of signaling that are responsible for epigenetic programming. As a result, we will be able to understand, and perhaps manipulate, the ways in which the genome learns from experience.

Indeed, our central interest precisely regards the manner in which the asymptotic limit theorems of information theory constrain such chains of signaling, in the same sense that the Central Limit Theorem constrains sums of stochastic variates.

Crews et al. [21, 22] provide a broad overview of induced epigenetic change in phenotype, as do Guerrero-Bosagna et al. [39], who focus particularly on early development. They propose that changes arising because of alterations in early development processes, in some cases environmentally induced, can appear whether or not such changes could become fixed and prosper in a population. They recognize two ways for this to occur, first by dramatically modifying DNA aspects in the germ line with transgenerational consequences – mutations or persistent epigenetic modifications of the genome – or by inducing ontogenetical variation in every generation, although not inheritance via the germ line. From their perspective inductive environmental forces can act to create, through these means, new conformations of organisms which also implies new possibilities within the surrounding environment.

Foley et al. [32] take a very general perspective on the prospects for epigenetic epidemiology. They argue that epimutation is estimated to be 100 times more frequent than genetic mutation and may occur randomly or in response to the environment. Periods of rapid cell division and epigenetic remodeling are likely to be most sensitive to stochastic or environmentally mediated epimutation. Disruption of epigenetic profile is a feature of most cancers and is speculated to play a role in the etiology of other complex diseases including asthma, allergy, obesity, type 2 diabetes, coronary heart disease, autism spectrum disorders, bipolar disorders, and schizophrenia.

They find evidence that a small change in the level of DNA methylation, especially in the lower range in an animal model, can dramatically alter expression for some genes. The timing of nutritional insufficiency or other environmental exposures may also be critical. In particular low-level maternal care was associated with developmental dysfunction and altered stress response in the young. Foley et al. emphasize the potential implications of such findings, given how widely stress is implicated in disease onset and relapse.

They especially note that when epigenetic status or change in status over time is the outcome, then models for either threshold-based dichotomies or proportional data will be required. Threshold models, defined by a given level or pattern of methylation or a degree of change in methylation over time, will, in their view, benefit from relevant functional data to identify meaningful thresholds.

A special contribution of the approach taken here is that just such threshold behavior leads 'naturally' to a language-like 'dual information source' constrained by the necessary conditions imposed by information theory's asymptotic limit theorems, allowing development of statistical models of complicated cognitive phenomena, including but not limited to cognitive gene expression.

A recent review by Weaver [77] focuses specifically on the epigenetic effects of glucocorticoids – stress hormones. In mammals, Weaver argues, the closeness or degree of positive attachment in parent-infant bonding and parental investment during early life has long-term consequences on development of interindividual differences in cognitive and emotional development in the offspring. The long-term effects of the early social experience, he continues, particularly of the mother-offspring interaction, have been widely investigated. The nature of that interaction influences gene expression and the development of behavioral responses in the offspring that remain stable from early development to the later stages of life. Although enhancing the offspring's ability to respond according to environmental clues early in life can have immediate adaptive value, the cost, Weaver says, is that these adaptations serve as predictors of ill health in later life. He concludes that maternal influences on the development of neuroendocrine systems that underlie hypothalamic-pituitary-adrenal (HPA) axis and behavioral responses to stress mediate the relation between early environment and health in the adult offspring. In particular, he argues, exposure of the mother to environmental adversity alters the nature of mother-offspring interaction, which, in turn, influences the development of defensive responses to threat and reproductive strategies in the progeny.

In an updated review of epigenetic epidemiology, Jablonka [43] finds it clear that the health and general physiology of animals and people can be affected not only by the interplay of their own genes and conditions of life, but also by the inherited effects of the interplay of genes and environment in their ancestors. These ancestral influences on health, Jablonka says, depend neither on inheriting particular genes, nor on the persistence of the ancestral environment.

Significantly, Bossdorf et al. [11] invoke 'contexts' much like Baars' model of consciousness [68], and infer a need to expand the concept of variation and evolution in natural populations, taking into account several likely interacting ecologically relevant inheritance systems. Potentially, this may result in a significant expansion, though by all means not a negation, of the Modern Evolutionary Synthesis as well as in more conceptual and empirical integration between ecology and evolution.

More formally, Scherrer and Jost [62, 63] use information theory arguments to extend the definition of the gene to include the local epigenetic machinery, something they characterize as the 'genon'. Their central point is that coding information is not simply contained in the coded sequence, but is, in their terms, *provided by* the genon that accompanies it on the expression pathway and controls in which peptide it will end up. In their view the information that counts is not about the identity of a nucleotide or an amino acid derived from it, but about the relative frequency of the transcription and generation of a particular type of coding sequence that then contributes to the determination of the types and numbers of functional products derived from the DNA coding region under consideration.

From our perspective the formal tools for understanding such phenomena involve asymptotic limit theorems affecting information sources – active systems that generate or 'provide' information – and these are respectively the Rate Distortion Theorem and its zero error limit, the Shannon-McMillan Theorem, described in the Mathematical Appendix.

We begin with a reconsideration of the current de-facto standard systems biology neural network-analog model of development, and proceed to its generalization.

2   Models of Development

2.1   The Spinglass Model

Ciliberti et al.[16, 17], culminating a long series of papers, apply the spinglass model from statistical physics to organisimal development in an evolutionary context. We summarize their formalism and look at some of the less obvious topological implications – in particular the mapping of disjoint directed homotopy classes of phenotype paths into interaction matrix space. We then extend the approach by applying a cognitive paradigm for gene expression first developed in [74]. Analogs to phase transition arguments in physical systems generate punctuated equilibrium evolutionary transitions in a 'highly natural' manner, even for the spinglass treatment, and a hierarchical extension permits incorporation of epigenetic effects as a kind of tunable catalysis.

The spinglass model of development assumes that N transcriptional regulators are represented by their expression patterns

$$\mathbf{S}(t) = [S_1(t), ..., S_N(t)]$$

at some time t during a developmental or cell-biological process and in one cell or domain of an embryo. The transcriptional regulators influence each other's expression through cross-regulatory and autoregulatory interactions described by a matrix $w = (w_{ij})$. For nonzero elements, if $w_{ij} > 0$ the interaction is activating, if $w_{ij} < 0$ it is repressing. w represents, in this model, the regulatory genotype of the system, while the expression state $\mathbf{S}(t)$ is the phenotype. These regulatory interactions change the expression of the network $\mathbf{S}(t)$ as time progresses according to a difference equation

$$S_i(t + \Delta t) = \sigma[\sum_{j=1}^{N} w_{ij} S_j(t)], \tag{1}$$

where Δt is a constant and σ a sigmodial function whose value lies in the interval $(-1, 1)$. In the spinglass limit σ is the sign function, taking only the values ± 1.

The networks of interest in the spinglass model are those whose expression state begins from a prespecified initial state $\mathbf{S}(0)$ at time $t = 0$ and converge to a prespecified stable equilibrium state $\mathbf{S}_\infty$. Such a network is termed *viable*, for obvious reasons.

After an elaborate and very difficult simulation exercise, a particular series of results emerges. Reference [16] finds that viable networks comprise a tiny fraction of possible ones. They could be widely scattered in the space of all possible networks and occupy disconnected islands in this space. However, direct computation indicates precisely the opposite. The metagraph of viable networks has one 'giant' connected component that comprises most or all viable networks. Any two networks in this component can be reached from one another through gradual changes of one regulatory interaction at a time, changes that never leave the space of viable networks, for this calculation.

In general, within the giant component, randomly chosen pairs of networks with the same phenotype will have vastly different organization, in terms of the matrix (w_{ij}).

Define $0 \leq d \leq 1$ as the the fraction of genes that differ in their expression state between $\mathbf{S}_0$ and $\mathbf{S}_\infty$. A typical result is that for $N = 5$ genes, $6 \leq M \leq 7$ total regulatory interactions, and $d = 0.4$, full enumeration finds a total of only 37,338 viable networks out of 6.3×10^7 possible ones [16]. Long random walks through the space of viable networks, however, visit all but a very small fraction of the nodes of the metagraph, and this missing fraction decreases as N increases. Large N require elaborate Monte Carlo sampling for simulation, a difficult and computationally intensive enterprise.

In w-space [16, 17] define a metric characterizing the distance between two network topologies as

$$D(w, w') = \frac{1}{2M_+} \sum_{i,j} |sign(w_{ij}) - sign(w'_{ij})|,$$

where M_+ is the maximum number of regulatory interactions, and sign(x)=± 1 depends on the sign of x, and is 0 for $x = 0$.

Several observations emerge directly.

1. This approach is formally similar to spinglass neural network models of learning by selection, e.g., as proposed by Toulouse et al. [66] nearly a generation ago. Subsequent work [4, 5], summarized in [23], suggests that such models are simply not sufficient to the task of understanding high level cognitive function, and these have been largely supplanted by complicated 'global workspace' concepts whose mathematical characterization is highly nontrivial [3].

2. What [16, 17] observe, in another idiom, is that in phenotype space, in $\mathbf{S}$-space, the set of all paths associated with viable networks forms an equivalence class, closely analogous to the directed homotopy equivalence classes in the sense of [36, 37]. Directed homotopy differs from simple homotopy (e.g., [50]) in that one uses paths from one point to another rather than loops, and seeks continuous deformations between them. See [74] for discussion in a biological context. Thus there is, in this spinglass model, a mapping from $\mathbf{S}$-space into (w_{ij}) space, characterized by the metric D, that associates a unique simply connected component with each dihomotopy-like equivalence class of paths connecting two particular phenotype points. Indeed, the w-space component might well be treated according to standard homotopy arguments, i.e., using loops.

3. What one does with homotopically simply connected components is patch them together to build larger, and more interesting, topological structures, using the Seifert-Van Kampen Theorem (SVKT) (e.g., [50], Ch. 10). If paths within S-space are not continuously transformable into one another, (if there are 'holes'), then several distinct dihomotopy classes will exist, e.g., as in figures 1 and 2 of [74], explored further below in terms of developmental critical periods and their 'shadows'. The obvious conjecture is that, under such a circumstance, very complex topological objects may lurk in w-space, not just the simply connected component discovered by by [16, 17]. These may, according to the SVKT, intersect as well as exist as isolated and disconnected sets.

In particular, if there are dihomotopy 'holes' in S-space, consequently reflected in disconnected patches in w-space, then punctuated transition events of various sorts may well become an evolutionary norm, as in [38], even for the spinglass model.

4. A large and increasing body of work surrounding coupled cell networks invokes groupoids, a natural generalization of symmetry groups. As [25] remarks, until recently the abstract theory of coupled cell systems has mainly focused on the effects of symmetry in the network and the consequent formation of spatial and spatiotemporal patterns. The formal setting for this theory centers upon the symmetry group of the network.

Reference [25] concludes that analysis of robust patterns of synchrony in general coupled cell systems – that is, dynamics in which sets of cells behave identically as a consequence of the network topology – leads to the fruitful notion of the 'symmetry groupoid' of a coupled cell network. A groupoid is a generalization of a group, in which products of elements are not always defined. The symmetry groupoid of a coupled cell network is a natural algebraic formalization of the 'local symmetries' that relate subsets of the network to each other. In particular 'admissible' vector fields – those specified by the network topology – are precisely those that are equivariant under the action of the symmetry groupoid.

The Appendix provides a summary of standard material on groupoids that will be of later use.

5. Both of these – analogous – approaches can apparently be coarse-grained into a symbolic dynamics associated with (simple) information sources having particular grammar and syntax. The method is straightforward (e.g., [7, 55]). One could, thus, probably translate the spinglass results of Ciliberti et al. into symbolic dynamics, using groupoid methods to study the underlying topological objects.

6. The spinglass model of development is abstracted from longstanding (if ultimately unsucessful) attempts at similar treatments of neural networks involved in high level cognition (e.g., [44, 56, 61, 64]). Thus and consequently [16, 17] are invoking an implicit cognitive paradigm for gene expression (e.g., [18, 19, 74]). Cognitive process, as the philosopher Fred Dretske eloquently argues (e.g., [26]), is constrained by the necessary conditions imposed by the asymptotic limit theorems of information theory. A little work produces a very general cognitive gene expression metanetwork structure recognizably similar to that found

in [16, 17]. The massively parallel computations are hidden, somewhat, in the required empirical fitting of regression model analogs based on the asymptotic limit theorems of information theory rather than on the central limit theorem.

7. A salient characteristic of high level cognitive process is precisely its inherent punctuation (e.g., [4, 5, 68]), and this emerges directly using an information theory approach via the famous homology between information and free energy (e.g., [31]). 'Simple' neural network analogs will inevitably have more difficulty replicating such behavior, but as discussed, the mapping of disjoint dihomotopy equivalence classes from phenotype sequence space to disjoint sets in interaction matrix space provides a straightforward example for spinglass models.

The next sections use information theory methods to make the transition from crossectional w-space into that of serially correlated sequences of phenotypes, expanding on the results of [74].

2.2 Shifting Perspective: Cognition as an Information Source

Atlan and Cohen [2], in the context of a study of the immune system, argue that the essence of cognition is the comparison of a perceived signal with an internal, learned picture of the world, and then choice of a single response from a large repertoire of possible responses.

Such choice inherently involves information and information transmission since it always generates a reduction in uncertainty, as explained in [1] (p. 21).

More formally, a pattern of incoming input – like the $\mathbf{S}(t)$ of equation (1) – is mixed in a systematic algorithmic manner with a pattern of internal ongoing activity – like the (w_{ij}) according to equation (1) – to create a path of combined signals $x = (a_0, a_1, ..., a_n, ...)$ – analogous to the sequence of $\mathbf{S}(t+\Delta t)$ of equation (1), with, say, $n = t/\Delta t$. Each a_k thus represents some functional composition of internal and external signals.

This path is fed into a highly nonlinear decision oscillator, h, a 'sudden threshold machine', in a sense, that generates an output $h(x)$ that is an element of one of two disjoint sets B_0 and B_1 of possible system responses. Let us define the sets B_k as

$$B_0 = \{b_0, ..., b_k\},$$

$$B_1 = \{b_{k+1}, ..., b_m\}.$$

Assume a graded response, supposing that if

$$h(x) \in B_0,$$

the pattern is not recognized, and if

$$h(x) \in B_1,$$

the pattern has been recognized, and some action $b_j, k+1 \leq j \leq m$ takes place.

The principal objects of formal interest are paths x triggering pattern recognition-and-response. That is, given a fixed initial state a_0, examine all possible subsequent paths x beginning with a_0 and leading to the event $h(x) \in B_1$. Thus $h(a_0, ..., a_j) \in B_0$ for all $0 < j < m$, but $h(a_0, ..., a_m) \in B_1$.

For each positive integer n, let $N(n)$ be the number of high probability grammatical and syntactical paths of length n which begin with some particular a_0 and lead to the condition $h(x) \in B_1$. Call such paths 'meaningful', assuming, not unreasonably, that $N(n)$ will be considerably less than the number of all possible paths of length n leading from a_0 to the condition $h(x) \in B_1$.

While the combining algorithm, the form of the nonlinear oscillator, and the details of grammar and syntax are all unspecified in this model, the critical assumption which permits inference of the necessary conditions constrained by the asymptotic limit theorems of information theory is that the finite limit

$$H = \lim_{n \to \infty} \frac{\log[N(n)]}{n} \qquad (2)$$

both exists and is independent of the path x.

Define such a pattern recognition-and-response cognitive process as *ergodic*. Not all cognitive processes are likely to be ergodic in this sense, implying that H, if it indeed exists at all, is path dependent, although extension to nearly ergodic processes seems possible [73].

Invoking the spirit of the Shannon-McMillan Theorem, whose content is described in more detail in the Appendix, as choice involves an inherent reduction in uncertainty, it is then possible to define an adiabatically, piecewise stationary, ergodic (APSE) information source $\mathbf{X}$ associated with stochastic variates X_j having joint and conditional probabilities $P(a_0, ..., a_n)$ and $P(a_n|a_0, ..., a_{n-1})$ such that appropriate conditional and joint Shannon uncertainties satisfy the classic relations

$$H[\mathbf{X}] = \lim_{n \to \infty} \frac{\log[N(n)]}{n} =$$

$$\lim_{n \to \infty} H(X_n|X_0, ..., X_{n-1}) =$$

$$\lim_{n \to \infty} \frac{H(X_0, ..., X_n)}{n+1}. \qquad (3)$$

See the Mathematical Appendix for a summary of basic information theory results.

This information source is defined as *dual* to the underlying ergodic cognitive process.

Adiabatic means that the source has been parametized according to some scheme, and that, over a certain range, along a particular piece, as the parameters vary, the source remains as close to stationary and ergodic as needed for information theory's central theorems to apply. *Stationary* means that the system's probabilities do not change in time, and *ergodic*, roughly, that the cross sectional means approximate long-time averages. Between pieces it is necessary to invoke various kinds of phase transition formalisms, as described more fully in [68, 74].

Using the developmental vernacular of [16, 17], we now examine paths in phenotype space that begins at some $\mathbf{S}_0$ and converges $n = t/\Delta t \to \infty$ to some other $\mathbf{S}_\infty$. Suppose the system is conceived at $\mathbf{S}_0$, and h represents (for example) reproduction when phenotype $\mathbf{S}_\infty$ is reached. Thus $h(x)$ can have two values, i.e., B_0 not able to reproduce, and B_1, mature enough to reproduce. Then $x = (\mathbf{S}_0, \mathbf{S}_{\Delta t}, ..., \mathbf{S}_{n\Delta t}, ...)$ until $h(x) = B_1$.

Structure is now subsumed *within the sequential grammar and syntax of the dual information source* rather than within the cross sectional internals of (w_{ij})-space, a simplifying shift in perspective.

This transformation carries heavy computational burdens, as well as providing deeper mathematical insight.

First, the fact that viable networks comprise a tiny fraction of all those possible emerges easily from the spinglass formulation simply because of the 'mechanical' limit that the number of paths from $\mathbf{S}_0$ to $\mathbf{S}_\infty$ will always be far smaller than the total number of possible paths, most of which simply do not end on the target configuration.

From the information source perspective, which inherently subsumes a far larger set of dynamical structures than possible in a spinglass model – not simply those of symbolic dynamics – the result is what [47] characterizes as the 'E-property' of a stationary, ergodic information source. This property is that, in the limit of infinitely long output, the classification of output strings into two sets:

1. A very large collection of gibberish which does not conform to underlying (sequential) rules of grammar and syntax, in a large sense, and which has near-zero probability, and

2. A relatively small 'meaningful' set, in conformity with underlying structural rules, having very high probability.

The essential content of the Shannon-McMillan Theorem is that, if $N(n)$ is the number of meaningful strings of length n, then the uncertainty of an information source X can be defined as $H[X] = \lim_{n\to\infty} \log[N(n)]/n$, that can be expressed in terms of joint and conditional probabilities as in equation (3) above. Proving these results for general stationary, ergodic information sources requires considerable mathematical machinery [20, 24, 47].

Second, information source uncertainty has an important heuristic interpretation that [1] describes as follows:

> ...[W]e may regard a portion of text in a particular language as being produced by an information source. The probabilities $P[X_n = a_n | X_0 = a_0, ... X_{n-1} = a_{n-1}]$ may be estimated from the available data about the language; in this way we can estimate the uncertainty associated with the language. A large uncertainty means, by the [Shannon-McMillan Theorem], a large number of 'meaningful' sequences. Thus given two languages with uncertainties H_1 and H_2 respectively, if $H_1 > H_2$, then in

the absence of noise it is easier to communicate in the first language; more can be said in the same amount of time. On the other hand, it will be easier to reconstruct a scrambled portion of text in the second language, since fewer of the possible sequences of length n are meaningful.

This will prove important below.

Third, information source uncertainty is homologous with free energy density in a physical system, a matter having implications across a broad class of dynamical behaviors.

The free energy density of a physical system having volume V and partition function $Z(K)$ derived from the system's Hamiltonian – the energy function – at inverse temperature K is (e.g., [49])

$$F[K] = \lim_{V \to \infty} -\frac{1}{K} \frac{\log[Z(K,V)]}{V} =$$

$$\lim_{V \to \infty} \frac{\log[\hat{Z}(K,V)]}{V}, \tag{4}$$

where $\hat{Z} = Z^{-1/K}$.

The partition function for a physical system is the normalizing sum in an equation having the form

$$P[E_i] = \frac{\exp[-E_i/kT]}{\sum_j \exp[-E_j/kT]}$$

where E_i is the energy of state i, k a constant, and T the system temperature, and $P[E_i]$ is the probability of state i.

Feynman [31], following the classic arguments of [9] that present idealized machines using information to do work, concludes *the information contained in a message is most simply measured by the free energy needed to erase it.* The arguments of [9] are clever indeed, and the Feynman treatment of them in [31] is well worth reading.

Thus, according to this argument, source uncertainty is homologous to free energy density as defined above, i.e., from the similarity with the relation $H = \lim_{n \to \infty} \log[N(n)]/n$.

Ash's comment above then has an important corollary: If, for a biological system, $H_1 > H_2$, source 1 will require more metabolic free energy than source 2.

3 Symmetry Arguments

A formal equivalence class algebra, in the sense of the groupoid section of the Appendix, can now be constructed by choosing different origin and end points S_0, S_∞ and defining equivalence of two states by the existence of a high probability meaningful path connecting them with the same origin and end. Disjoint partition by equivalence class, analogous to orbit equivalence classes for dynamical systems, defines the vertices of the proposed network of cognitive dual languages, much enlarged beyond the spinglass example. We thus envision a *network of metanetworks*, in the sense of [16]. Each vertex then represents a different equivalence class of information sources dual to a cognitive process. This is an abstract set of metanetwork 'languages' dual to the cognitive processes of gene expression and development.

This structure generates a groupoid, in the sense of [78]. States a_j, a_k in a set A are related by the groupoid morphism if and only if there exists a high probability grammatical path connecting them to the same base and end points, and tuning across the various possible ways in which that can happen – the different cognitive languages – parametizes the set of equivalence relations and creates the (very large) groupoid.

There is an implicit hierarchy. First, there is structure *within the system having the same base and end points*, as in [16]. Second, there is a complicated groupoid structure defined by sets of dual information sources surrounding the variation of base and end points. We do not need to know what that structure is in any detail, but can show that its existence has profound implications.

We begin with the simple case, the set of dual information sources associated with a fixed pair of beginning and end states.

3.1   The First Level

The spinglass model of [16, 17] produced a simply connected, but otherwise undifferentiated, metanetwork of gene expression dynamics that could be traversed continuously by single-gene transitions in the highly parallel w-space. Taking the serial grammar/syntax model above, we find that not all high probability meaningful paths from $\mathbf{S}_0$ to $\mathbf{S}_\infty$ are actually the same. They are structured by the uncertainty of the associated dual information source, and that has a homological relation with free energy density.

Let us index possible dual information sources connecting base and end points by some set $A = \cup \alpha$. Argument by abduction from statistical physics is direct: Given metabolic energy density available at a rate M, and an allowed development time τ, let $K = 1/\kappa M \tau$ for some appropriate scaling constant κ, so that $M\tau$ is total developmental free energy. Then the probability of a particular H_α will be determined by the standard expression (e.g., [49]),

$$P[H_\beta] = \frac{\exp[-H_\beta K]}{\sum_\alpha \exp[-H_\alpha K]}, \tag{5}$$

where the sum may, in fact, be a complicated abstract integral.

This is just a version of the fundamental probability relation from statistical mechanics, as above. The sum in the denominator, the partition function in statistical physics, is a crucial normalizing factor that allows the definition of of $P[H_\beta]$ as a probability.

A basic requirement, then, is that the sum/integral always converges. K is the inverse product of a scaling factor, a metabolic energy density rate term, and a characteristic development time τ. The developmental energy might be raised to some power, e.g., $K = 1/(\kappa(M\tau)^b)$, suggesting the possibility of allometric scaling.

Thus, in this formulation, there must be structure *within* a (cross sectional) connected component in the w-space of [16, 17], determined in no small measure by available energy. Some dual information sources will be 'richer'/smarter than others, but, conversely, must use more metabolic energy for their completion.

3.2   The Second Level

The next generalization is crucial:

While we might simply impose an equivalence class structure based on equal levels of energy/source uncertainty, producing a groupoid in the sense of the Appendix (and possibly allowing a Morse Theory approach in the sense of [52, 59]), we can do more *by now allowing both source and end points to vary*, as well as by imposing energy-level equivalence. This produces a far more highly structured groupoid that we now investigate.

Equivalence classes define groupoids, by standard mechanisms [13, 35, 78]. The basic equivalence classes – here involving both information source uncertainty level and the variation of $\mathbf{S}_0$ and $\mathbf{S}_\infty$, will define transitive groupoids, and higher order systems can be constructed by the union of transitive groupoids, having larger alphabets that allow more complicated statements in the sense of Ash above.

Again, given an appropriately scaled, dimensionless, fixed, inverse available metabolic energy density rate and development time, so that $K = 1/\kappa M\tau$, we propose that the metabolic-energy-constrained probability of an information source representing equivalence class D_i, H_{D_i}, will again be given by

$$P[H_{D_i}] = \frac{\exp[-H_{D_i}K]}{\sum_j \exp[-H_{D_j}K]}, \qquad (6)$$

where the sum/integral is over all possible elements of the largest available symmetry groupoid. By the arguments of Ash above, compound sources, formed by the union of underlying transitive groupoids, being more complex, generally having richer alphabets, as it were, will all have higher free-energy-density-equivalents than those of the base (transitive) groupoids.

Let

$$Z_D = \sum_j \exp[-H_{D_j}K]. \qquad (7)$$

We now define the *Groupoid free energy* of the system, F_D, at inverse normalized metabolic energy density K, as

$$F_D[K] = -\frac{1}{K}\log[Z_D[K]], \qquad (8)$$

again following the standard arguments from statistical physics [31, 49].

The groupoid free energy construct permits introduction of important ideas from statistical physics.

3.3   Spontaneous Symmetry Breaking

We have expressed the probability of an information source in terms of its relation to a fixed, scaled, available (inverse) metabolic free energy, seen as a kind of equivalent (inverse) system temperature. This gives a statistical thermodynamic path leading to definition of a 'higher' free energy construct – $F_D[K]$ – to which we now apply Landau's fundamental heuristic phase transition argument [49, 59, 65].

The essence of Landau's insight was that certain phase transitions were usually in the context of a significant symmetry change in the physical states of a system, with one phase being far more symmetric than the other. A symmetry is lost in the transition, a phenomenon called spontaneous symmetry breaking. The greatest possible set of symmetries in a physical system is that of the Hamiltonian describing its energy states. Usually states accessible at lower temperatures will lack the symmetries available at higher temperatures, so that the lower temperature phase is less symmetric: The randomization of higher temperatures – in this case limited by available metabolic free energy – ensures that higher symmetry/energy states – mixed transitive groupoid structures – will then be accessible to the system. Absent high metabolic free energy, however, only the simplest transitive groupoid structures can be manifest. A full treatment from this perspective requires invocation of groupoid representations, no small matter (e.g., [10, 14]).

Somewhat more rigorously, the biological renormalization schemes of the Appendix to [74] may now be imposed on $F_D[K]$ itself, leading to a spectrum of highly punctuated transitions in the overall system of developmental information sources.

Most deeply, however, an extended version of Pettini's Morse-Theory-based topological hypothesis [59] can now be invoked, i.e., that changes in underlying groupoid structure are a necessary (but not sufficient) consequence of phase changes in $F_D[K]$. Necessity, but not sufficiency, is important, as it, in theory, allows mixed groupoid symmetries.

The essential insight is that the single simply connected giant component of [16, 17] is unlikely to be the full story, and that more complete models will likely be plagued – or graced – by highly punctuated dynamics.

Several matters are worth noting. First, Landau's spontaneous symmetry breaking arguments are perhaps the simplest approach possible here. The formal mathematical development requires invoking holonomy groups and groupoids, as in [34].

Second, one need not be restricted to terms of the form $\exp[-H_jK]$, as any $f(H_j, K)$ such that the sum over j converges will serve, although the resulting 'thermodynamic' relations between variates of central interest may then be less elegant.

Third, there may be some allometric scaling tradeoff between metabolic energy rate and development time determined by a relation of the form $K \propto (\tau M)^\alpha$.

4   Tunable Epigenetic Catalysis

Incorporating the influence of embedding contexts – epigenetic effects – is most elegantly done by invoking the Joint Asymptotic Equipartition Theorem (JAEPT) [20]. For example, given an embedding contextual information source, say Z, that affects development, then the dual cognitive source uncertainty H_{D_i} is replaced by a joint uncertainty $H(X_{D_i}, Z)$. The objects of interest then become the jointly typical dual sequences $y^n = (x^n, z^n)$, where x is associated with cognitive gene expression and z with the embedding context. Restricting consideration of x and z to those sequences that are in fact jointly typical allows use of the information transmitted from Z to X as the splitting criterion.

One important inference is that, from the information theory 'chain rule' [20],

$$H(X, Y) = H(X) + H(Y|X) \leq H(X) + H(Y),$$

while there are approximately $\exp[nH(X)]$ typical X sequences, and $\exp[nH(Z)]$ typical Z sequences, and hence $\exp[n(H(x) + H(Y))]$ independent joint sequences, there are only about $\exp[nH(X, Z)] \leq \exp[n(H(X) + H(Y))]$ jointly typical sequences, so that the effect of the embedding context, in this model, is to lower the *relative* free energy of a particular developmental channel.

Thus the effect of epigenetic regulation is to channel development into pathways that might otherwise be inhibited by an energy barrier. Hence the epigenetic information source Z acts as a *tunable catalyst*, a kind of second order cognitive enzyme, to enable and direct developmental pathways. This result permits hierarchical models similar to those of higher order cognitive neural function that incorporate Baars' contexts in a natural way [73, 74].

It is worth emphasizing that this is indeed a relative energy argument, since, metabolically, two systems must now be supported, i.e., that of the 'reaction' itself and that of its catalytic regulator. 'Programming' and stabilizing inevitably intertwined, as it were.

This elaboration allows a spectrum of possible 'final' phenotypes, what [33] calls developmental or phenotype plasticity. Thus gene expression is seen as, in part, responding to environmental or other, internal, developmental signals.

West-Eberhard [79] argues that any new input, whether it comes from the genome, like a mutation, or from the external environment, like a temperature change, a pathogen, or a parental opinion, has a developmental effect only if the preexisting phenotype is responsive to it. A new input causes a reorganization of the phenotype, or 'developmental recombination.' In developmental recombination, phenotypic traits are expressed in new or distinctive combinations during ontogeny, or undergo correlated quantitative change in dimensions. Developmental recombination can result in evolutionary divergence at all levels of organization.

Individual development can be visualized as a series of branching pathways. Each branch point, according to [79], is a developmental decision, or switch point, governed by some regulatory apparatus, and each switch point defines a modular trait. Developmental recombination implies the origin or deletion of a branch and a new or lost modular trait. It is important to realize that the novel regulatory response and the novel trait originate simultaneously. Their origins are, in fact, inseparable events. There cannot, [79] concludes, be a change in the phenotype, a novel phenotypic state, without an altered developmental pathway.

These mechanisms are accomplished in our formulation by allowing the set B_1 in section 2.2 to span a distribution of possible 'final' states $\mathbf{S}_\infty$. Then the groupoid arguments merely expand to permit traverse of both initial states and possible final sets, recognizing that there can now be a possible overlap in the latter, and the epigenetic effects are realized through the joint uncertainties $H(X_{D_i}, Z)$, so that the epigenetic information source Z serves to direct as well the possible final states of X_{D_i}.

Again, [62, 63] use information theory arguments to suggest something similar to epigenetic catalysis, finding the information in a sequence is not contained in the sequence but has been provided by the machinery that accompanies it on the expression pathway. That work does not, however, invoke a cognitive paradigm, its attendant groupoid symmetries, or the homology between information source uncertainty and free energy density that drives dynamics.

The mechanics of channeling can be made more precise as follows.

5   Rate Distortion Dynamics

Real time problems, like the crosstalk between epigenetic and genetic structures, are inherently rate distortion problems, and the interaction between biological structures can be restated in communication theory terms. Suppose a sequence of signals is generated by a biological information source Y having output $y^n = y_1, y_2, \ldots$. This is 'digitized' in terms of the observed behavior of the system with which it communicates, say a sequence of observed behaviors $b^n = b_1, b_2, \ldots$. The b_i happen in real time. Assume each b^n is then deterministically retranslated back into a reproduction of the original biological signal,

$$b^n \to \hat{y}^n = \hat{y}_1, \hat{y}_2, \ldots.$$

Here the information source Y is the epigenetic Z, and B is X_{D_i}, but the terminology used here is more standard [20].

Define a distortion measure $d(y, \hat{y})$ which compares the original to the retranslated path. Many distortion measures are possible, as described in the Mathematical Appendix.

The distortion between *paths* y^n and $\hat{y}^n$ is defined as

$$d(y^n, \hat{y}^n) = \frac{1}{n} \sum_{j=1}^{n} d(y_j, \hat{y}_j).$$

A remarkable fact of the Rate Distortion Theorem is that *the basic result is independent of the exact distortion measure chosen* [20, 24].

Suppose that with each path y^n and b^n-path retranslation into the y-language, denoted $\hat{y}^n$, there are associated individual, joint, and conditional probability distributions

$$p(y^n), p(\hat{y}^n), p(y^n, \hat{y}^n), p(y^n|\hat{y}^n).$$

The average distortion is defined as

$$D = \sum_{y^n} p(y^n) d(y^n, \hat{y}^n). \tag{9}$$

It is possible, using the distributions given above, to define the information transmitted from the Y to the $\hat{Y}$ process using the Shannon source uncertainty of the strings:

$$I(Y, \hat{Y}) = H(Y) - H(Y|\hat{Y}) = H(Y) + H(\hat{Y}) - H(Y, \hat{Y}), \tag{10}$$

where $H(..., ...)$ is the joint and $H(...|...)$ the conditional uncertainty [1, 20].

If there is no uncertainty in Y given the retranslation $\hat{Y}$, then no information is lost, and the systems are in perfect synchrony.

In general, of course, this will not be true.

The *rate distortion function* $R(D)$ for a source Y with a distortion measure $d(y, \hat{y})$ is defined as

$$R(D) = \min_{p(y,\hat{y}); \sum_{(y,\hat{y})} p(y)p(y|\hat{y})d(y,\hat{y}) \leq D} I(Y, \hat{Y}). \tag{11}$$

The minimization is over all conditional distributions $p(y|\hat{y})$ for which the joint distribution $p(y, \hat{y}) = p(y)p(y|\hat{y})$ satisfies the average distortion constraint (i.e., average distortion $\leq D$).

The *Rate Distortion Theorem* states that $R(D)$ is the minimum necessary rate of information transmission which ensures communication does not exceed average distortion D. Thus $R(D)$ defines a minimum necessary channel capacity. References [20, 24] provide details. The rate distortion function has been explicitly calculated for a number of simple systems.

Recall, now, the relation between information source uncertainty and channel capacity [1, 20]:

$$H[\mathbf{X}] \leq C, \tag{12}$$

where H is the uncertainty of the source X and C the channel capacity, defined according to the relation [1, 20]

$$C = \max_{P(X)} I(X|Y). \tag{13}$$

X is the message, Y the channel, and the probability distribution $P(X)$ is chosen so as to maximize the rate of information transmission along a Y.

Finally, recall the analogous definition of the rate distortion function above, again an extremum over a probability distribution.

Recall, again, equations (4-8), i.e., that the free energy of a physical system at a normalized inverse temperature-analog $K = 1/\kappa T$ is defined as $F(K) = -\log[Z(K)]/K$ where $Z(K)$ the partition function defined by the system Hamiltonian. More precisely, if the possible energy states of the system are a set $E_i, i = 1, 2, \ldots$ then, at normalized inverse temperature K, the probability of a state E_i is determined by the relation $P[E_i] = \exp[-E_i K]/\sum_j \exp[-E_j K]$. The partition function is simply the normalizing factor.

Applying this formalism, it is possible to extend the rate distortion model by describing a probability distribution for D across an ensemble of possible rate distortion functions in terms of available free metabolic energy, $K = 1/\kappa M \tau$.

The key is to take the $R(D)$ as representing energy as a function of the average distortion. Assume a fixed K, so that the probability density function of an average distortion D, given a fixed K, is then

$$P[D, K] = \frac{\exp[-R(D)K]}{\int_{D_{min}}^{D_{max}} \exp[-R(D)K]dD}. \tag{14}$$

Thus lowering K in this model rapidly raises the possibility of low distortion communication between linked systems.

We define the *rate distortion partition function* as just the normalizing factor in this equation:

$$Z_R[K] = \int_{D_{min}}^{D_{max}} \exp[-R(D)K]dD, \tag{15}$$

again taking $K = 1/\kappa M \tau$.

We now define a new free energy-analog, the *rate distortion free-energy*, as

$$F_R[K] = -\frac{1}{K} \log[Z_R[K]], \tag{16}$$

and apply Landau's spontaneous symmetry breaking argument to generate punctuated changes in the linkage between the genetic information source X_{D_i} and the embedding epigenetic information source Z. Recall that Landau's insight was that certain phase transitions were usually in the context of a significant symmetry change in the physical states of a system.

Again, the biological renormalization schemes of the Appendix to [74] may now be imposed on $F_R[K]$ itself, leading to a spectrum of highly punctuated transitions in the overall system of interacting biological substructures.

Since $1/K$ is proportional to the embedding metabolic free energy, we assert that

1. The greatest possible set of symmetries will be realized for high developmental metabolic free energies, and

2. Phase transitions, related to total available developmental metabolic free energy, will be accompanied by fundamental changes in the final topology of the

system of interest – phenotype changes – recognizing that evolutionary selection acts on phenotypes, not genotypes.

The relation $1/K = \kappa M \tau$ suggests the possibility of evolutionary tradeoffs between development time and the rate of available metabolic free energy.

6   More Topology

It seems possible to extend this treatment using standard topological arguments.

Taking $T = 1/K$ in equations (6) and (14) *as a product of eigenvalues*, we can define it as the determinant of a Hessian matrix representing a Morse Function, f, on some underlying, background, manifold, $\mathcal{M}$, characterized in terms of (as yet unspecified) variables $\mathcal{X} = (x^1, ..., x^n)$, so that

$$1/K = \det(\mathcal{H}_{i,j}),$$

$$\mathcal{H}_{i,j} = \partial^2 f / \partial x^i \partial x^j. \tag{17}$$

Again, see the Appendix for a brief outline of Morse Theory.

Thus κ, M, and the development time τ are seen as eigenvalues of $\mathcal{H}$ on the manifold $\mathcal{M}$ in an abstract space defined by some set of variables $\mathcal{X}$.

By construction $\mathcal{H}$ has everywhere only nonzero, and indeed, positive, eigenvalues, whose product thereby defines T as a generalized volume. Thus, and accordingly, all critical points of f have index zero, that is, no eigenvalues of $\mathcal{H}$ are ever negative at any point, and hence at any critical point $\mathcal{X}_c$ where $df(\mathcal{X}_c) = 0$.

This defines a particularly simple topological structure for $\mathcal{M}$: If the interval $[a, b]$ contains a critical value of f with a single critical point $\mathcal{X}_c$, then the topology of the set $\mathcal{M}_b$ defined above differs from that of $\mathcal{M}_a$ in a manner determined by the index i of the critical point. $\mathcal{M}_b$ is then homeomorphic to the manifold obtained from attaching to $\mathcal{M}_a$ an i-handle, the direct product of an i-disk and an $(m - i)$-disk.

One obtains, in this case, since $i = 0$, the two halves of a sphere with critical points at the top and bottom [52, 59]. This is, as in [16], a simply connected object. What one does then is to invoke the Seifert-Van Kampen Theorem (SVKT, [50]) and patch together the various simply connected subcomponents to construct the larger, complicated, topological object representing the full range of possibilities.

The physical natures of κ, M, and τ thus impose constraints on the possible complexity of this system, in the sense of the SVKT.

7   Inherited Epigenetic Memory

The cognitive paradigm for gene expression invoked here requires an internal picture of the world against which incoming signals are compared – algorithmically combined according to the rules of Section 2.2 – and then fed into a sharply step-wise decision oscillator that chooses one (or a few) action(s) from a much large repertoire of possibilities. Memory is inherent, and much recent work, as described in the introduction, suggests that epigenetic memory is indeed heritable.

The abduction of spinglass and other models from neural network studies to the analysis of development and its evolution carries with it the possibility of more than one system of memory. What Baars called 'contexts' channeling high level animal cognition may often be the influence of cultural inheritance, in a large sense. Our formalism suggests a class of statistical models that indeed greatly generalize those used for measuring the effects of cultural inheritance on human behavior in populations.

Epigenetic machinery, as a dual information source to a cognitive process, serves as a heritable system, intermediate between (relatively) hard-wired classical genetics, and a (usually) highly Larmarckian embedding cultural context. In particular, the three heritable systems interact, in our model, through a crosstalk in which the epigenetic machinery acts as a kind of intelligent catalyst for gene expression.

8 Multiple Processes

The argument to this point has, in large measure, been directly abducted from recent formal studies of high level cognition – consciousness – based on a Dretske-style information theoretic treatment of Bernard Baars' global workspace model [3, 68]. A defining and grossly simplifying characteristic of that phenomenon is its rapidity: typically the global broadcasts of consciousness occur in a matter of a few hundred milliseconds, limiting the number of processes that can operate simultaneously. Slower cognitive dynamics can, therefore, be far more complex than individual consciousness. One well known example is institutional distributed cognition that encompasses both individual and group cognition in a hierarchical structure typically operating on timescales ranging from a few seconds or minutes in combat or hunting groups, to years at the level of major governmental structures, commercial enterprises, religious organizations, or other analogous large scale cultural artifacts. Reference [73] provides the first formal mathematical analysis of institutional distributed cognition.

Clearly cognitive gene expression is not generally limited to a few hundred milliseconds, and something much like the distributed cognition analysis may be applied here as well. Extending the analysis requires recognizing an individual cognitive actor can participate in more than one 'task', synchronously, asynchronously, or strictly sequentially. Again, the analogy is with institutional function whereby many individuals often work together on several distinct projects: Envision a multiplicity of possible cognitive gene expression dual 'languages' that themselves form a higher order network linked by crosstalk.

Next, describe crosstalk measures linking different dual languages on that meta-meta (MM) network by some characteristic magnitude ω, and *define a topology on the MM network by renormalizing the network structure to zero if*

the crosstalk is less than ω *and set it equal to one if greater or equal to it.*
A particular ω, of sufficient magnitude, defines a giant component of network
elements linked by mutual information greater or equal to it, in the sense of [29],
as more fully described in [73] (Section 3.4).

The fundamental trick is, in the Morse Theory sense [52], to invert the argument so that a given topology for the giant component will, in turn, define some
critical value, ω_C, so that network elements interacting by mutual information
less than that value will be unable to participate, will be locked out and not
active. ω becomes an epigenetically syntactically-dependent detection limit, and
depends critically on the instantaneous topology of the giant component defining
the interaction between possible gene interaction MM networks.

Suppose, now, that a set of such giant components exists at some generalized
system 'time' k and is characterized by a set of parameters $\Omega_k = \omega_1^k, ..., \omega_m^k$.
Fixed parameter values define a particular giant component set having a par-
ticular set of topological structures. Suppose that, over a sequence of times the
set of giant components can be characterized by a possibly coarse-grained path
$\gamma_n = \Omega_0, \Omega_1, ..., \Omega_{n-1}$ having significant serial correlations that, in fact, permit
definition of an adiabatically, piecewise stationary, ergodic (APSE) information
source Γ.

Suppose that a set of (external or internal) epigenetic signals impinging on
the set of such giant components can also be characterized by another APSE
information source Z that interacts not only with the system of interest globally,
but with the tuning parameters of the set of giant components characterized by
Γ. Pair the paths (γ_n, z_n) and apply the joint information argument above,
generating a splitting criterion between high and low probability sets of pairs of
paths. We now have a multiple workspace cognitive genetic expression structure
driven by epigenetic catalysis.

9   'Coevolutionary' Development

The model can be applied to multiple interacting information sources representing simultaneous gene expression processes, for example across a spatially
differentiating organism as it develops. This is, in a broad sense, a 'coevolutionary' phenomenon in that the development of one segment may affect that of
others.

Most generally we assume that different cognitive developmental subprocesses
of gene expression characterized by information sources H_m interact through
chemical or other signals and assume that *different processes become each other's
principal environments*, a broadly coevolutionary phenomenon.

We write

$$H_m = H_m(K_1...K_s, ...H_j...), \tag{18}$$

where the K_s represent other relevant parameters and $j \neq m$.

The dynamics of such a system is driven by a recursive network of stochastic
differential equations, similar to those used to study many other highly parallel
dynamic structures (e.g., [83]).

Letting the K_j and H_m all be represented as parameters Q_j, (with the caveat that H_m not depend on itself), one can define, according to the generalized Onsager development of the Appendix,

$$S^m = H_m - \sum_i Q_i \partial H_m / \partial Q_i$$

to obtain a complicated recursive system of phenomenological 'Onsager relations' stochastic differential equations,

$$dQ_t^j = \sum_i [L_{j,i}(t, ...\partial S^m/\partial Q^i...)dt + \sigma_{j,i}(t, ...\partial S^m/\partial Q^i...)dB_t^i], \qquad (19)$$

where, again, for notational simplicity only, we have expressed both the H_j and the external K's in terms of the same symbols Q_j.

m ranges over the H_m and we could allow different kinds of 'noise' dB_t^i, having particular forms of quadratic variation that may, in fact, represent a projection of environmental factors under something like a rate distortion manifold [73, 74].

As usual for such systems, there will be multiple quasi-stable points within a given system's H_m, representing a class of generalized resilience modes accessible via punctuation.

Second, however, there may well be analogs to fragmentation when the system exceeds the critical values of K_c according to the approach of [74]. That is, the K-parameter structure will represent full-scale fragmentation of the entire structure, and not just punctuation within it.

We thus infer two classes of punctuation possible for this kind of structure.

There are other possible patterns:

1. Setting equation (19) equal to zero and solving for stationary points again gives attractor states since the noise terms preclude unstable equilibria.

2. This system may converge to limit cycle or 'strange attractor' behaviors in which the system seems to chase its tail endlessly, e.g., the cycle of climate-driven phenotype changes in persistent temperate region plants.

3. What is converged to in both cases is not a simple state or limit cycle of states. Rather it is an equivalence class, or set of them, of highly dynamic information sources coupled by mutual interaction through crosstalk. Thus 'stability' in this extended model represents particular patterns of ongoing dynamics rather than some identifiable 'state', although such dynamics may be indexed by a 'stable' set of phenotypes.

Here we become enmeshed in a system of highly recursive phenomenological stochastic differential equations, but at a deeper level than the standard stochastic chemical reaction model (e.g., [84]), and in a dynamic rather than static manner: the objects of this system are equivalence classes of information sources and their crosstalk, rather than simple final states of a chemical system.

10   Multiple Models

Recent work [75] argues that consciousness may have undergone the characteristic branching and pruning of evolutionary development, particularly in view
of the rapidity of currently surviving conscious mechanisms. According to that
study, evolution is littered with polyphyletic parallelisms: many roads lead to
functional Romes, and consciousness, as a particular form of high order cognitive process operating in real time, embodies one such example, represented by
an equivalence class structure that factors the broad realm of necessary conditions information theoretic realizations of Baars' global workspace model. Many
different physiological systems, then, can support rapidly shifting, highly tunable, and even simultaneous assemblages of interacting unconscious cognitive
modules. Thus [75] concludes the variety of possibilities suggests minds today
may be only a small surviving fraction of ancient evolutionary radiations – bush
phylogenies of consciousness pruned by selection and chance extinction.

Even in the realms of rapid global broadcast inherent to real time cognition,
[75] speculates, following a long tradition, that ancient backbrain structures instantiate rapid emotional responses, while the newer forebrain harbors rapid
'reasoned' responses in animal consciousness. The cooperation and competition
of these two rapid phenomena produces, of course, a plethora of systematic
behaviors.

Since consciousness is necessarily restricted to realms of a few hundred milliseconds, evolutionary pruning may well have resulted in only a small surviving fraction of previous evolutionary radiations. Processes operating on longer
timescales may well be spared such draconian evolutionary selection. That is,
the vast spectrum of mathematical models of cognitive gene expression inherent
to our analysis here, in the context of development times much longer than a
few hundred milliseconds, implies current organisms may simultaneously harbor
several, possibly many, quite different cognitive gene expression mechanisms.

It seems likely, then, that, with some generality, slow phenomena, ranging
from institutional distributed cognition to cognitive gene expression, permit the
operation of very many quite different cognitive processes simultaneously or in
rapid succession.

One inference is, then, that gene expression and its epigenetic regulation are, for
even very simple organisms, far more complex than individual human consciousness, currently regarded as one of the 'really big' unsolved scientific problems.

Neural network models adapted or abducted from inadequate cognitive studies
of a generation ago are unlikely to cleave the Gordian Knot of scientific inference
surrounding gene expression.

11   Epigenetic Focus

The Tuning Theorem analysis of the Appendix permits an inattentional blind-
ness/concentrated focus perspective on the famous computational 'no free lunch'
theorem of [81, 82]. Following closely the arguments of [28], [81, 82] have established that there exists no generally superior function optimizer. There is no 'free

lunch' in the sense that an optimizer 'pays' for superior performance on some functions with inferior performance on others. If the distribution of functions is uniform, then gains and losses balance precisely, and all optimizers have identical average performance. The formal demonstration depends primarily upon a theorem that describes how information is conserved in optimization. This Conservation Lemma states that when an optimizer evaluates points, the posterior joint distribution of values for those points is exactly the prior joint distribution. Put simply, observing the values of a randomly selected function does not change the distribution: An optimizer has to 'pay' for its superiority on one subset of functions with inferiority on the complementary subset.

As [28] describes, anyone slightly familiar with the evolutionary computing literature recognizes the paper template 'Algorithm X was treated with modification Y to obtain the best known results for problems P_1 and P_2.' Anyone who has tried to find subsequent reports on 'promising' algorithms knows that they are extremely rare. Why should this be?

A claim that an algorithm is the very best for two functions is a claim that it is the very worst, on average, for all but two functions. It is due to the diversity of the benchmark set of test problems that the 'promise' is rarely realized. Boosting performance for one subset of the problems usually detracts from performance for the complement.

Reference [28] argues that hammers contain information about the distribution of nail-driving problems. Screwdrivers contain information about the distribution of screw-driving problems. Swiss army knives contain information about a broad distribution of survival problems. Swiss army knives do many jobs, but none particularly well. When the many jobs must be done under primitive conditions, Swiss army knives are ideal.

Thus, according to [28], the tool literally carries information about the task optimizers are literally tools-an algorithm implemented by a computing device is a physical entity.

Another way of looking at this is to recognize that a computed solution is simply the product of the information processing of a problem, and, by a very famous argument, information can never be gained simply by processing. Thus a problem X is transmitted as a message by an information processing channel, Y, a computing device, and recoded as an answer. By the Tuning Theorem argument of the Appendix there will be a channel coding of Y which, when properly tuned, is most efficiently transmitted by the problem. In general, then, the most efficient coding of the transmission channel, that is, the best algorithm turning a problem into a solution, will necessarily be highly problem-specific. Thus there can be no best algorithm for all equivalence classes of problems, although there may well be an optimal algorithm for any given class. The tuning theorem form of the No Free Lunch theorem will apply quite generally to cognitive biological and social structures, as well as to massively parallel machines.

Rate distortion, however, occurs when the problem is collapsed into a smaller, simplified, version and then solved. Then there must be a tradeoff between allowed average distortion and the rate of solution: the retina effect. In a very fundamental sense – particularly for real time systems – rate distortion manifolds present a generalization of the converse of the no free lunch arguments. The neural corollary is known as inattentional blindness [69].

We are led to suggest that there may well be a considerable set of no free lunch-like conundrums confronting highly parallel real-time structures, including epigenetic control of gene expression, and that they may interact in distinctly complicated ways.

12   Developmental Disorders

12.1   Network Information Theory

Let U be an information source representing a systematic embedding environmental 'program' interacting with the process of cognitive gene expression, here defined as a complicated information set of sources having source joint uncertainty $H(Z_1, ..., Z_n)$ that guides the system into a particular equivalence class of desired developmental behaviors and trajectories.

To model the effect of U on development one can, most simply, invoke results from network information theory, ([20], p. 388). Given three interacting information sources, say Y_1, Y_2, Z, the splitting criterion between high and low probability sets of states, taking Z as the external context, is given by

$$I(Y_1, Y_2|Z) = H(Z) + H(Y_1|Z) + H(Y_2|Z) - H(Y_1, Y_2, Z),$$

where, again, $H(...|...)$ and $H(..., ..., ...)$ represent conditional and joint uncertainties. This generalizes to the relation

$$I(Y_1, ..., Y_n|Z) = H(Z) + \sum_{j=1}^{n} H(Y_j|Z) - H(Y_1, ..., Y_n, Z).$$

Thus the fundamental splitting criterion between low and high probability sets of joint developmental paths becomes

$$I(Z_1, ..., Z_n|U) = H(U) + \sum_{j=1}^{n} H(Z_j|U) - H(Z_1, ..., Z_n, U). \tag{20}$$

Again, the Z_i represent internal information sources and U that of the embedding environmental context.

The central point is that a one step extension of that system via the results of network information theory [20] allows incorporating the effect of an external environmental 'farmer' in guiding cognitive developmental gene expression.

12.2   Embedding Ecosystems as Information Sources

The principal farmer for a developing organism is the ecosystem in which it is embedded, in a large sense. Summarizing briefly the arguments of [74], ecosystems, under appropriate coarse graining, often have reconizable grammar and syntax. For example, the turn-of-the-seasons in a temperate climate, for most natural communities, is remarkably similar from year to year in the sense that the ice melts, migrating birds return, trees bud, flowers and grass grow, plants and animals reproduce, the foliage turns, birds migrate, frost, snow, the rivers freeze, and so on in a predictable manner from year to year.

Suppose, then, that we can coarse grain an ecosystem at time t according to some appropriate partition of the phase space in which each division A_j represents a particular range of numbers for each possible species in the ecosystem, along with associated parameters such as temperature, rainfall, humidity, insolation, and so on. We examine longitudinal paths, statements of the form

$$x(n) = A_0, A_1, ..., A_n$$

defined in terms of some 'natural' time unit characteristic of the system. Then n corresponds to a time unit T, so that $t = T, 2T, ..., nT$. Our interest is in the serial correlation along paths. If $N(n)$ is the number of possible paths of length n that are consistent with the underlying grammar and syntax of the appropriately coarse grained ecosystem, for example, spring leads to summer, autumn, winter, back to spring, etc., but never spring to autumn to summer to winter in a temperate climate.

The essential assumption is that, for appropriate coarse graining, $N(n)$, the number of possible grammatical paths, is much smaller than the total conceivable number of paths, and that, in the limit of large n,

$$H = \lim_{n \to \infty} \frac{\log[N(n)]}{n}$$

both exists and is independent of path.

Not all possible ecosystem coarse grainings are likely to lead to this result, as is sometimes the case with Markov models. Reference [40] in particular emphasizes that mesoscale ecosystem processes are most likely to entrain dynamics at larger and smaller scales, a process [74] characterizes as *mesoscale resonance*, a generalization of the Baldwin effect. See that reference for details, broadly based on the Tuning Theorem.

12.3 Ecosystems Farm Organismal Development

The environmental and ecosystem farming of development may not always be benign.

Suppose we can operationalize and quantify degrees of both overfocus or inattentional blindness (IAB) and of overall structure or environment distortion (D) in the actions of a highly parallel cognitive epigenetic regulatory system. The essential assumption is that the (internal) dual information source of a cognitive structure that has low levels of both IAB overfocus and structure/environment distortion will tend to be richer than that of one having greater levels. This is shown in figure 1a, where H is the source uncertainty dual to internal cognitive process, $X = IAB$, and $Y = D$. Regions of low X, Y, near the origin, have greater source uncertainty than those nearby, so $H(X, Y)$ shows a (relatively gentle) peak at the origin, taken here as simply the product of two error functions.

We are, then, particularly interested in the internal cognitive capacity of the structure itself, as paramatized by degree of overfocus and by the (large scale) distortion between implementation and impact. That capacity, a purely internal quantity, need not be convex in the parameter D, which is taken to characterize interaction with an external environment, and thus becomes a context for internal measures. Such measures need not themselves be convex in D.

The generalized Onsager argument, based on the homology between information source uncertainty and free energy, as explained more fully in the Appendix, is shown in figure 1b. $S = H(X, Y) - X dH/dX - Y dH/dY$, the 'disorder' analog to entropy in a physical system, is graphed on the Z axis against the $X - Y$ plane, assuming a gentle peak in H at the origin. Peaks in S, according to theory, constitute repulsive system barriers, which must be overcome by external forces. In figure 1b there are three quasi-stable topological resilience modes, in the sense of [71], marked as A, B, and C. The A region is locked in to low levels of both overfocus and distortion, as it sits in a pocket. Forcing the system in either direction, that is, increasing either IAB or D, will, initially, be met by homeostatic attempts to return to the resilience state A, according to this model.

If overall distortion becomes severe in spite of homeostatic developmental mechanisms, the system will then jump to the quasi-stable state B, a second pocket. According to the model, however, once that transition takes place, there will be a tendency for the system to remain in a condition of high distortion. That is, the system will become locked-in to a structure with high distortion in the match between structure implementation and structure impact, but one having lower overall cognitive capacity, i.e., a lower value of H in figure 1a.

The third pocket, marked C, is a broad plain in which both IAB and D remain high, a highly overfocused, poorly linked pattern of behavior which will require significant intervention to alter once it reaches such a quasi-stable resilience mode. The structure's cognitive capacity, measured by H in figure 1a, is the lowest of all for this condition of pathological resilience, and attempts to correct the problem – to return to condition A, will be met with very high barriers in S, according to figure 1b. That is, mode C is very highly resilient, although pathologically so, much like the eutrophication of a pure lake by sewage outflow. See [70, 71] for discussions of ecological resilience and literature references.

We can argue that the three quasi-equilibrium configurations of figure 1b represent different dynamical states of the system, and that the possibility of transition between them represents the breaking of the associated symmetry groupoid by external forcing mechanisms. That is, three manifolds representing three different kinds of system dynamics have been patched together to create a more complicated topological structure. For cognitive phenomena, such behavior is likely to be the rule rather than the exception. 'Pure' groupoids are abstractions, and the fundamental questions will involve linkages which break the underlying symmetry.

S=H-XdH/dX-YdS/dY

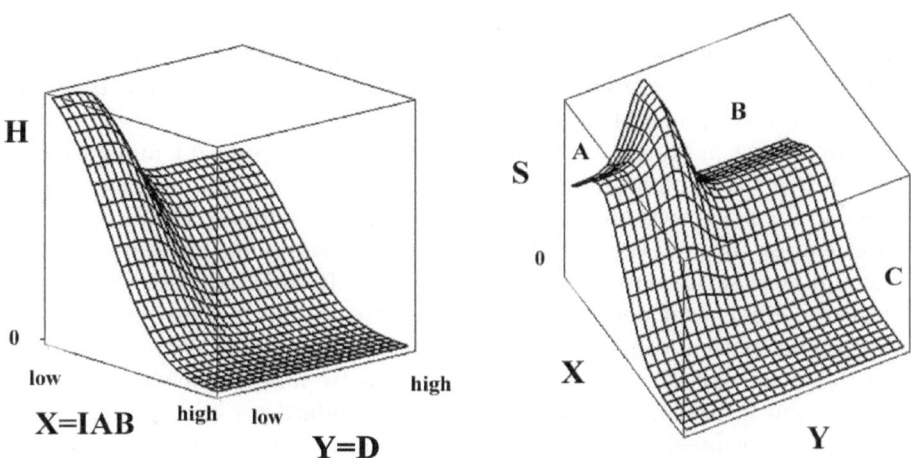

Fig. 1. a. Source uncertainty, H, of the dual information source of epigenetic cognition, as parametized by degrees of focus, $X = IAB$ and distortion, $Y = D$, between implementation and actual impact. Note the relatively gentle peak at low values of X, Y. Here H is generated as the product of two error functions. b. Generalized Onsager treatment of figure 1a. $S = H(X, Y) - XdH/dX - YdH/dY$. The regions marked $A, B,$ and C represent realms of resilient quasi-stability, divided by barriers defined by the relative peaks in S. Transition among them requires a forcing mechanism. From another perspective, limiting energy or other resources, or imposing stress from the outside, driving down H in figure 1a, would force the system into the lower plain of C, in which the system would then become trapped in states having high levels of distortion and inattentional blindness/overfocus.

In all of this, as in equation (19), system convergence is not to some fixed state, limit cycle, or pseudorandom strange attractor, but rather to some appropriate set of highly dynamic information sources, i.e., behavior patterns constituting, here, developmental trajectories, rather than to some fixed 'answer to a computing problem' [72].

What this model suggests is that sufficiently strong external perturbation can force a highly parallel real-time cognitive epigenetic structure from a normal, almost homeostatic, developmental path into one involving a widespread, comorbid, developmental disorder. This is a well studied pattern for humans and their institutions, reviewed at some length elsewhere [71, 73]. Indeed, this argument provides the foundation of a fairly comprehensive model of chronic developmental dysfunction across a broad class of cognitive systems, including, but not limited to, cognitive epigenetic control of gene expression. One approach might be as follows:

A developmental process can be viewed as involving a sequence of surfaces like figure 1, having, for example, 'critical periods' when the barriers between the normal state A and the pathological states B and C are relatively low. This might particularly occur under circumstances of rapid growth or long-term energy demand, since the peaks of figure 1 are inherently energy maxima by the duality between information source uncertainty and free energy density. During such a time the peaks of figure 1 might be relatively suppressed, and the system would become highly sensitive to perturbation, and to the onset of a subsequent pathological developmental trajectory.

To reiterate, then, during times of rapid growth, embryonic de- and re- methylation, and/or other high system demand, metabolic energy limitation imposes the need to focus via something like a rate distortion manifold. Cognitive process requires energy through the homologies with free energy density, and more focus at one end necessarily implies less at some other. In a distributed zero sum developmental game, as it were, some cognitive or metabolic processes must receive more free energy than others, and these may then be more easily affected by external chemical, biological, or social stressors, or by simple stochastic variation. Something much like this has indeed become a standard perspective (e.g., [76]).

A structure trapped in region C might be said to suffer something much like what [80] describes as the loss of gradient problem, in which one part of a multiple population coevolutionary system comes to dominate the others, creating an impossible situation in which the other participants do not have enough information from which to learn. That is, the cliff just becomes too steep to climb. Reference [80] also characterizes focusing problems in which a two-population coevolutionary process becomes overspecialized on the opponent's weaknesses, effectively a kind of inattentional blindness.

Thus there seems some consonance between our asymptotic analysis of cognitive structural function and current studies of pathologies affecting coevolutionary algorithms (e.g. [30, 72]). In particular the possibility of historic trajectory, of path dependence, in producing individualized failure modes, suggests there can be no one-size-fits-all amelioration strategy.

Equation (20) basically enables a kind of environmental catalysis to cognitive gene expression, in a sense closely similar to the arguments of Section 4. This is analogous to, but more general than, the 'mesoscale resonance' invoked by [74]: during critical periods, according to these models, environmental signals can have vast impact on developmental trajectory.

12.4   A Simple Probability Argument

Again, critical periods of rapid growth require energy, and by the homology between free energy density and cognitive information source uncertainty, that energy requirement may be in the context of a zero-sum game so that the barriers of figure 1 may be lowered by metabolic energy constraints or high energy demand. In particular the groupoid structure of equation (5) changes progressively as the organism develops, with new equivalence classes being added to $A = \cup \alpha$. If metabolic energy remains capped, then

$$P[H_\beta] = \frac{\exp[-H_\beta K]}{\sum_\alpha \exp[-H_\alpha K]}$$

must decrease with increase in α, i.e., with increase in the cardinality of A. Thus, for restricted K, barriers between different developmental paths must fall as the system becomes more complicated.

A precis of these results can be more formally captured using methods closely similar to recent algebraic geometry approaches to concurrent, i.e., highly parallel, computing [26, 37, 60].

13   Reconsidering Directed Homotopy: Shadows

Here we reconsider directed homotopy in a developmental context, as shadowed by critical developmental periods. First, we restrict the analysis to a two dimensional phenotype space, and begin development at some $\mathbf{S}_0$ as in figure 2.

If one requires temporal path dependence – no reverse development – then figure 2 shows two possible final states, $\mathbf{S}_1$ and $\mathbf{S}_2$, separated by a critical point $\mathbf{C}$ *that casts a path-dependent developmental shadow* in time. There are, consequently, two separate 'ways' of reaching a final state in this model. The $\mathbf{S}_i$ thus represent (relatively) static phenotypic expressions of the solutions to equation (19) that are, of themselves, highly dynamic information sources.

Elements of each 'way' can be transformed into each other by continuous deformation without crossing the impenetrable shadow cast by the critical period $\mathbf{C}$.

These ways are the equivalence classes defining the system's topological structure, a groupoid analogous to the fundamental homotopy group in spaces that admit of loops [50] rather than time-driven, one-way paths. That is, the closed loops needed for classical homotopy theory are impossible for this kind of system because of the 'flow of time' defining the output of an information source – one goes from $\mathbf{S}_0$ to some final state. The theory is thus one of *directed homotopy*, dihomotopy, and the central question revolves around the continuous deformation of paths in development space into one another, without crossing the shadow cast by the critical period $\mathbf{C}$. Reference [36] provides another introduction to the formalism.

Thus the external signals U of equation (20), as a catalytic mechanism, can define quite different developmental dihomotopies.

Such considerations suggest that a multitasking developmental process that becomes trapped in a particular pattern cannot, in general, expect to emerge from it in the absence of external forcing mechanisms or the stochastic resonance/mutational action of 'noise'. Emerging from such a trap involves large-scale topological changes, and this is the functional equivalent of a first order phase transition in a physical systems and requires energy.

The fundamental topological insight is that environmental context – the U in equation (20) – can be imposed on the 'natural' groupoids underlying massively parallel gene expression. This sort of behavior is, as noted in [71], central to ecosystem resilience theory.

Apparently the set of developmental manifolds, and its subsets of directed homotopy equivalence classes, formally classifies quasi-equilibrium states, and

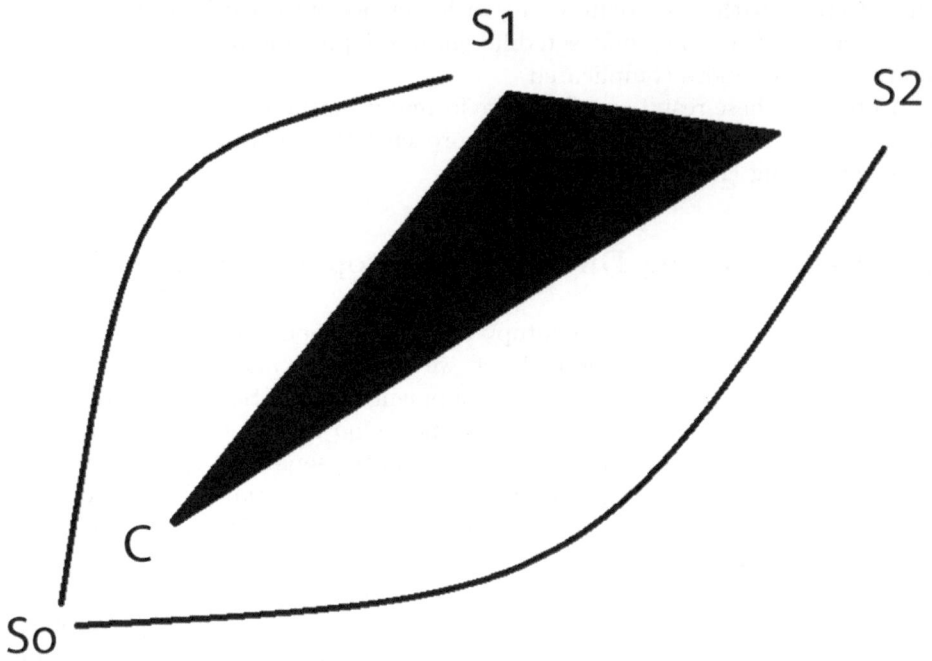

Fig. 2. Given an initial developmental state S_0 and a critical period C casting a path-dependent developmental shadow, there are two directed homotopy equivalence classes of deformable paths leading, respectively, to final phenotype states S_1 and S_2 that are expressions of the highly dynamic information source solutions to equation (19). These equivalence classes define a topological groupoid on the developmental system.

thus characterizes the different possible developmental resilience modes. Some of these may be highly pathological.

Shifts between markedly different topological modes appear to be necessary effects of phase transitions, involving analogs to phase changes in physical systems.

It seems clear that both 'normal development' and possible pathological states can be represented as topological resilience/phase modes in this model, suggesting a real equivalence between difficulties in carrying out gene expression and its stabilization. This mirrors recent results on the relation between programming difficulty and system stability in highly parallel computing devices [70].

14   Epigenetic Programming of Artificial Systems for Biotechnology

Reference [72] examines how highly parallel 'Self-X' computing machines – self-programming, protecting, repairing, etc. – are inevitably coevolutionary in the sense of Section 9 above, since elements of a dynamic structural hierarchy always interact, an effect that will asymptotically dominate system behavior at great scale. The 'farming' paradigm provides a model for programming such devices, that, while broadly similar to the liquid state machines of [51], differs in that convergence is to an information source, a systematic dynamic behavior pattern, rather than to a computed fixed 'answer'. As the farming metaphor suggests, stabilizing complex coevolutionary mechanisms appears as difficult as programming them. Sufficiently large networks of even the most dimly cognitive modules will become emergently coevolutionary, suggesting the necessity of 'second order' evolutionary programming that generalizes the conventional Nix/Vose models.

Although we cannot pursue the argument in detail here, very clearly such an approach to programming highly parallel coevolutionary machines – equivalent to deliberate epigenetic farming – should be applicable to a broad class of artificial biological systems/machines for which some particular ongoing behavior is to be required, rather than some final state 'answer'. Examples might include the manufacture, in a large sense, of a dynamic product, e.g., a chemical substance, anti-cancer or artificial immune search-and-destroy strategy, biological signal detection/transduction process, and so on.

Tunable epigenetic catalysis lowers an 'effective energy' associated with the convergence of a highly coevolutionary cognitive system to a final dynamic behavioral strategy. Given a particular 'farming' information source acting as the program, the behavior of the final state of interest will become associated with the lowest value of the free energy-analog, possibly calculable by optimization methods. If the retina-like rate distortion manifold has been properly implemented, a kind of converse to the no free lunch theorem, then this optimization procedure should converge to an appropriate solution, fixed or dynamic. Thus we invoke a synergism between the focusing theorem and a 'tunable epigenetic catalysis theorem' to raise the probability of an acceptable solution, particularly for a real-time system whose dynamics will be dominated by rate distortion theorem constraints.

The degree of catalysis needed for convergence in a real time system would seem critically dependent on the rate distortion function $R(D)$ or on its product with an acceptable reaction time, τ, that is, on there being sufficient bandwidth in the communication between a cognitive biological 'machine' and its embedding environment. If that bandwidth is too limited, or the available reaction time too short, then the system will inevitably freeze out into what amounts to a highly dysfunctional 'ground state'.

The essential point would seem to be a convergence between emerging needs in biotechnology and general strategies for programming coevolutionary computing devices.

15   Discussion and Conclusions

We have hidden the kind of massive calculations made explicit in [16, 17], burying them as 'fitting regression-model analogs to data', possibly at a second order epigenetic hierarchical level. In the real world such calculations would be quite difficult, particularly given the introduction of punctuated transitions that must be fitted using elaborate renormalization calculations, typically requiring such exotic objects as Lambert W-functions (e.g., [68, 73, 74]).

Analogies with neural network studies suggest, however, intractable conceptual difficulties for spinglass-type models of gene expression and development dynamics, much as claimed by [57]. In spite of nearly a century of sophisticated neural network model studies – including elegant treatments like [66] – Atmanspacher [3] claims that to formulate a serious, clear-cut and transparent formal framework for cognitive neuroscience is a challenge comparable to the early stage of physics four centuries ago. Only a very few contemporary approaches, including that of [68], are worth mentioning, in his view.

Furthermore, [48] has identified what might well be described as the sufficiency failing of neural network models, that is, neural networks can be constructed as Turing machines that can replicate any known dynamic behavior in the same sense that the Ptolemaic Theory of planetary motion, as a Fourier expansion in epicycles, can, to sufficient order, mimic any observed orbit. Keplerian central motion provides an essential reduction. The particular characterization of [48] is that 'neural possibility is not neural plausibility'.

Likewise, [8] concludes that neural-centered explanations of high order mental function commit the mereological fallacy, that is, the fundamental logical error of attributing what is in fact a property of an entirety to a limited part of the whole system. 'The brain' does not exist in isolation, but as part of a complete biological individual who is most often deeply embedded in social and cultural contexts.

Neural network-like models of gene expression and development applied to complex living things inherently commit both errors, particularly in a social, cultural, or environmental milieu. This suggests a particular necessity for the formal inclusion of the effects of embedding contexts – the epigenetic Z and the environmental U – in the sense of [4, 5]. That is, gene expression and development are conditioned by signals from embedding physiological, social, and for humans, cultural, environments. As described above, our formulation can include such influences in a highly natural manner, as they influence epigenetic catalysis. In addition, multiple, and quite different, cognitive gene expression mechanisms may operate simultaneously, or in appropriate sequence, given sufficient development time.

Although epigenetic catalysis, as we have explored it here, might seem worthy of special focus, this would be a kind of intellectual optical illusion akin to inattentional blindness. Epigenetic catalysis is only one aspect of a general cognitive paradigm for gene expression, and this larger, and very complicated 'perceptual field' should remain the center of intellectual attention, rather than any single element of that field. This is to take, perhaps, an 'East Asian' rather than 'Western' perspective on the matter [69].

Developmental disorders, in a broad sense that must include comorbid mental and physical characteristics, emerge as pathological 'resilience' modes, in the sense of [71], a viewpoint from ecosystem theory quite similar to that of epigenetic epidemiology [32, 76]. Environmental farming through an embedding information source affecting internal epigenetic regulation of gene expression, can, as a kind of programming of a highly parallel cognitive system, place the organism into a quasi-stable pathological developmental pattern converging on a dysfunctional phenotype.

The probability models of cognitive process presented here will lead, most fundamentally, to statistical tools based on the asymptotic limit theorems of information theory, in the same sense that the usual parametric statistics are based on the Central Limit Theorem. We have not, then, given 'a' model of development and its disorders in cognitive gene expression, but, rather, outlined a possible general strategy for fitting empirically-determined statistical models to real data, in precisely the sense that one would fit the usual parametric statistical models to normally distributed data.

The fitting of statistical models does not, of itself, perform scientific inference. That is done by comparing fitted models for similar systems under different, or different systems under similar, conditions, and by examining the structure of residuals.

One implication of this work, then, is that understanding complicated processes of gene expression and development – and their pathologies – will require construction of data analysis tools considerably more sophisticated than now available, including the present crop of simple models abducted from neural network studies or stochastic chemical reaction theory. Most centrally, however, currently popular (and fundable) reductionist approaches to understanding gene expression must eventually exhaust themselves in the same desert of sand-grain hyperparticularity that appears to have driven James Crick from molecular biology into consciousness studies, a field now mature enough to provide tools for use in the other direction.

Acknowledgments

The author thanks Dr. C. Guerrero-Bosagna and two anonymous reviewers for comments useful in revision.

References

1. Ash, R.: Information Theory. Dover Publications, New York (1990)
2. Atlan, H., Cohen, I.: Immune information, self-organization, and meaning. International Immunology 10, 711–717 (1998)
3. Atmanspacher, H.: Toward an information theoretical implementation of contextual conditions for consciousness. Acta Biotheoretica 54, 157–160 (2006)
4. Baars, B.: A Cognitive Theory of Consciousness. Cambridge University Press, New York (1988)
5. Baars, B.: Global workspace theory of consciousness: toward a cognitive neuroscience of human experience. Progress in Brain Research 150, 45–53 (2005)
6. Backdahl, L., Bushell, A., Beck, S.: Inflammatory signalling as mediator of epigenetic modulation in tissue-specific chronic inflammation. The International Journal of Biochemistry and Cell Biology (2009), doi:10.1016/j.biocel.2008.08.023

7. Beck, C., Schlogl, F.: Thermodynamics of Chaotic Systems. Cambridge University Press, Cambridge (1995)
8. Bennett, M., Hacker, P.: Philosophical Foundations of Neuroscience. Blackwell Publishing, Malden (2003)
9. Bennett, C.: Logical depth and physical complexity. In: Herkin, R. (ed.) The Universal Turing Machine: A Half-Century Survey, pp. 227–257. Oxford University Press, Oxford (1988)
10. Bos, R.: Continuous representations of groupoids. arXiv:math/0612639 (2007)
11. Bossdorf, O., Richards, C., Pigliucci, M.: Epigenetics for ecologists. Ecology Letters 11, 106–115 (2008)
12. Britten, R., Davidson, E.: Gene regulation for higher cells: a theory. Science 165, 349–357 (1969)
13. Brown, R.: From groups to groupoids: a brief survey. Bulletin of the London Mathematical Society 19, 113–134 (1987)
14. Buneci, M.: Representare de Groupoizi. Editura Mirton, Timisoara (2003)
15. Cannas Da Silva, A., Weinstein, A.: Geometric Models for Noncommutative Algebras. American Mathematical Society, RI (1999)
16. Ciliberti, S., Martin, O., Wagner, A.: Robustness can evolve gradually in complex regulatory networks with varying topology. PLoS Computational Biology 3(2), e15 (2007)
17. Ciliberti, S., Martin, O., Wagner, A.: Innovation and robustness in complex regulatory gene networks. Proceedings of the National Academy of Sciences 104, 13591–13596 (2007)
18. Cohen, I.: Immune system computation and the immunological homunculus. In: Nierstrasz, O., Whittle, J., Harel, D., Reggio, G. (eds.) MoDELS 2006. LNCS, vol. 4199, pp. 499–512. Springer, Heidelberg (2006)
19. Cohen, I., Harel, D.: Explaining a complex living system: dynamics, multi-scaling, and emergence. Journal of the Royal Society: Interface 4, 175–182 (2007)
20. Cover, T., Thomas, J.: Elements of Information Theory. John Wiley and Sons, New York (1991)
21. Crews, D., McLachlan, J.A.: Epigenetics, evolution, endocrine disruption, health, and disease. Endocrinology 147, S4–S10 (2006)
22. Crews, D., Gore, A., Hsu, T., Dangleben, N., Spinetta, M., Schallert, T., Anway, M., Skinner, M.: Transgenerational epigenetic imprints on mate preference. Proceedings of the National Academy of Sciences 104, 5942–5946 (2007)
23. Dehaene, S., Naccache, L.: Towards a cognitive neuroscience of consciousness: basic evidence and a workspace framework. Cognition 79, 1–37 (2001)
24. Dembo, A., Zeitouni, O.: Large Deviations: Techniques and Applications, 2nd edn. Springer, New York (1998)
25. Dias, A., Stewart, I.: Symmetry groupoids and admissible vector fields for coupled cell networks. Journal of the London Mathematical Society 69, 707–736 (2004)
26. Dretske, F.: The explanatory role of information. Philosophical Transactions of the Royal Society A 349, 59–70 (1994)
27. Emery, M.: Stochastic Calculus on Manifolds. Springer, New York (1989)
28. English, T.: Evaluation of evolutionary and genetic optimizers: no free lunch. In: Fogel, L., Angeline, P., Back, T. (eds.) Evolutionary Programming V: Proceedings of the Fifth Annual Conference on Evolutionary Programming, pp. 163–169. MIT Press, Cambridge (1996)
29. Erdos, P., Renyi, A.: On the evolution of random graphs (1960); reprinted in The Art of Counting, pp. 574–618 (1973), and in Selected Papers of Alfred Renyi, pp. 482–525 (1976)

30. Ficici, S., Milnik, O., Pollak, J.: A game-theoretic and dynamical systems analysis of selection methods in coevolution. IEEE Transactions on Evolutionary Computation 9, 580–602 (2005)
31. Feynman, R.: Lectures on Computation. Westview Press, New York (2000)
32. Foley, D., Craid, J., Morley, R., Olsson, C., Dwyer, T., Smith, K., Saffery, R.: Prospects for epigenetic epidemiology. American Journal of Epidemiology 169, 389–400 (2009)
33. Gilbert, S.: Mechanisms for the environmental regulation of gene expression: ecological aspects of animal development. Journal of Bioscience 30, 65–74 (2001)
34. Glazebrook, J.F., Wallace, R.: Small worlds and red queens in the global workspace: an information-theoretic approach. Cognitive Systems Reserch (2009), doi:10.1016/j.cogsys.2009.01.002
35. Golubitsky, M., Stewart, I.: Nonlinear dynamics and networks: the groupoid formalism. Bulletin of the American Mathematical Society 43, 305–364 (2006)
36. Goubault, E., Raussen, M.: Dihomotopy as a tool in state space analysis. In: Rajsbaum, S. (ed.) LATIN 2002. LNCS, vol. 2286, pp. 16–37. Springer, Heidelberg (2002)
37. Goubault, E.: Some geometric perspectives on concurrency theory. Homology, Homotopy, and Applications 5, 95–136 (2003)
38. Gould, S.: The Structure of Evolutionary Theory. Harvard University Press, Cambridge (2002)
39. Guerrero-Bosagna, C., Sabat, P., Valladares, L.: Environmental signaling and evolutionary change: can exposure of pregnant mammals to environmental estrogens lead to epigenetically induced evolutionary changes in embryos? Evolution and Development 7, 341–350 (2005)
40. Holling, C.: Cross-scale morphology, geometry and dynamicsl of ecosystems. Ecological Monographs 41, 1–50 (1992)
41. Jablonka, E., Lamb, M.: Epigenetic Inheritance and Evolution: The Lamarckian Dimension. Oxford University Press, Oxford (1995)
42. Jablonka, E., Lamb, M.J.: Epigenetic inheritance in evolution. Journal of Evolutionary Biology 11, 159–183 (1998)
43. Jablonka, E.: Epigenetic epidemiology. International Journal of Epidemiology 33, 929–935 (2004)
44. Jaeger, J., Surkova, S., Blagov, M., Janssens, H., Kosman, D., Kozlov, K., Manu, M., Myasnikova, E., Vanario-Alonso, C., Samsonova, M., Sharp, D., Reintiz, J.: Dynamic control of positional information in the early Drosophila embryo. Nature 430, 368–371 (2004)
45. Jaenisch, R., Bird, A.: Epigenetic regulation of gene expression: how the genome integrates intrinsic and environmental signals. Nature Genetics Supplement 33, 245–254 (2003)
46. Kastner, M.: Phase transitions and configuration space topology. ArXiv cond-mat/0703401 (2006)
47. Khinchin, A.: Mathematical Foundations of Information Theory. Dover, New York (1957)
48. Krebs, P.: Models of cognition: neurological possibility does not indicate neurological plausibility. In: Bara, B., Barsalou, L., Bucciarelli, M. (eds.) Proceedings of CogSci 2005, Stresa, Italy, pp. 1184–1189 (2005), http://cogprints.org/4498/
49. Landau, L., Lifshitz, E.: Statistical Physics, Part I, 3rd edn., Part I. Elsevier, New York (2007)
50. Lee, J.: Introduction to topological manifolds. Springer, New York (2000)

51. Maas, W., Natschlager, T., Markram, H.: Real-time computing without stable states: a new framework for neural computation based on perturbations. Neural Computation 14, 2531–2560 (2002)
52. Matsumoto, Y.: An Introduction to Morse Theory. American Mathematical Society, Providence (2002)
53. Maturana, H.R., Varela, F.J.: Autopoiesis and Cognition. Reidel Publishing Company, Dordrecht (1980)
54. Maturana, H.R., Varela, F.J.: The Tree of Knowledge. Shambhala Publications, Boston (1992)
55. McCauly, J.: Chaos, Dynamics, and Fractals. Cambridge Nonlinear Science Series, Cambridge, UK (1994)
56. Mjolsness, E., Sharp, D., Reinitz, J.: A connectionist model of development. Journal of Theoretical Biology 152, 429–458 (1991)
57. O'Nuallain, S.: Code and context in gene expression, cognition, and consciousness. In: Barbiere, M. (ed.) The Codes of Life: The Rules of Macroevolution, ch. 15, pp. 347–356. Springer, New York (2008)
58. O'Nuallain, S., Strohman, R.: Genome and natural language: how far can the analogy be extended? In: Witzany, G. (ed.) Proceedings of Biosemiotics. Tartu University Press, Umweb (2007)
59. Pettini, M.: Geometry and Topology in Hamiltonian Dynamics and Statistical Mechanics. Springer, New York (2007)
60. Pratt, V.: Modeling concurrency with geometry. In: Proceedings of the 18th ACM SIGPLAN-SIGACT Symposium on Principles of Programming Languages, pp. 311–322 (1991)
61. Reinitz, J., Sharp, D.: Mechanisms of even stripe formation. Mechanics of Development 49, 133–158 (1995)
62. Scherrer, K., Jost, J.: The gene and the genon concept: a functional and information-theoretic analysis. Molecular Systems Biology 3, 87–95 (2007)
63. Scherrer, K., Jost, J.: Gene and genon concept: coding versus regulation. Theory in Bioscience 126, 65–113 (2007)
64. Sharp, D., Reinitz, J.: Prediction of mutant expression patterns using gene circuits. BioSystems 47, 79–90 (1998)
65. Skierski, M., Grundland, A., Tuszynski, J.: Analysis of the three-dimensional time-dependent Landau-Ginzburg equation and its solutions. Journal of Physics A (Math. Gen.) 22, 3789–3808 (1989)
66. Toulouse, G., Dehaene, S., Changeux, J.: Spin glass model of learning by selection. Proceedings of the National Academy of Sciences 83, 1695–1698 (1986)
67. Turner, B.: Histone acetylation and an epigeneticv code. Bioessays 22, 836–845 (2000)
68. Wallace, R.: Consciousness: A Mathematical Treatment of the Global Neuronal Workspace Model. Springer, New York (2005)
69. Wallace, R.: Culture and inattentional blindness. Journal of Theoretical Biology 245, 378–390 (2007)
70. Wallace, R.: Toward formal models of biologically inspired, highly parallel machine cognition. International Journal of Parallel, Emergent, and Distributed Systems 23, 367–408 (2008)
71. Wallace, R.: Developmental disorders as pathological resilience domains. Ecology and Society 13(1), 29 (2008), http://www.ecologyandsociety.org/vol13/iss1/art29/
72. Wallace, R.: Programming coevolutionary machines: the emerging conundrum. International Journal of Parallel, Emergent, and Distributed Systems (in press, 2009)

73. Wallace, R., Fullilove, M.: Collective Consciousness and its Discontents: Institutional Distributed Cognition, Racial Policy, and Public Health in the United States. Springer, New York (2008)
74. Wallace, R., Wallace, D.: Punctuated equilibrium in statistical models of generalized coevolutionary resilience: how sudden ecosystem transitions can entrain both phenotype expression and Darwinian selection. In: Priami, C. (ed.) Transactions on Computational Systems Biology IX. LNCS (LNBI), vol. 5121, pp. 23–85. Springer, Heidelberg (2008)
75. Wallace, R.G., Wallace, R.: Evolutionary radiation and the spectrum of consciousness. Consciousness and Cognition (2009), doi:10.1016/j.concog.2008.12.002
76. Waterland, R., Michels, K.: Epigenetic epidemiology of the developmental origins hypothesis. Annual Reviews of Nutrition 27, 363–388 (2007)
77. Weaver, I.: Epigenetic effects of glucocorticoids. Seminars in Fetal and Neonatal Medicine (2009), doi:10.1016/j.siny.2008.12.002
78. Weinstein, A.: Groupoids: unifying internal and external symmetry. Notices of the American Mathematical Association 43, 744–752 (1996)
79. West-Eberhard, M.: Developmental plasticity and the origin of species differences. Proceedings of the National Academy of Sciences 102, 6543–6549 (2005)
80. Wiegand, R.: An analysis of cooperative coevolutionary algorithms. PhD Thesis, George Mason University (2003)
81. Wolpert, D., Macready, W.: No free lunch theorems for search. Santa Fe Institute, SFI-TR-02-010 (1995)
82. Wolpert, D., Macready, W.: No free lunch theorems for optimization. IEEE Transactions on Evolutionary Computation 1, 67–82 (1997)
83. Wymer, C.R.: Structural nonlinear continuous-time models in econometrics. Macroeconomic Dynamics 1, 518–548 (1997)
84. Zhu, R., Rebirio, A., Salahub, D., Kaufmann, S.: Studying genetic regulatory networks at the molecular level: delayed reaction stochastic models. Journal of Theoretical Biology 246, 725–745 (2007)

16 Mathematical Appendix

16.1 The Shannon-McMillan Theorem

According to the structure of the underlying language of which a message is a particular expression, some messages are more 'meaningful' than others, that is, are in accord with the grammar and syntax of the language. The Shannon-McMillan or Asymptotic Equipartition Theorem, describes how messages themselves are to be classified.

Suppose a long sequence of symbols is chosen, using the output of a random variable X, so that an output sequence of length n, with the form

$$x_n = (\alpha_0, \alpha_1, ..., \alpha_{n-1})$$

has joint and conditional probabilities

$$P(X_0 = \alpha_0, X_1 = \alpha_1, ..., X_{n-1} = \alpha_{n-1})$$

$$P(X_n = \alpha_n | X_0 = \alpha_0, ..., X_{n-1} = \alpha_{n-1}).$$

Using these probabilities we may calculate the conditional uncertainty

$$H(X_n|X_0, X_1, ..., X_{n-1}).$$

The uncertainty of the *information source*, $H[\mathbf{X}]$, is defined as

$$H[\mathbf{X}] = \lim_{n \to \infty} H(X_n|X_0, X_1, ..., X_{n-1}). \tag{21}$$

In general

$$H(X_n|X_0, X_1, ..., X_{n-1}) \leq H(X_n).$$

Only if the random variables X_j are all stochastically independent does equality hold. If there is a maximum n such that, for all $m > 0$

$$H(X_{n+m}|X_0, ..., X_{n+m-1}) = H(X_n|X_0, ..., X_{n-1}),$$

then the source is said to be of *order* n. It is easy to show that

$$H[\mathbf{X}] = \lim_{n \to \infty} \frac{H(X_0, ...X_n)}{n+1}.$$

In general the outputs of the $X_j, j = 0, 1, ..., n$ are *dependent*. That is, the output of the communication process at step n depends on previous steps. Such serial correlation, in fact, is the very structure which enables most of what is done in this paper.

Here, however, the processes are all assumed statble in time, that is, the probabilities and serial correlations do not change in time, and the system is *stationary*.

A very broad class of such self-correlated, stationary, information sources, the so-called *ergodic* sources for which the long-run relative frequency of a sequence converges stochastically to the probability assigned to it, have a particularly interesting property:

It is possible, in the limit of large n, to divide all sequences of outputs of an ergodic information source into two distinct sets, S_1 and S_2, having, respectively, very high and very low probabilities of occurrence, with the source uncertainty providing the splitting criterion. In particular the Shannon-McMillan Theorem states that, for a (long) sequence having n (serially correlated) elements, the number of 'meaningful' sequences, $N(n)$ – those belonging to set S_1 – will satisfy the relation

$$\frac{\log[N(n)]}{n} \approx H[\mathbf{X}]. \tag{22}$$

More formally,

$$\lim_{n \to \infty} \frac{\log[N(n)]}{n} = H[\mathbf{X}]$$

$$= \lim_{n \to \infty} H(X_n|X_0, ..., X_{n-1})$$

$$= \lim_{n \to \infty} \frac{H(X_0, ..., X_n)}{n+1}. \tag{23}$$

Using the internal structures of the information source permits *limiting attention only to high probability 'meaningful' sequences of symbols.*

16.2   The Rate Distortion Theorem

The Shannon-McMillan Theorem can be expressed as the 'zero error limit' of the Rate Distortion Theorem [20, 24] which defines a splitting criterion that identifies high probability pairs of sequences. We follow closely the treatment of [20].

The origin of the problem is the question of representing one information source by a simpler one in such a way that the least information is lost. For example we might have a continuous variate between 0 and 100, and wish to represent it in terms of a small set of integers in a way that minimizes the inevitable distortion that process creates. Typically, for example, an analog audio signal will be replaced by a 'digital' one. The problem is to do this in a way which least distorts the *reconstructed* audio waveform.

Suppose the original stationary, ergodic information source Y with output from a particular alphabet generates sequences of the form

$$y^n = y_1, ..., y_n.$$

These are 'digitized,' in some sense, producing a chain of 'digitized values'

$$b^n = b_1, ..., b_n,$$

where the b-alphabet is much more restricted than the y-alphabet.

b^n is, in turn, *deterministically retranslated* into a reproduction of the original signal y^n. That is, each b^m is mapped on to a unique n-length y-sequence in the alphabet of the information source Y:

$$b^m \to \hat{y}^n = \hat{y}_1, ..., \hat{y}_n.$$

Note, however, that many y^n sequences may be mapped onto the *same* retranslation sequence $\hat{y}^n$, so that information will, in general, be lost.

The central problem is to explicitly minimize that loss.

The retranslation process defines a new stationary, ergodic information source, $\hat{Y}$.

The next step is to define a *distortion measure*, $d(y, \hat{y})$, which compares the original to the retranslated path. For example the *Hamming distortion* is

$$d(y, \hat{y}) = 1, y \neq \hat{y}$$

$$d(y, \hat{y}) = 0, y = \hat{y}. \tag{24}$$

For continuous variates the *Squared error distortion* is

$$d(y, \hat{y}) = (y - \hat{y})^2. \tag{25}$$

There are many possibilities.

The distortion between paths y^n and $\hat{y}^n$ is defined as

$$d(y^n, \hat{y}^n) = \frac{1}{n} \sum_{j=1}^{n} d(y_j, \hat{y}_j). \tag{26}$$

Suppose that with each path y^n and b^n-path retranslation into the y-language and denoted y^n, there are associated individual, joint, and conditional probability distributions

$$p(y^n), p(\hat{y}^n), p(y^n | \hat{y}^n).$$

The *average distortion* is defined as

$$D = \sum_{y^n} p(y^n) d(y^n, \hat{y}^n). \tag{27}$$

It is possible, using the distributions given above, to define the information transmitted from the incoming Y to the outgoing $\hat{Y}$ process in the usual manner, using the Shannon source uncertainty of the strings:

$$I(Y, \hat{Y}) = H(Y) - H(Y | \hat{Y}) = H(Y) + H(\hat{Y}) - H(Y, \hat{Y}).$$

If there is no uncertainty in Y given the retranslation $\hat{Y}$, then no information is lost.

In general, this will not be true.

The *information rate distortion function* $R(D)$ for a source Y with a distortion measure $d(y, \hat{y})$ is defined as

$$R(D) = \min_{p(y, \hat{y}); \sum_{(y, \hat{y})} p(y) p(y | \hat{y}) d(y, \hat{y}) \leq D} I(Y, \hat{Y}). \tag{28}$$

The minimization is over all conditional distributions $p(y | \hat{y})$ for which the joint distribution $p(y, \hat{y}) = p(y) p(y | \hat{y})$ satisfies the average distortion constraint (i.e., average distortion $\leq D$).

The *Rate Distortion Theorem* states that $R(D)$ *is the maximum achievable rate of information transmission which does not exceed the distortion* D. See [20, 24] details.

More to the point, however, is the following: Pairs of sequences $(y^n, \hat{y}^n)$ can be defined as *distortion typical*; that is, for a given average distortion D, defined in terms of a particular measure, pairs of sequences can be divided into two sets, a high probability one containing a relatively small number of (matched) pairs with $d(y^n, \hat{y}^n) \leq D$, and a low probability one containing most pairs. As $n \to \infty$, the smaller set approaches unit probability, and, for those pairs,

$$p(y^n) \geq p(\hat{y}^n | y^n) \exp[-nI(Y, \hat{Y})]. \tag{29}$$

Thus, roughly speaking, $I(Y, \hat{Y})$ embodies the splitting criterion between high and low probability pairs of paths.

For the theory of interacting information sources, then, $I(Y, \hat{Y})$ can play the role of H in the dynamic treatment above.

The rate distortion function can actually be calculated in many cases by using a Lagrange multiplier method – see Section 13.7 of [20].

16.3 Groupoids

Basic ideas. Following [78] closely, a groupoid, G, is defined by a base set A upon which some mapping – a morphism – can be defined. Note that not all possible pairs of states (a_j, a_k) in the base set A can be connected by such a morphism. Those that can define the groupoid element, a morphism $g = (a_j, a_k)$ having the natural inverse $g^{-1} = (a_k, a_j)$. Given such a pairing, it is possible to define 'natural' end-point maps $\alpha(g) = a_j, \beta(g) = a_k$ from the set of morphisms G into A, and a formally associative product in the groupoid $g_1 g_2$ provided $\alpha(g_1 g_2) = \alpha(g_1), \beta(g_1 g_2) = \beta(g_2)$, and $\beta(g_1) = \alpha(g_2)$. Then the product is defined, and associative, $(g_1 g_2) g_3 = g_1 (g_2 g_3)$.

In addition, there are natural left and right identity elements λ_g, ρ_g such that $\lambda_g g = g = g \rho_g$ [78].

An orbit of the groupoid G over A is an equivalence class for the relation $a_j \sim Ga_k$ if and only if there is a groupoid element g with $\alpha(g) = a_j$ and $\beta(g) = a_k$. Following [15], we note that a groupoid is called transitive if it has just one orbit. The transitive groupoids are the building blocks of groupoids in that there is a natural decomposition of the base space of a general groupoid into orbits. Over each orbit there is a transitive groupoid, and the disjoint union of these transitive groupoids is the original groupoid. Conversely, the disjoint union of groupoids is itself a groupoid.

The isotropy group of $a \in X$ consists of those g in G with $\alpha(g) = a = \beta(g)$. These groups prove fundamental to classifying groupoids.

If G is any groupoid over A, the map $(\alpha, \beta) : G \to A \times A$ is a morphism from G to the pair groupoid of A. The image of (α, β) is the orbit equivalence relation $\sim G$, and the functional kernel is the union of the isotropy groups. If $f : X \to Y$ is a function, then the kernel of f, $ker(f) = [(x_1, x_2) \in X \times X : f(x_1) = f(x_2)]$ defines an equivalence relation.

Groupoids may have additional structure. As [78] explains, a groupoid G is a topological groupoid over a base space X if G and X are topological spaces and α, β and multiplication are continuous maps. A criticism sometimes applied to groupoid theory is that their classification up to isomorphism is nothing other than the classification of equivalence relations via the orbit equivalence relation and groups via the isotropy groups. The imposition of a compatible topological structure produces a nontrivial interaction between the two structures. It is possible to introduce a metric structure on manifolds of related information sources, producing such interaction.

In essence, a groupoid is a category in which all morphisms have an inverse, here defined in terms of connection to a base point by a meaningful path of an information source dual to a cognitive process.

As [78] points out, the morphism (α, β) suggests another way of looking at groupoids. A groupoid over A identifies not only which elements of A are equivalent to one another (isomorphic), but *it also parametizes the different ways (isomorphisms) in which two elements can be equivalent*, i.e., all possible information sources dual to some cognitive process. Given the information theoretic characterization of cognition presented above, this produces a full modular cognitive network in a highly natural manner.

Brown [13] describes the fundamental structure as follows:

> A groupoid should be thought of as a group with many objects, or with many identities... A groupoid with one object is essentially just a group. So the notion of groupoid is an extension of that of groups. It gives an additional convenience, flexibility and range of applications...
>
> EXAMPLE 1. A disjoint union [of groups] $G = \cup_\lambda G_\lambda, \lambda \in \Lambda$, is a groupoid: the product ab is defined if and only if a, b belong to the same G_λ, and ab is then just the product in the group G_λ. There is an identity 1_λ for each $\lambda \in \Lambda$. The maps α, β coincide and map G_λ to λ, $\lambda \in \Lambda$.
>
> EXAMPLE 2. An equivalence relation R on [a set] X becomes a groupoid with $\alpha, \beta : R \to X$ the two projections, and product $(x, y)(y, z) = (x, z)$ whenever $(x, y), (y, z) \in R$. There is an identity, namely (x, x), for each $x \in X$...

[78] makes the following fundamental point:

> Almost every interesting equivalence relation on a space B arises in a natural way as the orbit equivalence relation of some groupoid G over B. Instead of dealing directly with the orbit space B/G as an object in the category S_{map} of sets and mappings, one should consider instead the groupoid G itself as an object in the category G_{htp} of groupoids and homotopy classes of morphisms.

The groupoid approach has become quite popular in the study of networks of coupled dynamical systems which can be defined by differential equation models, [35].

Global and local symmetry groupoids. Here we follow [78] fairly closely, using the example of a finite tiling.

Consider a tiling of the euclidean plane R^2 by identical 2 by 1 rectangles, specified by the set X (one dimensional) where the grout between tiles is $X = H \cup V$, having $H = R \times Z$ and $V = 2Z \times R$, where R is the set of real numbers and Z the integers. Call each connected component of $R^2 \backslash X$, that is, the complement of the two dimensional real plane intersecting X, a tile.

Let Γ be the group of those rigid motions of R^2 which leave X invariant, i.e., the normal subgroup of translations by elements of the lattice $\Lambda = H \cap V =$

$2Z \times Z$ (corresponding to corner points of the tiles), together with reflections through each of the points $1/2\Lambda = Z \times 1/2Z$, and across the horizontal and vertical lines through those points. As noted in [78], much is lost in this coarse-graining, in particular the same symmetry group would arise if we replaced X entirely by the lattice Λ of corner points. Γ retains no information about the local structure of the tiled plane. In the case of a real tiling, restricted to the finite set $B = [0, 2m] \times [0, n]$ the symmetry group shrinks drastically: The subgroup leaving $X \cap B$ invariant contains just four elements even though a repetitive pattern is clearly visible. A two-stage groupoid approach recovers the lost structure.

We define the transformation groupoid of the action of Γ on R^2 to be the set

$$G(\Gamma, R^2) = \{(x, \gamma, y | x \in R^2, y \in R^2, \gamma \in \Gamma, x = \gamma y\},$$

with the partially defined binary operation

$$(x, \gamma, y)(y, \nu, z) = (x, \gamma\nu, z).$$

Here $\alpha(x, \gamma, y) = x$, and $\beta(x, \gamma, y) = y$, and the inverses are natural.

We can form the restriction of G to B (or any other subset of R^2) by defining

$$G(\Gamma, R^2)|_B = \{g \in G(\Gamma, R^2) | \alpha(g), \beta(g) \in B\}$$

1. An orbit of the groupoid G over B is an equivalence class for the relation $x \sim_G y$ if and only if there is a groupoid element g with $\alpha(g) = x$ and $\beta(g) = y$.

Two points are in the same orbit if they are similarly placed within their tiles or within the grout pattern.

2. The isotropy group of $x \in B$ consists of those g in G with $\alpha(g) = x = \beta(g)$. It is trivial for every point except those in $1/2\Lambda \cap B$, for which it is $Z_2 \times Z_2$, the direct product of integers modulo two with itself.

By contrast, embedding the tiled structure within a larger context permits definition of a much richer structure, i.e., the identification of local symmetries.

We construct a second groupoid as follows. Consider the plane R^2 as being decomposed as the disjoint union of $P_1 = B \cap X$ (the grout), $P_2 = B \backslash P_1$ (the complement of P_1 in B, which is the tiles), and $P_3 = R^2 \backslash B$ (the exterior of the tiled room). Let E be the group of all euclidean motions of the plane, and define the local symmetry groupoid G_{loc} as the set of triples (x, γ, y) in $B \times E \times B$ for which $x = \gamma y$, and for which y has a neighborhood $\mathcal{U}$ in R^2 such that $\gamma(\mathcal{U} \cap P_i) \subseteq P_i$ for $i = 1, 2, 3$. The composition is given by the same formula as for $G(\Gamma, R^2)$.

For this groupoid-in-context there are only a finite number of orbits:

$\mathcal{O}_1$ = interior points of the tiles.
$\mathcal{O}_2$ = interior edges of the tiles.
$\mathcal{O}_3$ = interior crossing points of the grout.
$\mathcal{O}_4$ = exterior boundary edge points of the tile grout.

$\mathcal{O}_5$ = boundary 'T' points.

$\mathcal{O}_6$ = boundary corner points.

The isotropy group structure is, however, now very rich indeed:

The isotropy group of a point in $\mathcal{O}_1$ is now isomorphic to the entire rotation group O_2.

It is $Z_2 \times Z_2$ for $\mathcal{O}_2$.

For $\mathcal{O}_3$ it is the eight-element dihedral group D_4.

For $\mathcal{O}_4, \mathcal{O}_5$ and $\mathcal{O}_6$ it is simply Z_2.

These are the 'local symmetries' of the tile-in-context.

16.4   Morse Theory

Morse theory examines relations between analytic behavior of a function – the location and character of its critical points – and the underlying topology of the manifold on which the function is defined. We are interested in a number of such functions, for example information source uncertainty on a parameter space and 'second order' iterations involving parameter manifolds determining critical behavior, for example sudden onset of a giant component in the mean number model [74], and universality class tuning in the mean field model of the next section. These can be reformulated from a Morse theory perspective. Here we follow closely the elegant treatments of [46, 59].

The essential idea of Morse theory is to examine an n-dimensional manifold M as decomposed into level sets of some function $f : M \rightarrow \mathbf{R}$ where $\mathbf{R}$ is the set of real numbers. The a-level set of f is defined as

$$f^{-1}(a) = \{x \in M : f(x) = a\},$$

the set of all points in M with $f(x) = a$. If M is compact, then the whole manifold can be decomposed into such slices in a canonical fashion between two limits, defined by the minimum and maximum of f on M. Let the part of M below a be defined as

$$M_a = f^{-1}(-\infty, a] = \{x \in M : f(x) \leq a\}.$$

These sets describe the whole manifold as a varies between the minimum and maximum of f.

Morse functions are defined as a particular set of smooth functions $f : M \rightarrow \mathbf{R}$ as follows. Suppose a function f has a critical point x_c, so that the derivative $df(x_c) = 0$, with critical value $f(x_c)$. Then f is a Morse function if its critical points are nondegenerate in the sense that the Hessian matrix of second derivatives at x_c, whose elements, in terms of local coordinates are

$$\mathcal{H}_{i,j} = \partial^2 f / \partial x^i \partial x^j,$$

has rank n, which means that it has only nonzero eigenvalues, so that there are no lines or surfaces of critical points and, ultimately, critical points are isolated.

The index of the critical point is the number of negative eigenvalues of $\mathcal{H}$ at x_c.

A level set $f^{-1}(a)$ of f is called a critical level if a is a critical value of f, that is, if there is at least one critical point $x_c \in f^{-1}(a)$.

Again following [59], the essential results of Morse theory are:

1. If an interval $[a, b]$ contains no critical values of f, then the topology of $f^{-1}[a, v]$ does not change for any $v \in (a, b)$. Importantly, the result is valid even if f is not a Morse function, but only a smooth function.

2. If the interval $[a, b]$ contains critical values, the topology of $f^{-1}[a, v]$ changes in a manner determined by the properties of the matrix H at the critical points.

3. If $f : M \to \mathbf{R}$ is a Morse function, the set of all the critical points of f is a discrete subset of M, i.e., critical points are isolated. This is Sard's Theorem.

4. If $f : M \to \mathbf{R}$ is a Morse function, with M compact, then on a finite interval $[a, b] \subset \mathbf{R}$, there is only a finite number of critical points p of f such that $f(p) \in [a, b]$. The set of critical values of f is a discrete set of $\mathbf{R}$.

5. For any differentiable manifold M, the set of Morse functions on M is an open dense set in the set of real functions of M of differentiability class r for $0 \le r \le \infty$.

6. Some topological invariants of M, that is, quantities that are the same for all the manifolds that have the same topology as M, can be estimated and sometimes computed exactly once all the critical points of f are known: Let the Morse numbers $\mu_i (i = 0, ..., m)$ of a function f on M be the number of critical points of f of index i, (the number of negative eigenvalues of H). The Euler characteristic of the complicated manifold M can be expressed as the alternating sum of the Morse numbers of any Morse function on M,

$$\chi = \sum_{i=1}^{m} (-1)^i \mu_i.$$

The Euler characteristic reduces, in the case of a simple polyhedron, to

$$\chi = V - E + F$$

where V, E, and F are the numbers of vertices, edges, and faces in the polyhedron.

7. Another important theorem states that, if the interval $[a, b]$ contains a critical value of f with a single critical point x_c, then the topology of the set M_b defined above differs from that of M_a in a way which is determined by the index, i, of the critical point. Then M_b is homeomorphic to the manifold obtained from attaching to M_a an i-handle, i.e., the direct product of an i-disk and an $(m - i)$-disk.

Again, see [52, 59] for details.

16.5   Generalized Onsager Theory

Understanding the time dynamics of groupoid-driven information systems away from phase transition critical points requires a phenomenology similar to the

Onsager relations of nonequilibrium thermodynamics. This also leads to a general theory involving large-scale topological changes in the sense of Morse theory.

If the Groupoid Free Energy (GFE) of a biological process is parametized by some vector of quantities $\mathbf{K} = (K_1, ..., K_m)$, then, in analogy with nonequilibrium thermodynamics, gradients in the K_j of the *disorder*, defined as

$$S_G = F_G(\mathbf{K}) - \sum_{j=1}^{m} K_j \partial F_G / \partial K_j \qquad (30)$$

become of central interest.

Equation (30) is similar to the definition of entropy in terms of the free energy of a physical system. Pursuing the homology further, the generalized Onsager relations defining temporal dynamics of systems having a GFE become

$$dK_j / dt = \sum_{i} L_{j,i} \partial S_G / \partial K_i, \qquad (31)$$

where the $L_{j,i}$ are, in first order, constants reflecting the nature of the underlying cognitive phenomena. The L-matrix is to be viewed empirically, in the same spirit as the slope and intercept of a regression model, and may have structure far different than familiar from more simple chemical or physical processes. The $\partial S_G / \partial K$ are analogous to thermodynamic forces in a chemical system, and may be subject to override by external physiological or other driving mechanisms: biological and cognitive phenomena, unlike simple physical systems, can make choices as to resource allocation.

That is, an essential contrast with simple physical systems driven by (say) entropy maximization is that complex biological or cognitive structures can make decisions about resource allocation, to the extent resources are available. Thus resource availability is a context, not a determinant, of behavior.

Equations (30) and (31) can be derived in a simple parameter-free covariant manner which relies on the underlying topology of the information source space implicit to the development [74]. We will not pursue that development here.

The dynamics, as we have presented them so far, have been noiseless, while biological systems are always very noisy. Equation (31) might be rewritten as

$$dK_j / dt = \sum_{i} L_{j,i} \partial S_G / \partial K_i + \sigma W(t)$$

where σ is a constant and $W(t)$ represents white noise. This leads directly to a family of classic stochastic differential equations having the form

$$dK_t^j = L^j(t, \mathbf{K})dt + \sigma^j(t, \mathbf{K})dB_t, \qquad (32)$$

where the L^j and σ^j are appropriately regular functions of t and $\mathbf{K}$, and dB_t represents the noise structure, and we have readjusted the indices.

Further progress in this direction requires introduction of methods from stochastic differential geometry and related topics in the sense of [27]. The obvious inference is that noise – not necessarily 'white' – can serve as a tool to shift the system between various topological modes, as a kind of crosstalk and the source of a generalized stochastic resonance.

Effectively, topological shifts between and within dynamic manifolds constitute another theory of phase transitions [59], and this phenomenological Onsager treatment would likely be much enriched by explicit adoption of a Morse theory perspective.

16.6   The Tuning Theorem

Messages from an information source, seen as symbols x_j from some alphabet, each having probabilities P_j associated with a random variable X, are 'encoded' into the language of a 'transmission channel', a random variable Y with symbols y_k, having probabilities P_k, possibly with error. Someone receiving the symbol y_k then retranslates it (without error) into some x_k, which may or may not be the same as the x_j that was sent.

More formally, the message sent along the channel is characterized by a random variable X having the distribution

$$P(X = x_j) = P_j, j = 1, ..., M.$$

The channel through which the message is sent is characterized by a second random variable Y having the distribution

$$P(Y = y_k) = P_k, k = 1, ..., L.$$

Let the joint probability distribution of X and Y be defined as

$$P(X = x_j, Y = y_k) = P(x_j, y_k) = P_{j,k}$$

and the conditional probability of Y given X as

$$P(Y = y_k | X = x_j) = P(y_k | x_j).$$

Then the Shannon uncertainty of X and Y independently and the joint uncertainty of X and Y together are defined respectively as

$$H(X) = -\sum_{j=1}^{M} P_j \log(P_j)$$

$$H(Y) = -\sum_{k=1}^{L} P_k \log(P_k)$$

$$H(X, Y) = -\sum_{j=1}^{M} \sum_{k=1}^{L} P_{j,k} \log(P_{j,k}). \tag{33}$$

The *conditional uncertainty* of Y given X is defined as

$$H(Y|X) = -\sum_{j=1}^{M}\sum_{k=1}^{L} P_{j,k} \log[P(y_k|x_j)]. \tag{34}$$

For any two stochastic variates X and Y, $H(Y) \geq H(Y|X)$, as knowledge of X generally gives some knowledge of Y. Equality occurs only in the case of stochastic independence.

Since $P(x_j, y_k) = P(x_j)P(y_k|x_j)$, we have

$$H(X|Y) = H(X,Y) - H(Y).$$

The information transmitted by translating the variable X into the channel transmission variable Y – possibly with error – and then retranslating without error the transmitted Y back into X is defined as

$$I(X|Y) = H(X) - H(X|Y) = H(X) + H(Y) - H(X,Y) \tag{35}$$

See, for example, [1, 20, 47] for details. The essential point is that if there is no uncertainty in X given the channel Y, then there is no loss of information through transmission. In general this will not be true, and herein lies the essence of the theory.

Given a fixed vocabulary for the transmitted variable X, and a fixed vocabulary and probability distribution for the channel Y, we may vary the probability distribution of X in such a way as to maximize the information sent. The capacity of the channel is defined as

$$C = \max_{P(X)} I(X|Y) \tag{36}$$

subject to the subsidiary condition that $\sum P(X) = 1$.

The critical trick of the Shannon Coding Theorem for sending a message with arbitrarily small error along the channel Y at any rate $R < C$ is to encode it in longer and longer 'typical' sequences of the variable X; that is, those sequences whose distribution of symbols approximates the probability distribution $P(X)$ above which maximizes C.

If $S(n)$ is the number of such 'typical' sequences of length n, then

$$\log[S(n)] \approx nH(X),$$

where $H(X)$ is the uncertainty of the stochastic variable defined above. Some consideration shows that $S(n)$ is much less than the total number of possible messages of length n. Thus, as $n \to \infty$, only a vanishingly small fraction of all possible messages is meaningful in this sense. This observation, after some considerable development, is what allows the Coding Theorem to work so well. In sum, the prescription is to encode messages in typical sequences, which are sent at very nearly the capacity of the channel. As the encoded messages become longer and longer, their maximum possible rate of transmission without error approaches channel capacity as a limit. Again, [1, 20, 47] provide details.

This approach can be, in a sense, inverted to give a tuning theorem which parsimoniously describes the essence of the Rate Distortion Manifold.

Telephone lines, optical wave, guides and the tenuous plasma through which a planetary probe transmits data to earth may all be viewed in traditional information-theoretic terms as a *noisy channel* around which we must structure a message so as to attain an optimal error-free transmission rate.

Telephone lines, wave guides, and interplanetary plasmas are, relatively speaking, fixed on the timescale of most messages, as are most other signaling networks. Indeed, the capacity of a channel, is defined by varying the probability distribution of the 'message' process X so as to maximize $I(X|Y)$.

Suppose there is some message X so critical that its probability distribution must remain fixed. The trick is to fix the distribution $P(x)$ but *modify the channel* – i.e., tune it – so as to maximize $I(X|Y)$. The *dual* channel capacity C^* can be defined as

$$C^* = \max_{P(Y),P(Y|X)} I(X|Y). \tag{37}$$

But

$$C^* = \max_{P(Y),P(Y|X)} I(Y|X)$$

since

$$I(X|Y) = H(X) + H(Y) - H(X,Y) = I(Y|X).$$

Thus, in a purely formal mathematical sense, *the message transmits the channel*, and there will indeed be, according to the Coding Theorem, a channel distribution $P(Y)$ which maximizes C^*.

One may do better than this, however, by modifying the channel matrix $P(Y|X)$. Since

$$P(y_j) = \sum_{i=1}^{M} P(x_i)P(y_j|x_i),$$

$P(Y)$ is entirely defined by the channel matrix $P(Y|X)$ for fixed $P(X)$ and

$$C^* = \max_{P(Y),P(Y|X)} I(Y|X) = \max_{P(Y|X)} I(Y|X).$$

Calculating C^* requires maximizing the complicated expression

$$I(X|Y) = H(X) + H(Y) - H(X,Y),$$

that contains products of terms and their logs, subject to constraints that the sums of probabilities are 1 and each probability is itself between 0 and 1. Maximization is done by varying the channel matrix terms $P(y_j|x_i)$ within the constraints. This is a difficult problem in nonlinear optimization. However, for the special case $M = L$, C^* may be found by inspection:

If $M = L$, then choose

$$P(y_j|x_i) = \delta_{j,i},$$

where $\delta_{i,j}$ is 1 if $i = j$ and 0 otherwise. For this special case

$$C^* = H(X),$$

with $P(y_k) = P(x_k)$ for all k. *Information is thus transmitted without error when the channel becomes 'typical' with respect to the fixed message distribution $P(X)$.*

If $M < L$, matters reduce to this case, but for $L < M$ information must be lost, leading to Rate Distortion limitations.

Thus modifying the channel may be a far more efficient means of ensuring transmission of an important message than encoding that message in a 'natural' language which maximizes the rate of transmission of information on a fixed channel.

We have examined the two limits in which either the distributions of $P(Y)$ or of $P(X)$ are kept fixed. The first provides the usual Shannon Coding Theorem, and the second a tuning theorem variant, a tunable retina-like Rate Distortion Manifold. It seems likely, however, than for many important systems $P(X)$ and $P(Y)$ will interpenetrate, to use Richard Levins' terminology. That is, $P(X)$ and $P(Y)$ will affect each other in characteristic ways, so that some form of mutual tuning may be the most effective strategy.

Author Index